REACTIONS OF ACIDS AND BASES
IN ANALYTICAL CHEMISTRY

ELLIS HORWOOD SERIES IN ANALYTICAL CHEMISTRY

Series Editors: Dr. R.A. CHALMERS and Dr. MARY MASSON, University of Aberdeen

Consultant Editor: Prof. J. N. MILLER, Loughborough University of Technology

REACTIONS OF ACIDS AND BASES IN ANALYTICAL CHEMISTRY

ADAM HULANICKI

Professor of Chemistry,
University of Warsaw

Translation Editor:
MARY R. MASSON
Department of Chemistry,
University of Aberdeen

ELLIS HORWOOD LIMITED
Publishers · Chichester
Halsted Press: a division of
JOHN WILEY & SONS
New York · Chichester · Brisbane · Toronto
PWN–Polish Scientific Publishers
Warsaw

English Edition first published in 1987 in coedition between
ELLIS HORWOOD LIMITED
Market Cross House, Cooper Street, Chichester, West Sussex, PO19 1EB, England
and
PWN–POLISH SCIENTIFIC PUBLISHERS
Warsaw, Poland

The publisher's colophon is reproduced from James Gillison's drawing of the ancient Market Cross, Chichester.

Translated from Polish by *Halina Galicka* (Chapters 1 and 2.1, 2.2—*Barbara Lipska*)

Revised and enlarged edition of A. Hulanicki, *Reakcje kwasów i zasad w chemii analitycznej*, published by Państwowe Wydawnictwo Naukowe, Warszawa 1980

Distributors:
Australia and New Zealand:
JACARANDA WILEY LIMITED
GPO Box 859, Brisbane, Queensland 4-001, Australia

Canada:
JOHN WILEY & SONS CANADA LIMITED
22 Worcester Road, Rexdale, Ontario, Canada

Europe and Africa:
JOHN WILEY & SONS LIMITED
Baffins Lane, Chichester, West Sussex, England

Albania, Bulgaria, Cuba, Czechoslovakia, German Democratic Republic, Hungary, Korean People's Democratic Republic, Mongolia, People's Republic of China, Poland, Romania, the U.S.S.R., Vietnam, Yugoslavia:
ARS POLONA—Foreign Trade Enterprise
Krakowskie Przedmieście 7, 00-068 Warszawa, Poland

North and South America and the rest of the world:
Halsted Press: a division of
JOHN WILEY & SONS
605 Third Avenue, New York, N.Y. 10158, U.S.A.

British Library Cataloguing in Publication Data
Hulanicki, A.
    Reactions of acids and bases in analytical
    chemistry. — (Ellis Horwood series in analytical
    chemistry)
    1. Acid–base equilibrium
    I. Title
    546'.24   QD477
    ISBN 0-85312-330-6 (Ellis Horwood Limited)
    ISBN 0-470-20246-7 (Halsted Press)
LIBRARY OF CONGRESS Card No. 85-16344

Table of Contents

Preface to the first Polish edition

The aim of this book is to present acid–base reactions primarily in the context of the Brønsted–Lowry theory. The general foundations of that theory can be found in many Polish textbooks of analytical chemistry. However, one constantly comes across remnants of the Arrhenius theory in those textbooks, particularly in the treatment of the reaction tradition- ally called hydrolysis, and in some other, more complicated, cases. The present author's guiding idea was to avoid this duality, which, incidentally, is also common in the foreign literature.

The study and applications of acid–base reactions in analytical chem- istry require knowledge of the states of thermodynamic equilibrium of the systems under consideration. The calculation of ionic equilibria, often elementary, presents difficulties in more complicated cases. Success can only be ensured by a systematic approach to problem solving, and a precise statement of those simplifying assumptions which make it possible to obtain the final result. Textbooks usually cover only the simplest cases of ionic equilibrium calculations. The next stage should be a full understand- ing of the equilibria, permitting results which give a better representation of the real state. The present book should be an introduction to such an approach, which is helpful in solving analytical problems.

The book was primarily written for junior students of chemistry. This can be seen most clearly in the initial chapters. Owing to their elementary character those chapters are a good introduction to the subject. The book also discusses some topics which might be considered either superfluous or too complicated for beginners. In order to be easily identified, the appro- priate sections are marked with an asterisk in the table of contents and in the text. They are included in the book to provide a more detailed descrip- tion of the basic problems concerning acid–base reactions.

Obviously, the book does not contain everything that might be rel- evant to its title. It deals with selected topics of a more fundamental character, and intentionally leaves out the practical side, the methodology of analytical determination based on acid–base titrations, and a vast range of theoretical and practical problems relating to the influence of acid–base reactions on redox reactions, complexation, precipitation, and also catalytic, kinetic and other effects. To discuss these problems, even superficially, the book would have to grow into several volumes. However, it seemed useful to devote one chapter to other acid and base theories, which are extensions of the Brønsted and Lowry theory.

The general plan of the book is not based on any particular textbook, although many of the books listed in the auxiliary bibliography have influenced the presentation of the problems, suggesting ways for their formulation or exemplification. The final opinion as to whether the design of the book is appropriate is left to the readers, and especially to those who will be able to test its usefulness for teaching purposes. The author will appreciate any critical remarks in this matter. A considerable role in the preparation of the book has been played by Professor Wiktor Kemula, who was my teacher in the field of research and teaching over many years. I wish to thank him for numerous stimulating discussions and words of encouragement. I also thank Professor Jerzy Minczewski and Professor Antoni Swinarski for their valuable and kind remarks, which beneficially influenced the content of the book.

ADAM HULANICKI

Warsaw, April 1971

Preface to the second Polish edition

The warm reception of my book by the readers of the first edition has stimulated the preparation of a subsequent edition, which does not differ in general outline from the original version. The additions introduced in the present edition include a slightly more extensive treatment of acid–base ionic equilibria involving metal ions and also of problems arising from the increasing interest in reactions in non-aqueous solvents. However, since I regard usefulness in teaching during the initial years of chemical studies as the main aim of the book, I have abstained from expanding those problems too much, for fear of losing the proportions which—I believe—were just right in the first edition. For the same reasons I have only slightly expanded the final chapters, e.g. by briefly discussing the hard and soft acid–base theory.

In eliminating errors and obscurities I was greatly helped by the kind remarks of my colleagues and friends, mainly from the Institute of Fundamental Problems in Chemistry, Warsaw University. I wish to express my thanks to all of them. I am particularly indebted to Dr Marek T. Krygowski for his remarks concerning non-aqueous solvents. I also thank Professor Zygmunt Marczenko for his numerous constructive remarks. I am very grateful to Dr B. J. Kapłan of the Institute of Analytical Chemistry and Geochemistry in Moscow, who pointed out several defects in the work while preparing its Russian translation, and who helped in eliminating them. I wish to thank Mrs Zofia Boglewska-Bareja for her assistance in preparing the present edition.

ADAM HULANICKI

Warsaw, September 1977

CHAPTER 1

Introduction

1.1 GENERAL REMARKS

The reactions of acids and bases, which we encounter in many chemical problems, particularly in analytical chemistry, are most familiar when they occur in aqueous solutions, although they can also take place in solution in other solvents, in molten salts and even in solids. The rules that govern these reactions are general rules for chemical processes. This introductory chapter reviews the elementary reactions and definitions, since they are necessary for a proper understanding of the subsequent chapters.

Reactions of acids and bases in solutions are fundamentally ionic reactions. Ions in a solution occur as a result of either dissociation of ionic compounds or ionization and dissociation of molecules with polarized covalent bonds. The crystal structure of an ionic compound (present in the solid state as an ionic crystal lattice) under the influence of molecules of water—a polar solvent—is broken down with formation of hydrated ions (aquo-ions). In an aquo-ion, the ion, which in a crystal lattice was surrounded by other ions, becomes surrounded instead by molecules of water. Such a process is observed for instance during dissolution of sodium chloride or sodium hydroxide.

Compounds with polarized bonds behave in a slightly different way, since under the influence of the dipoles of the solvent, their molecules become ionized as a result of a total displacement of an electron pair in the direction of a more electronegative atom, after which each ion becomes hydrated. In a dissociation process these ions become separated and exist in solution independently of one another. Such behaviour is typical of those covalent molecules which in solutions form strong acids, e.g. HCl. In many cases, despite the polarized character of the bond, many molecules are not completely ionized, and we do not observe complete dissociation in the solution. We then have solutions of weak, incompletely dissociated electrolytes. An example of such an electrolyte is hydrogen sulphide. Complete dissociation of this acid requires agents stronger than molecules of water, e.g. hydroxide ions.

Reactions of acids and bases are in most cases reversible—they can be made to proceed in either direction by changing the concentration. This applies also to dissociation reactions of acids and bases in cases where the solution contains undissociated molecules. On the other hand, dissociation of strong electrolytes, for which undissociated molecules do not exist in solutions, e.g. NaOH or HCl, cannot be reversed. This holds for processes occurring in dilute solutions. It is formally expressed by using equations with a single arrow to denote the process of dissolving strong electrolytes. A double arrow $\rightleftharpoons$ is used only when the reaction is reversible and when the molecules appearing in the equation are actually present in the solution.

It is characteristic of reversible reactions that a certain state of chemical equilibrium, which is a dynamic equilibrium, is reached. In a stage of dynamic equilibrium the rate of reaction from left to right is the same as that from right to left. As a result, the total concentrations of substrates and products do not change in this state, they characterize the state of equilibrium.

1.2 LAW OF MASS ACTION

The *rate of a chemical reaction* depends on the probability of collisions between the molecules present at a given instant in the space where the reaction takes place. Consider the reaction $A + B \rightleftharpoons C + D$; the rate of the reaction occurring between the molecules of A and B is proportional to their actual molar concentrations. If these concentrations, expressed in moles per litre, are denoted by the symbols in square brackets, then the rate of the reaction is given by

$$v_{A,B} = k_1 [A] [B] \tag{1.1}$$

where k_1 is the proportionality factor, called the reaction-rate constant. As the reaction progresses, the concentrations of reactants decrease, and this causes the reaction to slow down until it reaches the equilibrium state.

At the beginning of course of the reaction between A and B, the concentrations of the products C and D are negligible, but as more and more molecules of A and B react, the concentrations of the products increase. There is then, of course, a possibility of collisions between molecules C and D, and such collisions will result in formation of molecules of A and B, if the reaction is a reversible one. The rate of this reverse reaction depends on the probability of collisions between molecules C and D, which is proportional to the concentrations of these molecules. Consequently, the rate of the reaction between C and D is given by

$$v_{C,D} = k_2[C][D] \tag{1.2}$$

and is small at first, but gradually attains higher and higher values. Finally, the rate of the reverse reaction becomes equal to the rate of the reaction between A and B, for the given composition of the mixture. The number of molecules A and B reacting in unit time is then equal to the number of molecules of A and B formed as a result of the reaction between molecules C and D. This final rate determines the final composition of the reaction. It can be measured directly from measurements of changes in the concentrations of products (or reactants) in unit time.

The rates of the two reactions become equal at chemical equilibrium, which, as previously stated, is a state of dynamic equilibrium. At equilibrium, $v_{A,B} = v_{C,D}$, and hence

$$k_1[A][B] = k_2[C][D] \tag{1.3}$$

or

$$K = \frac{k_1}{k_2} = \frac{[C][D]}{[A][B]} \tag{1.4}$$

The statement that the rate of a chemical reaction is proportional to the actual concentrations of reactants is the essence of the law of mass action, formulated by Guldberg and Waage in 1863. This law implies that for the state of equilibrium the ratio of the product of the concentrations of species formed to the product of concentrations of reactants is a constant quantity, characterizing a given reaction at a given temperature. This quantity is called the *equilibrium constant*.

An analogous argument applies to the reaction $2A \rightleftharpoons C+D$. The equilibrium constant will then be expressed by

$$K = \frac{[C][D]}{[A][A]} = \frac{[C][D]}{[A]^2} \tag{1.5}$$

In the most general case,

$$aA+bB+cC+ \ldots \rightleftharpoons xX+yY+zZ+ \ldots$$

the equilibrium constant is expressed as:

$$K = \frac{[X]^x[Y]^y[Z]^z \ldots}{[A]^a[B]^b[C]^c \ldots} \tag{1.6}$$

The equilibrium constant derived on the assumption of a certain reaction mechanism remains valid whatever the mechanism of the reaction, provided that all the steps are reversible reactions. If, for instance, the product ABC_2 were formed from reactants A, B and C, then one of many possible mechanisms could be assumed, e.g.:

$$A + C \rightleftharpoons AC$$
$$B + C \rightleftharpoons BC$$
$$AC + BC \rightleftharpoons ABC_2$$

The equilibrium constant of each of these reactions would reflect the situation at the general equilibrium state. Hence the product of the equilibrium constants at each stage

$$K_1 = \frac{[AC]}{[A][C]}, \quad K_2 = \frac{[BC]}{[B][C]}, \quad K_3 = \frac{[ABC_2]}{[AC][BC]}$$

would be equal to the general equilibrium constant:

$$K = \frac{[ABC_2]}{[A][B][C]^2} \tag{1.7}$$

The law of mass action and the concept of the chemical equilibrium constant are applicable to all reactions which can be considered reversible, in particular to ionic reactions, e.g. to dissociation reactions. Consequently, if a molecule AB in an aqueous solution dissociates into ions A^+ and B^-, then the equilibrium constant for the reaction is equal to

$$K = \frac{[A^+][B^-]}{[AB]} \tag{1.8}$$

where the symbols of ions in square brackets denote the actual concentrations of ions at the state of equilibrium.

Equilibrium constants can characterize a variety of kinds of chemical reactions occurring between substances in various states of aggregation. Under certain conditions, the concentrations of some substances may be effectively constant; then they do not have to be taken into account in the equation of the equilibrium constant. For instance, for a reaction taking place in aqueous solution in which molecules of water are reactants

$$NH_3 + H_2O \rightleftharpoons NH_4^+ + OH^-$$

we can write the equilibrium constant as

$$K = \frac{[NH_4^+][OH^-]}{[NH_3]} \tag{1.9}$$

by omitting the concentration of H_2O, which is practically constant in an aqueous solution, and so can be included in the value of the constant.

(The concentration of water is about 55 moles per litre, so it changes very little in dilute solutions.)

Another example is the reaction taking place in water in a heterogeneous (multiphase) system:

$$CaCO_3 + 2H^+ \rightleftharpoons Ca^{2+} + H_2O + CO_2\uparrow$$

The correct equilibrium constant of this reaction is of the form

$$K = \frac{[Ca^{2+}][CO_2]}{[H^+]^2} \tag{1.10}$$

As was previously mentioned, the concentration of water is constant, and since the solution is at equilibrium with the precipitate of $CaCO_3$, the concentration of $CaCO_3$ actually dissolved is also constant. This means that $CaCO_3$ and H_2O can both be omitted from the expression for the equilibrium constant. Sometimes it is advantageous to replace the concentration of dissolved CO_2 by the partial pressure of CO_2 above the solution. Since these two quantities are unequivocally related to each other, Eq. (1.10) will not fundamentally change if $[CO_2]$ is replaced by $p(CO_2)$; only the numerical value of the constant will be changed.

One of the essential consequences of the law of mass action is that the position of the equilibrium for a reaction shifts if one of the reactants is removed or if an excess of one of the reacting substances is added. As an example, consider the reaction of ammonia with water. If OH^- ions are added to an equilibrium mixture, the equilibrium of the reaction will be shifted from right to left, so as to decrease the concentration of NH_4^+ and increase the concentration of NH_3, to compensate for addition of hydroxide and to keep constant the value of the equilibrium constant defined by Eq. (1.9).

If, in the same reaction, the concentration of OH^- ions is decreased (for example, by addition of hydrogen ions), then the equilibrium state will be shifted in the opposite direction, i.e. to the right. Again, the changes in concentrations are such as to maintain invariant the value of the equilibrium constant.

1.3 THE CONCEPT OF ACTIVITY

It was stated in the previous section that the rate of a reaction and the equilibrium state of a reaction both depend on the molar concentrations of the reagents. However, this is not strictly true. In reality, it is only in very dilute solutions that the ions of an electrolyte exist independently of one another. If the number of ions in a given volume increases, positive ions begin to interact electrostatically with negative ions, and vice versa. This interaction hampers the independent movement of ions, for example under the influence of an electric field. When the electrolyte concentration is high, ions of one sign are said to become surrounded by a kind of cloud of ions of the opposite charge. Measurements of certain properties of the solutions give results suggesting the presence of a smaller number of ions than would be produced from the complete dissociation of the amount of substance known to be dissolved. Consequently, the properties of solutions do not depend, strictly speaking, on the concentration of the dissolved compounds and hence on the concentration of the ions, but on another quantity, called the *activity* of the ions.

The activity, a_X of an ion or molecule is equal to its concentration, c_X multiplied by a number called the *activity coefficient f_X*.

$$a_X = f_X c_X \qquad (1.11)$$

In most cases, activity coefficients are < 1, but tend towards a value of unity as the solution is diluted. Only in very concentrated solutions, as a result of repulsion between ions, can activity coefficients be greater than 1.

It follows that the rate of a reaction and hence also the state of equilibrium do not depend on the concentrations of ions but on the corresponding activities. Consequently, it is better to express the equilibrium constant of a reaction $A + B \rightleftharpoons C + D$ as

$$_a K = \frac{a_C a_D}{a_A a_B} \qquad (1.12)$$

or as

$$_a K = \frac{[C][D]}{[A][B]} \frac{f_C f_D}{f_A f_B} = K \frac{f_C f_D}{f_A f_B} \qquad (1.13)$$

The equilibrium constant defined in this way is called the *thermodynamic constant* or the *activity constant*. It depends only on temperature and is independent of other parameters such as the composition of the solution.

In very concentrated solutions, where the activity coefficients differ considerably from 1, the value of the thermodynamic constant usually differs a great deal from the value of the *concentration constant*. However, if the solutions are very dilute and the activity coefficients are very close to 1, the value of the concentration constant approaches the value of the thermodynamic constant.

The data in Table 1.1 show how the concentration dissociation constant of acetic acid

$$K = \frac{[CH_3COO^-][H_3O^+]}{[CH_3COOH]}$$

i.e. the equilibrium constant of the reaction

$$CH_3COOH + H_2O \rightleftharpoons CH_3COO^- + H_3O^+$$

depends on the molar concentration of acetic acid. The activity coefficients of ions depend on two main factors—the charge on the ion and the *ionic strength*. Ionic strength is defined by the equation

$$I = \tfrac{1}{2}\sum C_i z_i^2 \tag{1.14}$$

where z_i is the charge on ion i, and C_i is the molar concentration of that ion.

Table 1.1

The concentration dissociation constant of acetic acid at various concentrations

Concentration, M	Dissociation constant $\times 10^5$	pK
1.00×10^{-5}	1.74	4.76
1.00×10^{-4}	1.75	4.76
1.00×10^{-3}	1.85	4.73
1.00×10^{-2}	2.10	4.68
1.00×10^{-1}	2.29	4.64
5.00×10^{-1}	2.64	4.58

Neutral molecules such as CH_3COOH and NH_3 present along with ions in a solution have no effect on the magnitude of electrostatic interactions. Hence, their presence is not taken into account in calculations of ionic strength.

Example 1.1. Calculate the ionic strength of a $0.20M$ solution of NaCl.

The solution contains the ions Na^+ and Cl^- at concentrations both equal to $0.20M$. Hence the ionic strength is

$$I = \tfrac{1}{2}(0.20 \times 1^2 + 0.20 \times 1^2) = 0.20$$

The ionic strength of an electrolyte dissociating into two univalent ions is numerically equal to its concentration.

Example 1.2. How does the ionic strength of an electrolyte of type A_2B_3, e.g. $La_2(SO_4)_3$, compare numerically with its concentration?

If we assume that the electrolyte in question dissociates completely and that its molar concentration is C, then

$$[La^{3+}] = 2C$$
$$[SO_4^{2-}] = 3C$$

so the ionic strength is

$$I = \tfrac{1}{2}(2C \times 3^2 + 3C \times 2^2) = 15C$$

The ionic strength is in this case 15 times the molar concentration of the electrolyte.

Example 1.3. Calculate the ionic strength of a solution containing $0.10M$ $MgSO_4$ and $0.10M$ Na_2SO_4.

We first calculate the concentration of each ion

$$[Mg^{2+}] = 0.10M, \quad [Na^+] = 0.20M, \quad [SO_4^{2-}] = 0.20M$$

Thus the ionic strength is

$$I = \tfrac{1}{2}(0.10 \times 2^2 + 0.20 \times 1^2 + 0.20 \times 2^2) = 0.70$$

Example 1.4. Calculate the ionic strength of a solution containing $0.10M$ CH_3COOH, $0.10M$ CH_3COONa and $0.20M$ $NaCl$.

Under the conditions given the dissociation of acetic acid is insignificant and may be neglected. The ionic strength of the solution will therefore depend only on the concentrations of the ions originating from sodium chloride and sodium acetate

$$[Na^+] = 0.30M, \quad [Cl^-] = 0.20M, \quad [CH_3COO^-] = 0.10M$$
$$I = \tfrac{1}{2}(0.30 \times 1^2 + 0.20 \times 1^2 + 0.10 \times 1^2) = 0.30$$

The calculations become more complicated in cases where it is necessary to take into account the actual degree of dissociation of a weak electrolyte.

1.3.1 The Debye–Hückel Theory

The dependence of the activity coefficient of an ion on ionic strength results from the state of equilibrium connected with the attraction forces between ions and their thermal movement which prevents the ions from coming too close to one another. According to the Debye–Hückel theory (1923), for an ion of charge z_i,

$$\log f_i = -\frac{Az_i^2 \sqrt{I}}{1 + Ba\sqrt{I}} \tag{1.15}$$

where A and B are constants related to the properties of the solvent and to the temperature of the measurement. In water at 25°C they are equal to 0.51 and 3.3×10^7, respectively. Parameter a is a function of the size of the ion and corresponds to the diameter of the hydrated ion; its value ranges from 3×10^{-8} to 11×10^{-8} cm (Table 1.2) and is expressed in cm in the equation. Equation (1.15) has been verified experimentally for solutions of ionic strength not exceeding 0.1. Since most of the solutions we will deal with also have $I < 0.1$ calculations made with Eq. (1.15) are satisfactory.

Table 1.2
Values of the parameter a for selected ions

Ion	$a \times 10^8$ cm
Sn^{4+}, Ce^{4+}, Th^{4+}, Zr^{4+}	11
H_3O^+, Al^{3+}, Fe^{3+}, Cr^{3+}	9
Mg^{2+}, Be^{2+}	8
Li^+, Ca^{2+}, Cu^{2+}, Zn^{2+}, Sn^{2+}, Mn^{2+}, Fe^{2+}, Ni^{2+}, Co^{2+}	6
Sr^{2+}, Ba^{2+}, Cd^{2+}, Hg^{2+}, S^{2-}, CH_3COO^-	5
Na^+, $H_2PO_4^-$, Pb^{2+}, CO_3^{2-}, SO_4^{2-}, CrO_4^{2-}, HPO_4^{2-}, PO_4^{3-}	4
OH^-, F^-, SCN^-, HS^-, ClO_4^-, Cl^-, Br^-, I^-, NO_3^-, K^+, NH_4^+, Ag^+	3

Example 1.5. Calculate the activity coefficient of the Sr^{2+} ion in a $0.01M$ solution of $SrCl_2$ in the absence and in presence of a $0.01M$ solution of KCl.

First, calculate the ionic strength of the solution for both cases, i.e.

$$I = \tfrac{1}{2}(0.01 \times 4 + 0.02 \times 1) = 0.03$$
$$I = \tfrac{1}{2}(0.01 \times 4 + 0.01 \times 1 + 0.03 \times 1) = 0.04$$

By applying the Debye–Hückel equation (Eq. (1.15)), next calculate the activity coefficient of the Sr^{2+} ion in the absence of KCl:

$$\log f_{Sr^{2+}} = -\frac{0.51 \times 4 \sqrt{0.03}}{1 + 3.3 \times 10^7 \times 5 \times 10^{-8} \sqrt{0.03}} = -\frac{2.04 \times 0.17}{1 + 1.65 \times 0.17}$$

$$= -\frac{0.35}{1.28} = -0.27 = \overline{1}.73$$

hence

$$f_{Sr^{2+}} = 0.54$$

Similar calculations for the solution with KCl added lead to

$$\log f_{Sr^{2+}} = -\frac{0.51 \times 4 \times 0.20}{1 + 1.65 \times 0.20} = -\frac{0.41}{1.33} = \overline{1}.69$$

and

$$f_{Sr^{2+}} = 0.49$$

These calculations indicate how the addition of an electrolyte influences the ionic strength and consequently the activity coefficients of all the ions present in the solution.

The Debye–Hückel equation may be simplified for very dilute solutions. If it is assumed that for many ions parameter a is equal or close to 3×10^{-8} cm, then the product $B \cdot a = 3.3 \times 10^7 \times 3 \times 10^{-8} \sim 1$, and hence

$$\log f_i = -\frac{A z_i^2 \sqrt{I}}{1 + \sqrt{I}} \tag{1.16}$$

This form of the equation is often used instead of Eq. (1.15). For very dilute solutions, when the ionic strength is not greater than 0.05, 0.014 and 0.005 for univalent, bivalent and tervalent ions, respectively, the term $\sqrt{I}$ in the denominator may be omitted as being considerably smaller than 1, and the activity coefficient may then be calculated according to the equation:

$$\log f_i = -0.5 z_i^2 \sqrt{I} \tag{1.17}$$

The values of the activity coefficients of three different ions calculated from Eqs. (1.15), (1.16) and (1.17) are given in Table 1.3. These data clearly indicate the range of application of the various equations for particular ion type.

Table 1.3

Activity coefficients of ions calculated from various forms of the Debye–Hückel equation

Ion	Equation	Activity coefficients f_i			
		$I = 0.005$	$I = 0.01$	$I = 0.05$	$I = 0.1$
	(1.15)	0.927	0.901	0.815	0.77
Na^+	(1.16)	0.93	0.90	0.81	0.76
	(1.17)	0.92	0.89	0.78	0.70
	(1.15)	0.749	0.675	0.485	0.405
Ca^{2+}	(1.16)	0.74	0.65	0.43	0.33
	(1.17)	0.63	0.63	0.36	0.23
	(1.15)	0.54	0.445	0.245	0.18
La^{3+}	(1.16)	0.50	0.39	0.15	0.083
	(1.17)	0.48	0.36	0.10	0.039

Sometimes a graph is sufficient for approximate calculations (Fig. 1.1). Naturally, we should always be fully aware of the attainable accuracy of calculation.

Data calculated from the Debye–Hückel equation are in good agreement with the results of experiment, for dilute solutions. However, if the ionic strength is greater than 0.1, experimentally obtained values should be used.

The major difficulty that arises in the determination of the activity coefficients of individual ions is that in a solution there are always at least

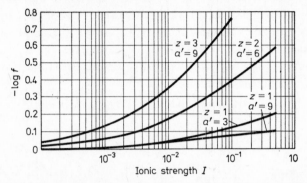

Fig. 1.1. The relationship between the activity coefficients and the ionic strength for H_3O^+ ($z = 1$, $a' = 9$) and for other ions with charges $z = 1$, $z = 2$, $z = 3$ on the basis of Eq. (1.15) on the assumption that $a' = a \times 10^8$.

two types of ions, having opposite charges. The activity coefficient then determined (e.g. on the basis of the measurement of conductance, electrode potentials, lowering of the freezing point) is the mean activity coefficient of the electrolyte

$$(f_\pm)^{m+n} = f_+^m f_-^n \tag{1.18}$$

where f_+ and f_- are the activity coefficients of the cation and the anion of the electrolyte $A_m B_n$, respectively. However, if certain assumptions are made, the activity coefficients of particular ions may be calculated from experimental data with fairly good accuracy.

For uncharged molecules of a dissolved substance present at low and moderate concentrations, the activity coefficients may be assumed to be equal to 1. (This follows from the character of the forces which have been taken into account in defining the notion of activity.)

The equations that follow from the Debye–Hückel theory allow derivation of the relationship between thermodynamic equilibrium constants and concentration constants. For the general case Eq. (1.13) can be presented in logarithmic form as

$$p_a K = pK - \log \frac{f_C^c f_D^d}{f_A^a f_B^b} \tag{1.19}$$

which after substitution of the values of the logarithms of the activity coefficients obtained from Eq. (1.16) gives

$$p_aK = pK - 0.5(az_A^2 + bz_B^2 - cz_C^2 - dz_D^2)\frac{\sqrt{I}}{1 + \sqrt{I}} \qquad (1.20)$$

If the expression in brackets is negative, the value of p_aK corresponding to the thermodynamic constant is greater than the value of pK corresponding to the concentration constant. In other words, the numerical value of the thermodynamic constant is then smaller than the value of the concentration constant. This equation is only true for those reactions that involve a separation of charge, e.g. for the reaction

$$H_3PO_4 + H_2O \rightleftharpoons H_3O^+ + H_2PO_4^- \qquad (1.21)$$

$$p_aK = pK + \frac{\sqrt{I}}{1 + \sqrt{I}}$$

for the reaction

$$H_2PO_4^- + H_2O \rightleftharpoons H_3O^+ + HPO_4^{2-} \qquad (1.22)$$

$$p_aK = pK + 2\frac{\sqrt{I}}{1 + \sqrt{I}}$$

and for the reaction

$$HPO_4^{2-} + H_2O \rightleftharpoons H_3O^+ + PO_4^{3-} \qquad (1.23)$$

$$p_aK = pK + 3\frac{\sqrt{I}}{1 + \sqrt{I}}$$

However, for a reaction in which on both sides of the equation there are ions with the same charge, e.g.

$$NH_4^+ + H_2O \rightleftharpoons H_3O^+ + NH_3$$

we have

$$p_aK = pK \qquad (1.24)$$

In most of the discussions of topics in this book, concentration constants will be used. These are quite adequate for introductory study of the basic types of equilibria and how to do calculations. However, the basic assumptions presented here can be utilized to convert systems based on concentrations into systems based on activities whenever necessary.

Example 1.6. The thermodynamic equilibrium constant for the reaction

$$H_2PO_4^- + H_2O \rightleftharpoons H_3O^+ + HPO_4^{2-}$$

is 6.3×10^{-8}. Calculate the concentration constant for a solution with ionic strength $I = 0.1$.

The value of pK is calculated from Eq. (1.20)

$$pK = p_aK + 0.5(1 \times 1^2 - 1 \times 1^2 - 1 \times 2^2)\frac{\sqrt{I}}{1 + \sqrt{I}}$$

$$pK = 7.20 - 0.5 \times 4\frac{\sqrt{0.1}}{1 + \sqrt{0.1}} = 7.20 - 0.48 = 6.72$$

This value is considerably smaller than the thermodynamic constant and the difference should be taken into account in accurate calculations, for instance in calculating the pH of a phosphate buffer.

The Debye–Hückel theory leads to the following conclusions, which are important for an analyst wanting to make qualitative estimates of interactions between the ions of an electrolyte.

(1) Neutral molecules in a solution are regarded as ideal molecules that do not interact with one another. Univalent ions (with a positive or a negative charge) behave less ideally, bivalent ions are still further from the ideal model, and so on. In each of these groups the activity coefficients increase with the radius of the hydrated ion.
(2) Ions with a definite charge have in a given solution approximately the same activity coefficients, independent of the individual concentrations.
(3) In solvents with a lower relative permittivity than water, deviations from the ideal behaviour of ions are greater than in solvents with a larger relative permittivity.

1.4 EQUILIBRIUM CONSTANTS AND OTHER THERMODYNAMIC QUANTITIES

The chemical equilibrium constant is a quantity often used in analytical chemistry. It is closely related to other thermodynamic quantities, and the relationships may be utilized for estimation of the properties of various systems, such as acid–base systems; moreover, they are helpful for determining the influence of temperature on reactions.

If the reaction takes place at constant pressure (which usually holds for analytical reactions in solution), then at a given temperature the tendency of the systems towards transition from the initial state (I) to the final state (II) is determined by the difference between the Gibbs free energy values for the two states.

$$\Delta G = G_{II} - G_I \tag{1.25}$$

A spontaneous chemical reaction is only possible when ΔG is negative, i.e. when the free energy decreases during the reaction. A positive value of ΔG would indicate a tendency of the reaction to go in the opposite direction, and in a system at equilibrium, $\Delta G = 0$.

The change of the free energy of a system in which a reaction takes place according to the equation

$$a\text{A} + b\text{B} \rightleftharpoons m\text{M} + n\text{N}$$

can be related to the values of the free energy for the reactants, and hence

$$\Delta G = (mG_M + nG_N) - (aG_A + bG_B) \tag{1.26}$$

For reactions occurring in solutions the values of the Gibbs free energy relate to unit concentration or activity of reactants (usually expressed in moles per litre of solution).

Hence for substance X,

$$G_X = G_X^0 + RT\ln C_X \tag{1.27}$$

or

$$G_X = G_X^0 + RT\ln a_X \tag{1.28}$$

The term G_X^0—the standard free energy—is equal to the free energy of component X in a solution of concentration or activity equal to 1.

For the chemical reaction above, the free energy change can be presented in the form:

$$\Delta G = [m(G_M^0 + RT\ln a_M) + n(G_N^0 + RT\ln a_N)]$$
$$- [a(G_A^0 + RT\ln a_A) + b(G_B^0 + RT\ln a_B)] \tag{1.29}$$

or

$$\Delta G = \Delta G^0 + RT\ln \frac{a_M^m a_N^n}{a_A^a a_B^b} \tag{1.30}$$

where

$$\Delta G^0 = mG_M^0 + nG_N^0 - aG_A^0 - bG_B^0 \tag{1.31}$$

The relationship between the standard free energy ΔG^0 of a reaction and the value of the equilibrium constant is evident from Eq. (1.30). At the equilibrium state determined by the given activities of the components, the free energy change is equal to zero, and hence

$$\Delta G^0 = -RT\ln \frac{a_M^m a_N^n}{a_A^a a_B^b} = -RT\ln K \tag{1.32}$$

By substituting common logarithms for natural logarithms we obtain an expression suitable for practical calculations:

$$\Delta G^0 = -2.303 RT \log K \qquad (1.33)$$

Example 1.7. Calculate the standard free energy change for the protonation of ammonia; the equilibrium constant is 1.74×10^9 in an aqueous solution at 25°C.

We first calculate the common logarithm of the equilibrium constant to be 9.24. Hence

$$\Delta G^0 = -2.303 RT \times 9.24$$

and by substituting $R = 8.314$ J·K^{-1}·mole^{-1} and $T = 298$ K (i.e. about 25°C) we obtain

$$\Delta G^0 = -2.303 \times 8.314 \times 298 \times 9.24 = -52.72 \text{ kJ/mole}$$

From these considerations it follows that a reaction is spontaneous if the free energy change is negative or if the corresponding value of $\log K$ is positive, i.e. $K > 1$. Values of K less than one always correspond to positive values of ΔG^0. It should be noted, however, that K can never be negative.

The ΔG^0 values for chemical reactions may be calculated from the value for the free energy of formation of a given compound from its elements. It is assumed that the value of G^0 for elements in the standard state and for a hydrogen ion is equal to zero. Values of the free energy of formation are given in tables of thermodynamic data of substances.

Example 1.8. Calculate the equilibrium constants for the dissociation of hydrogen sulphide at 25°C, if the free energies of formation of the reactants have the following values (in kJ/mole): $\Delta G^0(H_2S) = -27.4$, $\Delta G^0(HS^-) = +12.6$ and $\Delta G^0(S^{2-}) = +85.6$.

For the dissociation of the first hydrogen ion,

$$H_2S \rightleftharpoons H^+ + HS^-$$

the free energy change can be calculated in the following way:

$$\Delta G^0 = (0+12.6)-(-27.4) = 40.0 \text{ kJ/mole}$$

On the basis of this value it is easy to calculate the logarithm of the equilibrium constant

$$\log K = -\frac{\Delta G^0}{2.303 RT} = -\frac{40\,000}{2.303 \times 8.314 \times 298} = -7.01$$

hence the value of the equilibrium constant is 9.8×10^{-8}.

We proceed similarly for the second dissociation step:

$$HS^- \rightleftharpoons H^+ + S^{2-}$$

$$\Delta G^0 = (0+85.6) - (12.6) = 73.0 \text{ kJ/mole}$$

$$\log K = - \frac{73\ 000}{2.303 \times 8.314 \times 298} = -12.79$$

$$K = 10^{-12.79} = 1.6 \times 10^{-13}$$

These values are close to those obtained by direct measurements.

The relationship between the equilibrium constant and temperature follows from the van't Hoff equation

$$\frac{d\ln K}{dT} = \frac{\Delta H^0}{RT^2} \tag{1.34}$$

on the assumption that the standard enthalpy change ΔH^0 is independent of temperature. This assumption is not quite accurate, but it can serve as a basis for calculations over a limited range of temperatures. By integration of Eq. (1.34), the following expression for temperatures T_1 and T_2 is obtained:

$$\log K_{T_2} - \log K_{T_1} = \frac{\Delta H^0}{2.303R} \left(\frac{1}{T_1} - \frac{1}{T_2} \right) \tag{1.35}$$

If the equilibrium constant increases as the temperature changes from T_1 to T_2 (i.e. $(\log K_{T_2} - \log K_{T_1}) > 0$), then ΔH^0 is positive, which means that the reaction is endothermic, i.e. involves absorption of thermal energy. As the temperature increases, such reactions become more complete because of the increase in the equilibrium constant; the heat absorption tends to diminish the influence of the external factor of temperature. Conversely, if ΔH^0 is negative and, the reaction is exothermic, an increase of temperature inhibits the forward reaction, as a result of a decrease of the value of the equilibrium constant.

The values of ΔH^0 for many acid–base reactions are not high; for instance $\Delta H^0 = -0.4$ kJ/mole for acetic acid and $\Delta H^0 = -0.04$ kJ/mole for formic acid. However, for the dissociation of water ΔH^0 is 57.8 kJ/mole. This indicates that the dissociation constant of water, and hence also the ion-product of water, depend on temperature to a considerable degree.

1.5 METHODS OF PRESENTING THE EQUILIBRIUM CONSTANT

The content of this section requires some knowledge of the processes of acid dissociation and thus anticipates to some extent the material in subsequent chapters. Their strictly formal character, however, requires them to appear in the introductory part. Omitting this section should not affect the study of the subsequent chapters, and the reader can return to this section later.

Equilibrium constants, and thus also dissociation constants, mostly refer to molar concentrations of substrates and products. Such constants are usually given in tables and in chemical practice we are used to their approximate numerical values. They are usually presented as dimensionless quantities, and though this practice is accepted it is not strictly correct.

For a reaction

$$AB \rightleftharpoons A^+ + B^-$$

the equilibrium constant

$$K = \frac{[A^+][B^-]}{[AB]} \tag{1.36}$$

has the dimensions of concentration, i.e. mole per litre (of solution). Obviously, this refers only to the case where one molecule (concentration in the denominator) gives rise to two ions (concentrations in the numerator). For a more complex reaction, the equilibrium constant of the overall process should be given. For example, the stepwise dissociation of a polyprotic acid

$$H_2S + H_2O \rightleftharpoons H_3O^+ + HS^-$$
$$HS^- + H_2O \rightleftharpoons H_3O^+ + S^{2-}$$

can be presented as the overall equation

$$H_2S + 2H_2O \rightleftharpoons 2H_3O^+ + S^{2-}$$

and the equilibrium constant

$$K = \frac{[H_3O^+]^2[S^{2-}]}{[H_2S]} \tag{1.37}$$

has the dimensions of the square of concentration, i.e. mole^2/litre^2. These considerations can be extended to other types of reactions.

We only rarely use constants for which concentrations are expressed in other units. If the concentrations in Eq. (1.36) were expressed in millimoles, the constant would be numerically 10^3 times greater, and in the case of Eq. (1.37) 10^6 times greater. If the concentrations were expressed in moles per kilogram of the solvent (molal concentration), the dimensions of the constant would be mole per kilogram. If the concentrations were given in mole fractions, the constant would be dimensionless.

In the preceding section we considered thermodynamic activity constants and concentration constants. The relationships between these constants depend on the values of the activity coefficients (Eq. (1.13)). In the overwhelming majority of reactions encountered in analytical chemistry we use concentration constants determined from directly known concentrations of substances rather from their activities. Such an approach is usually fully justified and does not lead to any error which might substantially affect the final result.

The usual method of determination of the hydrogen ion, however, yields values in terms of activity rather than concentration (cf. Section 2.13). That is why there exist *mixed constants*, which are also called *Bjerrum constants*. The expression for the equilibrium constant includes the activities of hydrogen and hydroxide ions and the concentrations of the remaining ions. For instance, for the reaction

$$HA + H_2O \rightleftharpoons H_3O^+ + A^-$$

the mixed constant is defined as

$$K = \frac{a_{H_3O^+}[A^-]}{[HA]}$$

The applicability of mixed constants is limited to strictly defined analytical conditions, but many authors use them and confirm their usefulness.

Both the determination of the concentration constants and their use require maintenance of a fixed ionic strength. If the ionic strength of a solution obtained by addition of an electrolyte not participating in the reaction is large in comparison with the concentrations of the reactants, then it can be assumed that the activity coefficients of the reactants do not change significantly during the reaction. This is one of the reasons why concentration constants are usually quoted for fairly high values of ionic strength such as 0.1, 1.0 or even sometimes $3.0M$.

1.5.1 Conditional Constants

Conditional constants, also referred to as *apparent constants* or *effective constants* are unknown to physical chemists, but nevertheless they are extremely useful. They are particularly important for elucidation of complicated analytical reactions. From the analyst's point of view, it is not the actual concentration of a particular ion which is interesting but rather the total concentration of the various species which contain the ion to be determined. For instance, in the precipitation of ferric hydroxide from a ferric chloride solution:

$$Fe^{3+} + 3OH^- \rightleftharpoons Fe(OH)_3$$

the equilibrium constant is defined as

$$K = \frac{1}{[Fe^{3+}][OH^-]^3} \tag{1.38}$$

Thus it refers to a system in which there are only free Fe^{3+} ions (hydrated, but not bound in any other reaction). However, this is not the case in practice, since addition of Fe^{3+} ions to a solution containing, say, chloride ions, will result in the formation of complex ions such as $FeCl^{2+}$, $FeCl_2^+$ etc. The analyst is interested only in calculating the conditions necessary for complete precipitation of $Fe(OH)_3$. Therefore, instead of using Eq. (1.38) for the equilibrium constant, it is useful to define

$$K'_{Fe} = \frac{1}{[Fe'][OH^-]^3} \tag{1.39}$$

where K'_{Fe} is the conditional constant for precipitation of ferric hydroxide in a solution of specified composition (e.g. with a specified Cl^- concentration), and $[Fe']$ indicates the total concentration of iron species in solution from which $Fe(OH)_3$ may or should precipitate, and thus

$$[Fe'] = [Fe^{3+}] + [FeCl^{2+}] + [FeCl_2^+] + \ldots$$

The total concentration of non-precipitated species containing Fe(III) is given by

$$[Fe'] = [Fe^{3+}](1 + \beta_1[Cl^-] + \beta_2[Cl^-]^2 + \ldots)$$
$$= [Fe^{3+}]\alpha_{Fe} \tag{1.40}$$

where β_1, β_2 etc. are the overall stability constants for the complexes $FeCl^{2+}$, $FeCl_2^+$, etc. The coefficient $\alpha_{Fe(Cl)}$ is called the *side-reaction coefficient* for reaction of iron (III) with chloride.

The term "side-reactions" denotes additional reactions which involve the reactants and products of the main reaction. Binding of the reactants or products in side-reactions may cause a considerable shift in the equilibrium state, and hence would lead to a false value for the equilibrium constant if, in its calculation, side-reactions were disregarded.

In the reaction under consideration the relationship between the concentration equilibrium constant and the conditional constant is

$$K = K'_{Fe}\alpha_{Fe(Cl)} \tag{1.41}$$

where the subscript Fe following the symbol K' indicates that the particular side-reactions of Fe have been taken into account. The value of $\alpha_{Fe(Cl)}$ is unity if there are no side-reactions, and is greater than one if side-reactions do occur. Hence in this case the conditional constant is equal to or less than the concentration constant. If $[Cl^-] = 1.0M$, then

$\alpha_{Fe(Cl)} = 27$, and hence the equilibrium constant, which has a value of 3×10^{39} in the absence of Cl^- ions, will decrease to $3 \times 10^{39}/27 = 1 \times 10^{38}$, which is the conditional constant for the given conditions.

Conditional constants are also extremely useful for calculations concerned with the titration of weak acids in the presence of ions of metals which form complex compounds with the anions of the acids. The weak acids of greatest analytical important are complexing agents such as ethylenediaminetetra-acetic acid (EDTA), 1,2-diaminocyclohexanetetra-acetic acid (DCTA) etc. Usually, the analyst has a solution containing a mixture of the weak acid and its salts at a known total concentration (acid plus anions) and with known pH. The actual concentration of fully deprotonated anion can be calculated by use of side-reaction coefficients. Examples of calculations of this type are given in Sections 3.9 and 3.10, where equilibria in solutions of polyprotic acids are considered.

1.5.2 Protonation Constants

Properties of acids and bases are often specified by means of the values of the equilibrium constants of the following reactions:

$$HA + H_2O \rightleftharpoons H_3O^+ + A^- \quad \text{for acids}$$
$$B + H_2O \rightleftharpoons BH^+ + OH^- \quad \text{for bases}$$

Such equilibrium constants are called *acid* and *base dissociation constants* (Section 2.6).

Nowadays these constants are more and more often replaced by *protonation constants*, which are the equilibrium constants for the reaction of proton addition

$$A^- + H_3O^+ \rightleftharpoons HA + H_2O$$
$$B + H_3O^+ \rightleftharpoons BH^+ + H_2O$$

It is obvious that for acids the protonation constant is the reciprocal of the acid dissociation constant. For bases, the protonation constant is equal to the base dissociation constant divided by the ion-product of water, which is equal to 10^{-14}. The use of protonation constants leads to a more consistent labelling of reactions occurring in acid–base systems, and particularly in complexing systems. Subsequent chapters illustrate this assertion.

1.5.3 Compilations of Data

Since knowledge of the equilibrium constants for acid–base reactions and of complex-formation reactions is very important for an analyst, there exist many listings of equilibrium constants. Particular attention should be paid to studies that follow the guidelines of the International Union of Pure and Applied Chemistry (IUPAC), whose authors use uniform terminology and notation. Since equilibria are investigated in many scientific centres, the amount of information increases very rapidly and the collections of constants require continuous supplementing and updating. The following are the most useful compilations.

REFERENCES

L. G. Sillén and A. E. Martell, *Stability Constants of Metal–Ion Complexes*, The Chemical Society, London, 1964; *Supplement* No. **1**, 1971.

D. D. Perrin, *Dissociation Constants of Inorganic Acids and Bases in Aqueous Solution*, Butterworths, London, 1969.

G. Kortum, W. Vogel and K. Andrussow, *Dissociation of Organic Acids in Aqueous Solution*, Butterworths, London, 1961.

D. D. Perrin, *Dissociation of Organic Bases in Aqueous Solution*, Butterworths, London, 1965.

D. D. Perrin, *Stability Constants of Metal–Ion Complexes, Part B, Organic Ligands*, Pergamon Press, Oxford, 1979.

E. Högfeldt, *Stability Constants of Metal–Ion Complexes, Part A, Inorganic Ligands*, Pergamon Press, Oxford, 1982.

A. E. Martell and R. M. Smith, *Critical Stability Constants*, Vol. 1, *Amino Acids*, 1974; Vol. 2, *Amines*, 1975; Vol. 3, *Other Organic Ligands*; Vol. 4, *Inorganic Complexes*, 1976, Vol. 5, *First Supplement*, 1982, Plenum Press, New York.

PROBLEMS

1. Calculate the ionic strength of the following solutions:
  a. $0.05M$ Na_2SO_4;
  b. $0.08M$ $NaCH_3COO$ + $0.02M$ KCH_3COO;
  c. $0.20M$ $ZnSO_4$;
  d. $0.20M$ $ZnSO_4$ + $0.20M$ K_2SO_4;
  e. $0.20M$ $ZnSO_4$ + $0.20M$ KCl;
  f. $0.20M$ $ZnSO_4$ + $0.40M$ KCl;
  g. $0.10M$ NH_4Cl;
  h. $0.10M$ NH_4Cl + $0.10M$ NH_3;
  i. $0.05M$ HCl + $0.05M$ KCl;
  j. $0.05M$ HNO_3 + $0.05M$ $AgNO_3$ + $0.05M$ KCl.

2. Calculate the concentration needed to give a solution having an ionic strength of $0.30M$ when the electrolyte is:
  a. sodium nitrate;
  b. barium nitrate;

c. copper sulphate;

d. lanthanum sulphate;

e. potassium sulphate;

f. acetic acid.

3. The ionic strength in solutions containing a simple electrolyte is proportional to the electrolyte concentration. Calculate the proportionality factor for the following electrolytes:

a. 1:1 (univalent ions);

b. 1:1 (bivalent ions);

c. 1:1 (tervalent ions);

d. 1:2 (bi- and univalent ions);

e. 2:3 (ter- and bivalent ions).

4. Calculate the activity coefficient of the chloride ion in each of the following solutions:

a. $0.01M$ NaCl;

b. $0.01M$ NaCl + $0.01M$ KCl;

c. $0.01M$ NaCl + $0.01M$ K_2SO_4;

d. $0.10M$ NaCl;

e. $1.00M$ NaCl + $1.00M$ HCl.

5. Calculate the activity coefficients of the hydrogen ion in a solution which is:

a. $0.01M$ in HCl and $0.1M$ KCl;

b. $0.01M$ in HCl and $0.01M$ KCl;

c. $0.01M$ in HCl alone.

6. Calculate the concentration equilibrium constant at $I = 0.1$ for the following reactions:

a. $CH_3COOH + H_2O \rightleftharpoons H_3O^+ + CH_3COO^-$, when the thermodynamic constant is 1.75×10^{-5};

b. $NH_4^+ + H_2O \rightleftharpoons H_3O^+ + NH_3$, when the thermodynamic constant is 5.75×10^{-10};

c. $H_2C_2O_4 + H_2O \rightleftharpoons H_3O^+ + HC_2O_4^-$, when the thermodynamic constant is 5.37×10^{-2};

d. $HC_2O_4^- + H_2O \rightleftharpoons H_3O^+ + C_2O_4^{2-}$, when the thermodynamic constant is 5.37×10^{-5};

e. $H_2O + H_2O \rightleftharpoons H_3O^+ + OH^-$, when the thermodynamic constant is 1.00×10^{-14}.

Note: In these calculations, the approximate Eq. (1.20) may be used.

7. Derive the relationship between the thermodynamic and concentration equilibrium constant for the following reactions (assume a value for the ion size parameter of 3×10^{-8}):

a. $H_2PO_4^- + H_2O \rightleftharpoons OH^- + H_3PO_4$;

b. $HPO_4^{2-} + H_2O \rightleftharpoons OH^- + H_2PO_4^-$;

c. $PO_4^{3-} + H_2O \rightleftharpoons OH^- + HPO_4^-$;

d. $NH_4^+ + H_2O \rightleftharpoons OH^+ + NH_3$.

8. Calculate the standard free energy change for the following reactions:

a. $HF + H_2O \rightleftharpoons H_3O^+ + F^-$, $K = 6.76 \times 10^{-4}$;

b. $H_3O^+ + CN^- \rightleftharpoons H_2O + HCN$, $K = 2.5 \times 10^9$;

c. $2H_2O \rightleftharpoons H_3O^+ + OH^-$, $K = 1.0 \times 10^{-14}$;

d. $H_3AsO_4 + H_2O \rightleftharpoons H_3O^+ + H_2AsO_4^-$, $K = 6.03 \times 10^{-3}$;

e. $H_3O^+ + H_2AsO_4^- \rightleftharpoons H_2O + H_3AsO_4$, $K = 1.66 \times 10^2$.

9. Calculate the equilibrium constant for the following reactions:

a. $SO_4^{2-} + H_3O^+ \rightleftharpoons HSO_4^- + H_2O$, $\Delta G^0 = -10.8$ kJ/mole;

b. $HCOOH + H_2O \rightleftharpoons HCOO^- + H_3O^+$, $\Delta G^0 = +21.4$ kJ/mole;

c. $(COO)_2^{2-} + H_3O^+ \rightleftharpoons COOH.COO^- + H_2O$, $\Delta G^0 = -24.5$ kJ/mole;

d. $COOH.COO^- + H_2O^+ \rightleftharpoons (COOH)_2 + H_2O$, $\Delta G^0 = -7.4$ kJ/mole;

e. $C_6H_5NH_2 + H_2O \rightleftharpoons C_6H_5NH_3^+ + OH^-$, $\Delta G^0 = +53.6$ kJ/mole;

f. $H_2S + NH_3 \rightleftharpoons NH_4^+ + HS^-$, $\Delta G^0 = -11.6$ kJ/mole;

g. $H_2O_2 + OH^- \rightleftharpoons H_2O + HO_2^-$, $\Delta G^0 = -12.6$ kJ/mole;

h. $H_3BO_3 + H_2O \rightleftharpoons H_3O^+ + H_2BO_3^-$, $\Delta G^0 = +52.5$ kJ/mole.

CHAPTER 2

Properties of acids and bases in solution

2.1 DISSOCIATION OF WATER

Water is a solvent commonly found in nature and the usual solvent in analytical chemistry. Because of the covalent character of water molecules, pure water is almost a non-conductor of electric current. However, the oxygen–hydrogen bonds have a distinctly polar character, resulting from the differences in the electronegativities of the two elements, i.e. from the fact that an oxygen atom interacts more strongly than a hydrogen atom with the electron pair of the bond. The spatial structure of a water molecule, in which the bond angle is about 105°, causes the water molecule to have a polar character; the hydrogen atoms have a partial positive charge and the oxygen atom a partial negative charge.

Very accurate electrical measurements show that pure water conducts the electric current to a degree that is extremely low but still measurable. At 18°C the specific conductance of very pure water is 4.4×10^{-8} S·cm. This indicates very slight but still significant dissociation of water, according to the equation

$$2H_2O \rightleftharpoons H_3O^+ + OH^-$$

This equation can be presented in a simpler form, *viz.*

$$H_2O \rightleftharpoons H^+ + OH^-$$

However, neither equation is strictly correct. The simple equation suggests the existence of free protons, but it is most improbable that such a small charged particle would not interact with non-dissociated but polar water molecules. Moreover, molecules of liquid water also interact with each other, and water is known to contain aggregates composed of several molecules. Thus it is more accurate to represent the cation formed in the dissociation of water as H_3O^+ rather than H^+, but perhaps $H_9O_4^+$ would be nearer to the truth.

However, the size and composition of hydrated protons are not actually fixed, and may vary with the conditions, so by convention, we write H_3O^+ and speak of a *hydronium ion*. If there is no danger of confusion, and if the ion in question is not involved in any acid–base reaction, we often simply write H^+, remembering that, in aqueous solutions, this symbol always stands for a hydrated proton.

The equation for dissociation of water which shows the formation of the positive ion H_3O^+ and the negative ion OH^- expresses the essential nature of the process. One of the water molecules loses a proton to another molecule, which forms an H_3O^+ ion. The molecule which loses a proton is called the *proton donor* and the one which receives a proton is called the *acceptor*. The statement that some water molecules are donors and others are acceptors is of course a very artificial one. Generally, we refer to the *donor–acceptor character* of water molecules with regard to protons.

The donor–acceptor character of water molecules manifests itself not only in their interaction but also in their reactions with other molecules. These donor–acceptor properties are the basis of acid and base reactions.

2.2 THE ION-PRODUCT OF WATER

For the dissociation of water, expressed by the simplest equation

$$H_2O \rightleftharpoons H^+ + OH^-$$

just as for any other reversible chemical reaction, we can write an expression defining the equilibrium constant

$$K = \frac{[H^+][OH^-]}{[H_2O]} \tag{2.1}$$

In an aqueous medium the dissociation of water occurs to only a very slight degree. In pure water there are one hydrated H^+ ion and one OH^- ion per 554×10^6 water molecules. These data indicate that the degree of dissociation of water, i.e. the ratio of dissociated molecules to the total number of molecules, is equal to

$$x = \frac{1}{5.54 \times 10^8} = 1.8 \times 10^{-9}$$ (2.2)

Since the degree of dissociation is so small, we can assume that the concentration of non-dissociated water molecules is constant in water and in dilute aqueous solutions. This constant concentration of water, expressed in moles per litre, is $997/18.0 = 55.4$, since 997 g is the weight of 1 litre of water at a temperature of 25°C and 18.0 g is the weight of 1 mole of water.

By substituting the concentration of water and the numerical value of the dissociation constant of water in the equation for the equilibrium constant, we obtain

$$K = \frac{[H^+][OH^-]}{55.4} = 1.80 \times 10^{-16}$$ (2.3)

and multiplying the numerical values gives

$$[H^+][OH^-] = 1.00 \times 10^{-14} = K_w$$

The quantity K_w, equal to the product of the concentrations of the ions of water, is called the *ion-product of water*. Its numerical value characterizes the equilibrium state between the hydrated hydrogen ion and hydroxide ion and is the same for pure water and for solutions of electrolytes (provided that they are sufficiently dilute for other processess to be negligible, and the concentration of water is not significantly decreased, as it is in very concentrated solutions).

The ion-product of water depends to a considerable degree on temperature. At 0°C it is 0.13×10^{-14}, at 25°C it is approximately 1.0×10^{-14} (the most accurate measurements to date yield the value 1.008×10^{-14}) and at a temperature of 100°C it is 48×10^{-14}.

2.3 DISSOCIATION OF OTHER SOLVENTS

The dissociation of water as a solvent is not an isolated phenomenon and many other solvents behave in a similar way, although the process may vary. The behaviour of protic solvents is very similar to that of water, in that the dissociation involves the transfer of a proton from one molecule to another. The extent of any such reaction will depend on the donor–acceptor properties of the solvent and the value of its relative permittivity which directly influences the degree of interaction between the positive and the negative ions. As the relative permittivity increases, the forces of attraction between the negative and the positive ions decrease, and hence the possibility of dissociation increases.

Consider the following solvents:

acetic acid $2CH_3COOH \rightleftharpoons CH_3COOH_2^+ + CH_3COO^-$

ammonia $2NH_3 \rightleftharpoons NH_4^+ + NH_2^-$

ethanol $2C_2H_5OH \rightleftharpoons C_2H_5OH_2^+ + C_2H_5O^-$

In these equations the cation formed during the dissociation is, by convention, written as if it were solvated by one molecule of the solvent, just as it is for water. Again, this is undoubtedly an oversimplification, since the exact number of molecules of a solvent which solvate the proton is not known very precisely.

The dissociation reactions quoted occur only in non-aqueous systems, e.g. in anhydrous liquid ammonia or in anhydrous ethanol. In the presence of small amounts of water, and hence particularly in aqueous solutions, such reactions cannot usually be observed, since the interactions between the molecules are prevented by their stronger interactions with the molecules of water.

The equilibrium of the dissociation reaction is determined, as in the case of water, by the ion product for the particular solvent. The ion products for the dissociation reactions already mentioned are $K_{CH_3COOH} = 3.5 \times 10^{-15}$, $K_{NH_3} = 2 \times 10^{-28}$, and $K_{C_2H_5OH} = 8 \times 10^{-20}$.

The properties and the behaviour of various solvents will be clearer and more obvious when the theory of acid–base reactions is explained.

2.4 THE DEVELOPMENT OF CONCEPTS OF ACIDS AND BASES

Ideas about the nature of acids and bases have greatly developed over the years to account for newly obtained experimental evidence. In the past the antonyms acid and base were defined from the interactions of non-metal and metal oxides with water. The theory of electrolytic dissociation led to an explanation rather nearer to the truth. According to Arrhenius's ionic acid–base theory, an acid was regarded as a substance from which hydrogen ions dissociated, and a base as a substance which in a solution dissociates with the formation of an OH^- ion; a salt is the product of reaction of an acid with a base. Both acids and bases are neutral molecules.

However, it has been found that this theory is not valid for numerous compounds which should be regarded as salts and yet show typical acidic or basic properties. For example, sodium carbonate, which might appear to be a typical salt, has in aqueous solution all the properties of a base, whereas $NaHSO_4$ behaves like an acid. The theory cannot begin to account for the phenomena occurring in other solvents. The phenomena can be largely explained by the Brønsted–Lowry theory of acids and bases, which is very useful for description of the reactions taking place in aqueous solutions or other water-like solvents. The desire to be able to correlate acid–base properties more with the structure and properties of molecules has led to more generalized theories, introduced by Lewis, Usanovich and others.

2.5 BRØNSTED–LOWRY THEORY

On the basis of their observations of the phenomena occurring in aqueous solutions and in protic solvents, Brønsted and Lowry, independently of each other, published in 1923 an acid–base theory which explained a large number of processes taking place in solutions and predicted the properties of numerous substances.

The Brønsted–Lowry theory was a great advance over earlier theories, and particularly over the ionic acid–base theory. It took advantage of new precise experimental data and explained some previously not clearly understood properties of systems; above all it permitted a universal explanation of a number of facts which in the older theories required separate explanations.

According to the Brønsted–Lowry theory, an *acid* is a substance which may transfer a proton to another substance; i.e. it acts as a *proton donor*. *Bases* have the opposite property; they may accept a proton, i.e. they are *proton acceptors*. These definitions permit us to draw the correct conclusion that all acid–base reactions consist of the transfer of a proton from an acid species to a base species. The reactions of proton dissociation and proton association are reversible, so that a species formed as the result of loss of a proton from an acid may accept a proton back. Thus the reversible reaction in which a proton is dissociated from a donor—an acid—gives rise to an acceptor—a base. This may be expressed by the reaction

$$acid \rightleftharpoons base + p$$

where the symbol "p" denotes the proton which is transferred in the acid–base reaction, as opposed to the symbols H^+ or H_3O^+, which denote the hydrated proton which may exist in aqueous solutions only. A system of this kind, consisting of an acid and the base obtained from it by proton dissociation, is called a *conjugate acid–base pair*.

In the ionic theory of acids and bases, the acids and bases were assumed to be neutral molecules, but in the Brønsted–Lowry theory there is no such limitation. An acid or a base may be a neutral molecule or a cation or anion. Naturally the electric charge of a base will always be smaller by one unit than the charge of the conjugate acid. *Molecular acids* that exist in aqueous solutions include, acetic acid CH_3COOH, hydrocyanic acid HCN and water H_2O. *Cationic acids* include the ammonium ion NH_4^+, the hydronium ion H_3O^+, hydrated cations of metals such as aluminium $[Al(H_2O)_6]^{3+}$, zinc $[Zn(H_2O)_6]^{2+}$ and others. Examples of *anionic acids* are the hydrogen sulphate ion HSO_4^- and the hydrogen carbonate ion HCO_3^-. All these acids show a tendency to donate a proton. Proton acceptors include *molecular bases*, such as ammonia NH_3, methylamine CH_3NH_2, water H_2O; *cationic bases* such as the monohydroxoaluminate ion $[Al(H_2O)_5OH]^{2+}$, and *anionic bases* such as acetate CH_3COO^-, cyanide CN^-, hydrogen carbonate HCO_3^- and hydroxide OH^-.

The relationships between the acids and bases forming conjugate pairs can be expressed by means of the following equations:

$$acid \rightleftharpoons p + base$$
$$H_3O^+ \rightleftharpoons p + H_2O$$
$$CH_3COOH \rightleftharpoons p + CH_3COO^-$$
$$NH_4^+ \rightleftharpoons p + NH_3$$
$$H_2CO_3 \rightleftharpoons p + HCO_3^-$$
$$HSO_4^- \rightleftharpoons p + SO_4^{2-}$$
$$[Al(H_2O)_6]^{3+} \rightleftharpoons p + [Al(H_2O)_5OH]^{2+}$$
$$CH_3NH_3^+ \rightleftharpoons p + CH_3NH_2$$
$$H_2O \rightleftharpoons p + OH^-$$

The essential characteristic of each of these reactions is the proton that is transferred. In solution, however, a free proton cannot exist alone because it is charged and very small. Instead, it will associate with a particle or ion that is a proton acceptor, *viz.* a base. Thus, the reactions which actually occur are not reactions of proton dissociation from an acid, but of proton transfer to another base. The accepting base must be a stronger proton acceptor than the conjugate base of the initial acid. The simultaneous reactions of the two conjugate pairs

$$acid\ 1 \rightleftharpoons p + base\ 1$$
$$base\ 2 + p \rightleftharpoons acid\ 2$$

make up the total reaction

$$\text{acid } 1 + \text{base } 2 \rightleftharpoons \text{base } 1 + \text{acid } 2$$

When acid–base reactions occur in solutions, the solvent usually participates. The molecules of water, when it is the solvent, can both accept protons to form the hydronium ion H_3O^+ and lose protons to form the hydroxide ion OH^-. Thus, if a substance has stronger donor properties than water, then a reaction of forming H_3O^+ ions will occur and the substance will behave in solution as an acid. Here are examples of such reactions

$$\text{acid } 1 + \text{base } 2 \rightleftharpoons \text{base } 1 + \text{acid } 2$$

$$H_2S + H_2O \quad \rightleftharpoons HS^- + H_3O^+$$

$$HSO_4^- + H_2O \quad \rightleftharpoons SO_4^{2-} + H_3O^+$$

$$NH_4^+ + H_2O \quad \rightleftharpoons NH_3 + H_3O^+$$

In these reactions water always appears as a base.

However, if water has stronger donor properties than the substance dissolved, hydroxide ions will be formed. Such a substance is behaving as a base:

$$\text{base } 1 + \text{acid } 2 \qquad \rightleftharpoons \text{acid } 1 + \text{base } 2$$

$$[Zn(H_2O)_5OH]^+ + H_2O \rightleftharpoons [Zn(H_2O)_6]^{2+} + OH^-$$

$$CH_3NH_2 + H_2O \qquad \rightleftharpoons CH_3NH_3^+ + OH^-$$

$$CN^- + H_2O \qquad \rightleftharpoons HCN + OH^-$$

Such reactions in which the solvent participates are common in aqueous medium. However, if there is present in an aqueous solution a substance which is a stronger proton acceptor than water or a stronger proton donor than water, then its reaction will predominate. For example, when ammonia is introduced into an aqueous solution of acetic acid, since it is a stronger acceptor (i.e. a stronger base) than water, the predominant reaction will be

$$CH_3COOH + NH_3 \rightleftharpoons CH_3COO^- + NH_4^+$$

Similarly, if a proton donor stronger than water, e.g. the H_3O^+ ion, is introduced into a solution of sulphate ions the equilibrium of the reaction

$$SO_4^{2-} + H_3O^+ \rightleftharpoons HSO_4^- + H_2O$$

will be considerably shifted to the right.

A general rule governing acid–base reactions is the following: under given conditions acids react first with the strongest base, and bases with the strongest acid. The stronger the donor properties of an acid, the more difficult it is for the conjugate base to accept a proton, i.e. the weaker is the base as an acceptor. Conversely, the stronger the acceptor properties of a base the weaker the conjugate acid. For example, if we compare the donor properties of acetic acid and hydrocyanic acid, we find that acetic acid is the stronger donor, so the acetate ion is a weaker acceptor than the cyanide ion, the conjugate base of hydrocyanic acid.

Among the substances that can participate in acid–base reactions are some that can behave both as proton donors and proton acceptors. An example is provided by the molecules of water, whose behaviour depends on whether they are in the presence of substances of donor or acceptor nature. The hydrogen carbonate ion HCO_3^- reacts as a base in the presence of acids and as an acid in the presence of bases. The monohydroxoaluminate ion $[Al(H_2O)_5OH]^{2+}$ behaves similarly.

$$acid \rightleftharpoons p + base$$
$$H_3O^+ \rightleftharpoons p + H_2O$$
$$H_2CO_3 \rightleftharpoons p + HCO_3^-$$
$$[Al(H_2O)_6]^{3+} \rightleftharpoons p + [Al(H_2O)_5OH]^{2+}$$
$$acid \rightleftharpoons p + base$$
$$H_2O \rightleftharpoons p + OH^-$$
$$HCO_3^- \rightleftharpoons p + CO_3^{2-}$$
$$[Al(H_2O)_5OH]^{2+} \rightleftharpoons p + [Al(H_2O)_4(OH)_2]^+$$

Substances which can both donate and accept protons (under the appropriate conditions) are called *amphiprotic substances*. The donor or acceptor character will be manifested when another substance which has the properties of a stronger acceptor or a stronger donor, respectively, is present in the reaction medium.

In discussing the Brønsted–Lowry theory we have so far considered only aqueous solutions. However, the theory may also be applied to systems in other protic solvents, e.g. to solutions in anhydrous acetic acid or liquid ammonia.

2.6 DISSOCIATION CONSTANTS OF CONJUGATE ACIDS AND BASES

It would be useful if the tendency of an acid to lose a proton and the tendency of a base to accept a proton could be determined from the equilibrium state established in a system containing an acid, its conjugate base and protons. However, free protons cannot exist in solutions, so it is not possible to give an absolute measure of the acid–base properties of a conjugate acid–base pair. Nevertheless, if a conjugate acid–base pair exists in solution in a given solvent, the equilibrium reached may be determined experimentally. This state not only gives a measure of the donor–acceptor properties of the conjugate pair but also characterizes the reaction proceeding between the acid (or the base) and the molecules of the solvent. Thus, for acetic acid in aqueous solution, the following reaction occurs between the acid and the water:

$$CH_3COOH + H_2O \rightleftharpoons CH_3COO^- + H_3O^+$$

The position of the equilibrium state of this reaction is a measure of the ability of acetic acid to lose protons to the water molecules, i.e. it describes the relative strength of acetic acid as a donor compared with water.

The equilibrium constant of the reaction of acetic acid with water

$$K_a = \frac{[CH_3COO^-][H_3O^+]}{[CH_3COOH]} \tag{2.4}$$

is called the *acid dissociation constant*. Its numerical value is an indication of the strength of acetic acid, but only in relation to its aqueous solutions.

Similarly, for the reaction of a base with water

$$NH_3 + H_2O \rightleftharpoons NH_4^+ + OH^-$$

the equilibrium constant

$$K_b = \frac{[NH_4^+][OH^-]}{[NH_3]} \tag{2.5}$$

is called the *basic dissociation constant*. It is the measure of the acceptor ability of a base, but also only in comparison with water.

A basic dissociation constant may be written for any base. Thus the acetate ion reacts with water according to the equation

$$CH_3COO^- + H_2O \rightleftharpoons CH_3COOH + OH^-$$

and the equilibrium constant, which is the basic dissociation constant of the acetate ion, has the form:

$$K_b = \frac{[CH_3COOH][OH^-]}{[CH_3COO^-]} \tag{2.6}$$

The product of the acid dissociation constant of an acid and the basic dissociation constant of its conjugate base is equal to the ion product of the solvent, in this case, water.

$$K_a K_b = \frac{[CH_3COO^-][H_3O^+]}{[CH_3COOH]} \frac{[CH_3COOH][OH^-]}{[CH_3COO^-]}$$

$$= [H_3O^+][OH^-] = K_w. \tag{2.7}$$

Thus $K_a K_b$ is independent of the individual properties of the conjugate acid–base pair and is a function only of the solvent dissociation. This expression has some important consequences.

If, in a solvent with known ion product, an acid-dissociation reaction takes place, then from the value of its dissociation constant, the dissociation constant of the conjugate base may be calculated. Conversely, if the dissociation constant of the base is known, we can immediately calculate the dissociation constant of the conjugate acid. That is why present-day handbooks, instead of listing the dissociation constants of acids and independently the dissociation constants of bases, list only the acid dissociation constants. For example, instead of the basic dissociation constant of ammonia, the acid dissociation constant of the ammonium ion is given. It is $K_a = 5.75 \times 10^{-10}$. The basic dissociation constant of ammonia, if needed, may be calculated from

$$K_b = \frac{K_w}{K_a} = \frac{10^{-14}}{5.75 \times 10^{-10}} = 1.74 \times 10^{-5} \qquad (2.8)$$

which of course corresponds to the often quoted value of the dissociation constant of ammonia. Similarly, if the acid dissociation constant of hydrogen cyanide in an aqueous solution is $K_a = 4.4 \times 10^{-10}$, then the basic dissociation constant of the cyanide ion is

$$K_b = \frac{K_w}{K_a} = \frac{10^{-14}}{4.4 \times 10^{-10}} = 2.3 \times 10^{-5} \qquad (2.9)$$

Calculations of this kind can be done much more easily if logarithmic exponents are used. When $pK_a = 9.36$, the value of the basic dissociation constant pK_b is calculated according to the formula

$$pK_b = pK_w - pK_a = 14 - 9.36 = 4.64 \qquad (2.10)$$

The basic assumptions of the Brønsted–Lowry theory have allowed us to ascertain that the stronger the donor properties of an acid the weaker the acceptor properties of the conjugate base.

The values of the dissociation constants are the quantitative expression of that interdependence. Stronger acids have larger values of the dissociation constants, and so the conjugate bases are weaker—their base dissociation constants are small because the product of the two constants is invariable for a given solvent. Thus acetic acid ($K_a = 1.32 \times 10^{-5}$) is a stronger acid than the ammonium ion ($K_a = 5.75 \times 10^{-10}$) which in turn is a little stronger than hydrogen cyanide ($K_a = 4.4 \times 10^{-10}$). The cyanide ion, on the other hand, is the strongest of the corresponding conjugate bases ($K_b = 2.3 \times 10^{-5}$); ammonia is slightly weaker ($K_b = 1.74 \times 10^{-5}$) and the acetate ion is the weakest ($K_b = 5.79 \times 10^{-10}$).

Tables listing the acid dissociation constants of inorganic and organic compounds in aqueous solutions at a temperature of 25°C are given in the Appendix.

The influence of temperature on values of dissociation constants is not very great but it should be taken into consideration when precise calculations are made. For example, pK for acetic acid at 0°C is 4.78, at 20°C 4.76 and at 40°C 4.77. This shows that these changes are not monotonic; a pK can have a minimum value at a certain temperature (e.g. phosphoric acid at -113°C and water at 191°C). For these examples, at temperatures close to 20°C the pK value increases or decreases steadily (cf. Section 2.2). These phenomena result from two opposing factors—the lowering of the relative permittivity of the solvent as temperature increases, along with an increased tendency for protons to be lost from the acid at high temperatures.

As mentioned earlier (see Section 1.5), the properties of acids are sometimes expressed in terms, not of the acid dissociation constant, but of the constant of the proton association reaction, i.e. the *protonation constant*. The protonation constant K_H of an acid HA corresponds to the reaction

$$A^- + H_3O^+ \rightleftharpoons HA + H_2O$$

and it is given by the expression

$$K_H = \frac{[HA]}{[H_3O^+][A^-]} = \frac{1}{K_a} \tag{2.11}$$

For example, the protonation constant of acetic acid is 7.59×10^4, i.e. $\log K_H = 4.88$. For a weak acid, such as the ammonium ion, the protonation constant is higher, namely 1.74×10^9 ($\log K_H = 9.24$). These values will be easier to understand if we remember that the more strongly the proton is bound to the base (i.e. the more stable is the acid which is being formed) the higher is the value of the protonation constant. The protonation constant is thus analogous to the stability constant of a complex and an acid can be regarded as a proton complex of the conjugate base.

In approximate, semiquantitative calculations we usually assume that the ion product of water equals 1×10^{-14}, which corresponds to 14.00 on the logarithmic scale. This is no longer valid when we consider the effect of ionic strength in concentrated solution. Thus at 0.10 ionic strength the activity correction calculated as $\sqrt{I}(1+\sqrt{I})^{-1}$ equals 0.24, at 0.20 ionic strength—0.31. As a consequence the ion product of water in these conditions is $10^{-13.76}$ or $10^{-13.69}$, respectively. Therefore the product of the acid and base dissociation constants of a conjugate acid–base pair does not equal exactly 10^{-14}.

Taking as an example acetic acid and its conjugate base, acetate ion, at zero ionic strength the pK_a and pK_b values are equal to 4.88 and 9.12, which gives as a sum 14.00. For the same species at ionic strength 0.1 the corresponding pK values are 4.76 and 9.00, giving the value of pK_w = 13.76.

2.7 STRONG AND WEAK ACIDS AND BASES

The dissociation constant of an acid dissolved in a given solvent character-izes the tendency for a proton to be lost from that acid in comparison with the tendency of the solvent molecule to lose one. If the various substances which can lose protons in an aqueous solution are classified in decreasing order of their acid dissociation constants, we obtain a series of acids with gradually decreasing strength. At one end of this series there are substances which in aqueous medium lose protons very easily, and at the other end are the substances which are the weakest acids, but which nevertheless do dissociate in aqueous solution to a measurable extent. The strongest donor in an aqueous solution is the H_3O^+ ion and the weakest acid which can still dissociate is the H_2O molecule. Since the acid dissociation constant is in fact the equilibrium constant of the reaction of an acid with the mol-ecule of a solvent, the dissociation reaction of H_3O^+ can be expressed by the equation

$$H_3O^+ + H_2O \rightleftharpoons H_3O^+ + H_2O$$

acid 1 + base 2 $\rightleftharpoons$ acid 2 + base 1

It is the reaction in which a proton is transferred from one water mol-ecule to the other. The expression for the acid dissociation constant of acid 1 for this reaction is

$$K_a = \frac{[H_3O^+][H_2O]}{[H_3O^+]} \tag{2.12}$$

In this formula the constant concentration of water as a reactant is included in the value of the equilibrium constant. The formula implies that the value of the acid dissociation constant of the H_3O^+ ion is numerically equal to the concentration of water

$$K_a = [H_2O] = 55.4 \tag{2.13}$$

From this dissociation constant of the acid H_3O^+ in water, we can cal-culate the dissociation constant of the conjugate base H_2O

$$K_b = \frac{K_w}{K_a} = \frac{1 \times 10^{-14}}{55.4} = 1.80 \times 10^{-16} \qquad (2.14)$$

Since H_3O^+ is the strongest proton donor (i.e. the strongest acid) in an aqueous solution, the conjugate base H_2O is the weakest base. That is why bases with base dissociation constants that are smaller than $1.80 \times \times 10^{-16}$ do not dissociate in water. The dissociation is impossible because protons will be more likely to be attached to the molecules of the solvent, which are in excess.

A similar argument leads to the conclusion that OH^- ions are the strongest proton acceptors in aqueous solution. For the OH^- ion as a base the value of the basic dissociation constant is 55.4. The conjugate acid is here represented by the molecule of water, which is the weakest experimentally confirmed acid, with an acid dissociation constant of 1.80×10^{-16}. Weaker acids will not dissociate in water because water as a solvent will be more likely to react with bases.

If we consider the strengths of acids and bases in terms of the protonation reactions, i.e. if we characterize the reactions by the values of the protonation constants, we can say that the strongest proton acceptor in water, namely the base OH^-, has the largest protonation constant, $5.54 \times \times 10^{15}$. The weakest proton acceptor, the base H_2O, has the protonation constant 1.80×10^{-2}. All the values of the protonation constants of acid–base systems existing in water lie within these limits.

If some of the commonest acid–base reactions are arranged according to the values of the dissociation constants, the following series is obtained.

K_a	Strong acids	Weak bases	K_b
55.4	↑ H_3O^+	H_2O	1.80×10^{-16}
1.25×10^{-2}	HSO_4^-	SO_4^{2-}	$8.7 \ \times 10^{-13}$
$1.3 \ \times 10^{-5}$	CH_3COOH	CH_3COO^-	$7.6 \ \times 10^{-10}$
$4.0 \ \times 10^{-7}$	H_2CO_3	HCO_3^-	$2.5 \ \times 10^{-8}$
$8.0 \ \times 10^{-8}$	H_2S	HS^-	1.25×10^{-7}
5.75×10^{-10}	NH_4^+	NH_3	1.75×10^{-5}
$4.4 \ \times 10^{-10}$	HCN	CN^-	$2.3 \ \times 10^{-5}$
1.25×10^{-13}	HS^-	S^{2-}	$8.0 \ \times 10^{-2}$
1.80×10^{-16}	H_2O	OH^-	↓ 55.4
	Weak acids	Strong bases	

The weakest acid and the weakest base in this series is water; the strongest acid is the H_3O^+ ion, and the strongest base the OH^- ion.

If we introduce into an aqueous solution a molecule (or an ion) with a tendency to lose a proton greater than that of the H_3O^+ ion, then it will react with the molecules of water and the product of the reaction will be the H_3O^+ ion. Because of the large excess of H_2O molecules with acceptor properties, this reaction will proceed quantitatively. Therefore the solutions of the substances in question will exhibit identical acid properties, those of the H_3O^+ ions. Solutions of such strong acids contain no detectable undissociated molecules, but only H_3O^+ ions and the corresponding anions, which do not show basic properties in water and have a weaker ability than water to accept protons. The common strong acids (in dilute aqueous solution) are nitric acid, hydrochloric acid, sulphuric acid and perchloric acid. The term acids when applied to these substances, refers, according to Brønsted–Lowry acid–base theory, to the donor properties of the molecules, but it should be remembered that their acid properties in aqueous solutions result only from the presence of hydrated hydrogen ions, H_3O^+. Because of this, the reactions of strong acids with water are written as irreversible reactions:

$$HNO_3 + H_2O \rightarrow H_3O^+ + NO_3^-$$
$$HCl + H_2O \rightarrow H_3O^+ + Cl^-$$
$$H_2SO_4 + H_2O \rightarrow H_3O^+ + HSO_4^-$$
$$HClO_4 + H_2O \rightarrow H_3O^+ + ClO_4^-$$

In other solvents, the donor properties of the strong acids mentioned above can be differentiated. However, in water their strength as proton

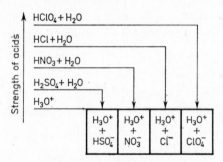

Fig. 2.1. Levelling properties of water in relation to strong acids.

donors reduces to the same level, which corresponds to the H_3O^+ ions. Such behaviour of a solvent is called the *levelling effect*. Water is thus a levelling solvent for the strong acids. This is presented schematically in Fig. 2.1.

In solvents other than water, it is possible to find that substances that are weak acids in aqueous solution are completely dissociated and therefore are "strong acids". Conversely, there are solvents with stronger donor properties than water (e.g. anhydrous acetic acid), in which at least some acids regarded as strong in an aqueous solution are only partially dissociated. In relation to the acids that are strong in aqueous medium, a solvent of this kind will be a differentiating solvent because it shows a *differentiating effect*. When the terms "differentiating solvent" and "levelling solvent" are used, it must be made clear which substance is being considered, because the same solvent shows different properties in relation to different substances.

Because water is an amphiprotic solvent, it shows a levelling action in relation to bases too. A substance which is a stronger proton acceptor than OH^- ions will give an immediate reaction with water molecules to yield the stoichiometric amount of OH^- ions. The compounds commonly regarded as strong bases in water have an ionic structure and one of the ions is hydroxide. For example, if we dissolve sodium hydroxide, NaOH, the crystal lattice is broken down and the ions become hydrated. The Na^+ ion takes no part in the acid–base reactions, its only function being to preserve electroneutrality both in the solid state and in the solution.

There exist bases stronger than the OH^- ion; but they can exist only in the absence of water. The anions of sodium ethoxide, sodium amide and sodium hydride are examples. In water these compounds react vigorously to produce OH^- ions and substances which, though they are potential proton donors (i.e. acids), cannot be regarded as acids in aqueous solutions. Consequently, such reactions (as in the case of strong acids) cannot be written as reversible and are presented as follows:

$$C_2H_5O^- + H_2O \rightarrow C_2H_5OH + OH^-$$
$$NH_2^- + H_2O \rightarrow NH_3 + OH^-$$
$$H^- + H_2O \rightarrow H_2 + OH^-$$

In all these reactions water as a solvent exhibits a levelling action; this is presented schematically in Fig. 2.2. In this diagram, just as in the

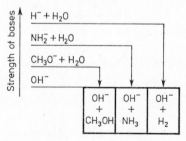

Fig. 2.2. Levelling properties of water in relation to strong bases.

equations, only the anions are shown, since they are the proton acceptors. The role of the cation, just as in the case of sodium hydroxide, is merely to preserve electroneutrality.

The acid–base reaction always takes place between the strongest acid and the strongest base under the relevant conditions, and the products of the reaction are the weakest conjugate base and the weakest conjugate acid, respectively. In other words, the reaction always proceeds towards the formation of the weakest base and the weakest acid of all those possible in a given system. The greater the difference in strength between the acid and base reactants and the base and acid products, the more the reaction proceeds towards the products. For the reaction of strong acids or strong bases with water the equilibrium lies very far to the right. For weak acids the reaction is not complete.

2.8 PROTOLYTIC REACTIONS

The term *protolytic reaction* is a synonym for an acid–base reaction proceeding in a protic solvent. The Brønsted–Lowry theory treats all such reactions uniformly, by introducing the generalized concept of an acid and a base. In the classical ionic acid–base theory there were several terms to describe various types of reactions now known to be essentially the same. Dissociation, hydrolysis and neutralization reactions are all examples of such protolysis reactions.

The concept of *dissociation* was formerly applied only to reactions of molecular acids (or bases) in water. According to the ionic theory, in a reaction of this kind, ions are produced from a neutral molecule, and the water serves only to create an advantageous medium for the reaction. In the case of strong acids and strong bases the dissociation reaction is almost complete, but weak acids and weak bases dissociate to only a small degree. When the ionic theory was applied to the most common weak base, namely ammonia, it was suggested that there existed NH_4OH molecules which could dissociate to yield OH^- ions and NH_4^+ ions.

In the Brønsted–Lowry theory, an important part is played in the dissociation reactions of molecular acids and molecular bases by the molecules of the solvent, which can act as proton acceptors or proton donors

$$HA + H_2O \rightleftharpoons H_3O^+ + A^-$$

$$B + H_2O \rightleftharpoons HB^+ + OH^-$$

According to the ionic theory the *hydrolysis reaction* proceeds between water and the salt of a weak acid and a strong base, the salt of a strong acid and a weak base, or the salt of a weak acid and a weak base. However, the cations of weak bases and the anions of weak acids are, in the light of the Brønsted–Lowry theory, cationic acids and anionic bases. Thus the only difference between hydrolysis reactions and the dissociation reactions of weak acids and weak bases is that the acids and the bases which take part in the hydrolysis reaction are electrically charged. Therefore the solution must also contain ions with an opposite sign, but these take no part in the protolysis reaction itself

$$HB^+ + H_2O \rightleftharpoons H_3O^+ + B$$
$$A^- + H_2O \rightleftharpoons HA + OH^-$$

Thus, there is no longer any need to use the term hydrolysis, because it only refers to a specific case of a protolysis reaction. Indeed, to calculate the pH of solutions of a weak molecular acid, or a cationic acid, we use the same mathematical expressions. The same holds for pH calculations of solutions of a weak molecular base or an ionic base.

According to the classical theories of acids and bases the *neutralization reaction* proceeds between a molecular acid and a molecular base, resulting in the formation of a salt and water. According to the ionic theory, the true product of a neutralization reaction is water, since the ions of the salt exist in the same form before and after the reaction. The formation of the salt can usually be observed only when the proper product of the reaction is removed, e.g. by evaporation.

According to the Brønsted–Lowry theory, the neutralization of a strong acid by a strong base consists of transfer of a proton from the H_3O^+ ion to the OH^- ion. In the reaction of a weak acid and a strong base the proton passes directly from the acid molecule to the OH^- ion. In the neutralization of a strong acid by a weak base proton-transfer takes place between the H_3O^+ ion and the base molecule. In the reaction between a weak acid and a weak base a proton is exchanged directly. The reactions may be represented as follows:

$$H_3O^+ + OH^- \rightleftharpoons 2H_2O$$
$$HA + OH^- \rightleftharpoons H_2O + A^-$$
$$H_3O^+ + B \rightleftharpoons HB^+ + H_2O$$
$$HA + B \rightleftharpoons HB^+ + A^-$$

All these cases involve a system of two conjugate acid–base pairs:

$$acid\ 1 + base\ 2 \rightleftharpoons acid\ 2 + base\ 1$$

The only important difference between a neutralization reaction and a dissociation reaction is that a dissociation reaction always involves neutral molecules of the solvent as a reactant whereas in a neutralization reaction, a molecule of the solvent is a product of the reaction; reactions between a weak acid and a weak base are an exception.

Table 2.1 presents various types of protolysis reactions, i.e. Brønsted–Lowry acid–base reactions, together with their traditional descriptions. Many of the reactions will be discussed in more detail later.

If a reaction proceeds in the solution with practically no participation of H_3O^+ ions or OH^- ions, then the total concentration of H_3O^+ and OH^- in the system does not change. Such a reaction however, is rare, because there are comparatively few ions and molecules that do not undergo protolysis reactions in solution. The main examples are the anions of strong acids, especially Cl^-, Br^-, NO_3^- and ClO_4^-, the alkali metal cations and some alkaline-earth metal cations.

The result of a protolysis reaction between acid 1 and base 2 is the formation of a new acid 2 and a new base 1. However, we need to be able to predict the direction of the reaction, and say why acid 1 reacts with base 2 and not acid 2 with base 1. There was some discussion of this in Section 2.7. The direction depends on the competitive character of the action of two acids of different strengths (acid 1 and acid 2) and two bases of different strengths (base 1 and base 2). A strong base, which has a strong affinity for protons, produces a weak conjugate acid. A strong acid releases protons more readily than a weak one and forms a weak conjugate base. Thus, as already mentioned, the general rule is that an acid–base reaction will always proceed towards formation of a weaker acid and a weaker base.

In the general case of an acid–base reaction the equilibrium constant has the form

$$K = \frac{[\text{acid 2}][\text{base 1}]}{[\text{acid 1}][\text{base 2}]} \tag{2.15}$$

Multiplying the numerator and denominator by $[H_3O^+]$ shows that this constant is equal to the quotient of the acid dissociation constants of acid 1 and acid 2

$$K = \frac{[\text{base 1}][H_3O^+]}{[\text{acid 1}]} \frac{[\text{acid 2}]}{[\text{base 2}][H_3O^+]} = \frac{K_{a1}}{K_{a2}} \tag{2.16}$$

The stronger acid 1 and the weaker acid 2 the greater will be the value of the equilibrium constant K and thus the more complete will be the corresponding overall reaction.

Table 2.1

Types of protolysis reactions

Reaction according to Arrhenius's terminology	Acid 1	+ Base 2	⇌ Acid 2	+ Base 1
Autodissociation of H_2O	H_2O	$+ H_2O$	$\rightleftharpoons H_3O^+$	$+ OH^-$
Dissociation of H_2S in H_2O	H_2S	$+ H_2O$	$\rightleftharpoons H_3O^+$	$+ HS^-$
Dissociation of $HClO_4$ in CH_3OH	$HClO_4$	$+ CH_3OH$	$\rightleftharpoons CH_3OH_2^+$	$+ ClO_4^-$
Dissociation of glycine in H_2O	$^+H_3NCH_2CO_2^-$	$+ H_2O$	$\rightleftharpoons H_3O^+$	$+ H_2NCH_2CO_2^-$
Hydrolysis of CH_3NH_3Cl	$CH_3NH_3^+$	$+ H_2O$	$\rightleftharpoons H_3O^+$	$+ CH_3NH_2$
Hydrolysis of $Zn(ClO_4)_2$	$Zn(H_2O)_4^{2+}$	$+ H_2O$	$\rightleftharpoons H_3O^+$	$+ Zn(H_2O)_3OH^+$
Hydrolysis of Na_2CO_3	H_2O	$+ CO_3^{2-}$	$\rightleftharpoons HCO_3^-$	$+ OH^-$
Neutralization of HCl by NaOH	H_3O^+	$+ OH^-$	$\rightleftharpoons H_2O$	$+ H_2O$
Neutralization of HCN by CH_3NH_2 in H_2O	HCN	$+ CH_3NH_2$	$\rightleftharpoons CH_3NH_3^+$	$+ CN^-$
Neutralization of NH_4Cl by $NaNH_2$ in liquid NH_3	NH_4^+	$+ NH_2^-$	$\rightleftharpoons NH_3$	$+ NH_3$
Dissolution of ZnO in HCl	$2H_3O^+$	$+ ZnO + 3H_2O$	$\rightleftharpoons Zn(H_2O)_4^{2+}$	$+ 2H_2O$
Dissolution of Ag_2CrO_4 in HNO_3	H_3O^+	$+ Ag_2CrO_4$	$\rightleftharpoons HCrO_4^-(+ 2Ag^+)$	$+ H_2O$
Displacement of HCN by $KHSO_4$	HSO_4^-	$+ CN^-$	$\rightleftharpoons HCN$	$+ SO_4^{2-}$
Displacement of NH_3 by NaOH	NH_4^+	$+ OH^-$	$\rightleftharpoons H_2O$	$+ NH_3$

For a protolysis reaction between a weak base B and an H_3O^+ ion, the equilibrium constant

$$K = \frac{[HB^+]}{[B][H_3O^+]} = \frac{1}{K_a} \tag{2.17}$$

is simply the reciprocal of the acid dissociation constant of the acid HB^+ formed in the reaction. Likewise, in the neutralization reaction of a weak acid HA with OH^- ions, the equilibrium constant is

$$K = \frac{[A^-]}{[HA][OH^-]} = \frac{K_a}{K_w} = \frac{1}{K_b} \tag{2.18}$$

i.e. it is the reciprocal of the dissociation constant of the base A^-. In the neutralization of a strong acid with a strong base, i.e. in the reaction of H_3O^+ ions and OH^- ions, the equilibrium constant

$$K = \frac{1}{[OH^-][H_3O^+]} = \frac{1}{K_w} = 10^{14} \tag{2.19}$$

is expressed in terms of the ionic product of water, i.e. by the product of the neutralization reaction. This is the highest value that the equilibrium constant of a protolysis reaction can reach in an aqueous solution. If at least one of the substrates is a weak electrolyte, the value of the equilibrium constant is smaller.

2.9 POLYPROTIC ACIDS AND BASES

So far we have considered only *monoprotic acids* (which can release one proton per molecule), and *monoprotic bases* (which can react with one proton per molecule). There are many substances that can lose several protons per molecule (*polyprotic acids*) and substances that can attach several protons per molecule (*polyprotic bases*). The stepwise dissociations of some typical polyprotic acids are given as examples:

orthophosphoric acid $H_3PO_4 \xrightarrow{-H_3O^+} H_2PO_4^- \xrightarrow{-H_3O^+} HPO_4^{2-}$
$\xrightarrow{-H_3O^+} PO_4^{3-}$

sulphuric acid $\qquad H_2SO_4 \xrightarrow{-H_3O^+} HSO_4^- \xrightarrow{-H_3O^+} SO_4^{2-}$

carbonic acid $\qquad H_2CO_3 \xrightarrow{-H_3O^+} HCO_3^- \xrightarrow{-H_3O^+} CO_3^{2-}$

hydrogen sulphide $\quad H_2S \xrightarrow{-H_3O^+} HS^- \xrightarrow{-H_3O^+} S^{2-}$

hydrated zinc ion $\quad [Zn(H_2O)_4]^{2+} \xrightarrow{-H_3O^+} [Zn(H_2O)_3OH]^+$
$\xrightarrow{-H_3O^+} [Zn(H_2O)_2(OH)_2] \xrightarrow{-H_3O^+}$

$$[Zn(H_2O)(OH)_3]^- \xrightarrow{-H_3O^+} [Zn(OH)_4]^{2-}$$

hydrazonium ion $N_2H_6^{2+} \xrightarrow{-H_3O^+} N_2H_5^+ \xrightarrow{-H_3O^+} N_2H_4$

protonated amino- $[NH_3CH_2COOH]^+ \xrightarrow{-H_3O^+}$

acetic acid $NH_3CH_2COO \xrightarrow{-H_3O^+} NH_2CH_2COO^-$

Since every acid has a corresponding conjugate base, the final result of the dissociation of a polyprotic acid is the formation of a polyprotic base. In the example given above the polyprotic bases are orthophosphate, sulphate, carbonate, sulphide, tetrahydroxozincate, and aminoacetate anions and the hydrazine molecule.

Every reaction of proton removal is characterized by a corresponding acid dissociation constant. For example for orthophosphoric acid there are three acid dissociation constants and three corresponding basic dissociation constants:

$$K_{a1} = \frac{[H_3O^+][H_2PO_4^-]}{[H_3PO_4]} = \frac{K_w}{K_{b3}} = 7.52 \times 10^{-3}$$

$$K_{a2} = \frac{[H_3O^+][HPO_4^{2-}]}{[H_2PO_4^-]} = \frac{K_w}{K_{b2}} = 6.31 \times 10^{-8} \qquad (2.20)$$

$$K_{a3} = \frac{[H_3O^+][PO_4^{3-}]}{[HPO_4^{2-}]} = \frac{K_w}{K_{b1}} = 1.26 \times 10^{-12}$$

The first acid dissociation constant corresponds to removal of the first proton and the first basic dissociation constant to the acceptance of the first proton by a polyprotic base. Thus for orthophosphoric acid the first acid dissociation constant corresponds to the third basic dissociation constant and the third acid dissociation constant corresponds to the first basic dissociation constant.

The values of the successive acid dissociation constants of polyprotic acids decrease. The constant for the second dissociation step is always smaller than the first and the third is smaller than the second. There are several reasons for the large difference between the values of the successive dissociation constants. First of all, the energy which is necessary for the removal of a proton, which is positively charged, increases with the negative charge on the base. The relevant charge on the base is the total ionic charge plus any partial charges resulting from inhomogeneous distribution of the electrons in the molecule or ion of the base. Another, comparatively small, contribution to differentiation of the successive dissociation constants is statistical in nature, and is a function of the probability of removal of one of two protons and the probability of proton attachment to one of two places in a base molecule. Also, it should not be forgotten that the presence in solution of the protons produced in the first step of any dissociation hinders all subsequent steps. Accordingly the main source of hydrogen ions is always the first step of a dissociation.

The equation for dissociation of a polyprotic acid is sometimes expressed as an overall equation, e.g.

$$H_2S \rightleftharpoons 2H^+ + S^{2-} \quad \text{or} \quad H_2S + 2H_2O \rightleftharpoons 2H_3O^+ + S^{2-}$$

which may lead to the erroneous conclusion that in a dissociation reaction the number of H_3O^+ ions produced is twice that of S^{2-} ions. The error is that such an overall equation does not represent the actual reaction, but instead shows the potential dissociation process for a polyprotic acid, i.e. a dissociation in the presence of acceptors sufficiently strong to shift the equilibrium state to the right. Therefore, overall equations should be avoided, because they do not correspond to real processes and are only the result of mathematical transformations. In reality, the first step of the dissociation $H_2S + H_2O \rightleftharpoons H_3O^+ + HS^-$ yields an equal number of H_3O^+ ions and HS^- ions. Of course the HS^- ions dissociate further, but only to a very small extent, so that only a tiny fraction of the total number of HS^- ions will dissociate according to the equation

$$HS^- + H_2O \rightleftharpoons H_3O^+ + S^{2-}$$

Consequently, the number of H_3O^+ ions in the H_2S solution is slightly higher than the number of HS^- ions, and the ratio of S^{2-} concentration to the total H_2S concentration is very small.

It becomes easier to understand the fact that the first dissociation step proceeds to a much greater extent than further dissociation steps if it is remembered that, when a second proton is removed from the parent acid, the negative charge on the conjugate base increases, and the proton removal has to overcome the increased electrostatic interaction, i.e. it requires greater energy.

In the Brønsted–Lowry theory the concept of polyprotic acids and bases is closely related to the concept of *amphiprotic substances*, i.e. *ampholytes*. The cited examples of acids and bases indicate that in intermediate stages of their reactions appear substances being able to attract as well as to donate protons; thus they are ampholytes.

An important group of compounds which can be treated as ampholytes is that of the amino acids. Each amino acid molecule contains the acid group —COOH and the basic group —NH_2. Formally, the protolysis reactions of the simplest amino acid, aminoacetic acid (glycine, NH_2CH_2COOH), may be described by the sequence

$$NH_3^+CH_2COOH \xrightarrow{-H_3O^+} NH_2CH_2COOH \xrightarrow{-H_3O^+} NH_2CH_2COO^-$$

for which the acid dissociation constants are $10^{-2.35}$ and $10^{-9.77}$ respectively. However, the glycine molecule, in spite of its overall neutrality, has charges with opposite signs localized at the two ends of the molecule. Thus, the formula of the neutral glycine is more correctly written as $NH_3^+CH_2COO^-$.

There is ample experimental evidence for this structure, which is called a *dipolar ion* or *zwitterion*. It results from the fact that the carboxyl group —COOH has more strongly acidic properties then the ammonium group —NH_3^+, and therefore, during acidification of an alkaline glycine solution, the —NH_2 group is protonated first, and —COO$^-$ group is protonated only when a larger amount of hydrogen ions has been added. Thus the equation of the reaction should be written as

$$NH_3^+CH_2COOH \xrightarrow{-H_3O^+} NH_3^+CH_2COO^- \xrightarrow{-H_3O^+} NH_2CH_2COO^-$$

and the properties of an aqueous solution of glycine depend on the basic character of the —COO$^-$ group and the acid character of the —NH_3^+ group; hence it is a typical case of an ampholyte or a diprotic acid of the form H_2R^+ (see Section 3.13). This is very important in view of the role of amino acids as the basic elements of the structure of proteins. Many substances which are of great importance in analytical chemistry behave in a similar way, e.g. ethylenediaminetetra-acetic acid, which may be described by a formula for a double dipolar ion

$$^-OOC-CH_2 \diagdown \qquad\qquad\qquad\qquad \diagup CH_2-COO^-$$
$$\qquad\qquad NH^+-CH_2-CH_2-NH^+$$
$$HOOC-CH_2 \diagup \qquad\qquad\qquad\qquad \diagdown CH_2-COOH$$

Thus it is a tetraprotic acid. However, in a very strongly acid medium it can attach two more protons, to form the double positively charged hexaprotic acid. Still, since such strongly acid solutions are not used in ordinary analytical conditions, it is a common practice to treat ethylenediamine-tetra-acetic acid as a tetraprotic acid.

Both for polyprotic acids and for monoprotic acids it is possible to describe the strength of the acid (or rather its stability) by means of the protonation constants. Table 2.2 lists the numerical values of the protonation constants of orthophosphoric acid in comparison with the acid dissociation constants and the basic dissociation constants of that acid.

Table 2.2

Various methods of presenting the equilibrium constants of the protolysis reactions of H_3PO_4

Name	Definition of equilibrium constant	Numerical value	Logarithmic expression
Acidic dissociation constants of H_3PO_4	$K_{a1} = \dfrac{[H^+][H_2PO_4^-]}{[H_3PO_4]}$	7.6×10^{-3}	$pK_{a1} = 2.12$
	$K_{a2} = \dfrac{[H^+][HPO_4^{2-}]}{[H_2PO_4^-]}$	6.3×10^{-8}	$pK_{a2} = 7.20$
	$K_{a3} = \dfrac{[H^+][PO_4^{3-}]}{[HPO_4^{2-}]}$	1.3×10^{-12}	$pK_{a3} = 11.90$
Basic dissociation constants of PO_4^{3-}	$K_{b1} = \dfrac{[OH^-][HPO_4^{2-}]}{[PO_4^{3-}]} = \dfrac{K_w}{K_{a3}}$	7.9×10^{-3}	$pK_{b1} = 2.10$
	$K_{b2} = \dfrac{[OH^-][H_2PO_4^-]}{[HPO_4^{2-}]} = \dfrac{K_w}{K_{a2}}$	1.6×10^{-7}	$pK_{b2} = 6.80$
	$K_{b3} = \dfrac{[OH^-][H_3PO_4]}{[H_2PO_4^-]} = \dfrac{K_w}{K_{a1}}$	1.3×10^{-12}	$pK_{b3} = 11.88$
Protonation constants of PO_4^{3-} (stability constants of proton complexes of PO_4^{3-} ions)	$K_{1H} = \dfrac{[HPO_4^{2-}]}{[H^+][PO_4^{3-}]} = \dfrac{1}{K_{a3}}$	7.9×10^{11}	$\log K_{1H} = 11.90$
	$K_{2H} = \dfrac{[H_2PO_4^-]}{[H^+][HPO_4^{2-}]} = \dfrac{1}{K_{a2}}$	1.6×10^{7}	$\log K_{2H} = 7.20$
	$K_{3H} = \dfrac{[H_3PO_4]}{[H^+][H_2PO_4^-]} = \dfrac{1}{K_{a1}}$	1.3×10^{2}	$\log K_{3H} = 2.12$

2.10 AMPHOTERIC SUBSTANCES

An *amphoteric substance* is one that can behave, according to the conditions, as either an acid or a base. This concept has long been of great importance in theories of acid–base reactions, and is included in the Brønsted–Lowry theory, according to which amphoteric properties are characteristic of substances which, in reactions in a solution, can both accept and lose protons, depending on the conditions; substances of this kind are called *amphiprotic substances* or *ampholytes*.

The most common ampholyte is water

$$H_2O + p \rightleftharpoons H_3O^+ \qquad H_2O - p \rightleftharpoons OH^-$$

The properties of water as an ampholyte are the basis of present theories of acids and bases. Similar properties are shown by many other protic solvents. As mentioned before, ampholytes include, intermediate products of dissociation of polyprotic acids and polyprotic bases.

Of particular interest are the hydrated ions (aquo-ions) of metals. An example is zinc, which may lose four protons from its aquo-ion. Zinc hydroxide $Zn(OH)_2$, which is one of the intermediate products of the stepwise dissociation, is a typical example of an amphoteric compound, according to both classical and Brønsted–Lowry theory. Classically an amphoteric compound can react with both acids and bases; zinc hydroxide, which is sparingly soluble in water, dissolves both in acids and bases.

$$Zn(OH)_2 + 2HCl \rightleftharpoons ZnCl_2 + 2H_2O$$
(according to classical theory)

$$Zn(H_2O)_2(OH)_2 + 2H_3O^+ \rightleftharpoons Zn(H_2O)_4^{2+} + 2H_2O$$
(according to Brønsted–Lowry theory)

$$Zn(OH)_2 + 2NaOH \rightleftharpoons Na_2ZnO_2 + 2H_2O$$
(according to classical theory)

$$Zn(H_2O)_2(OH)_2 + 2OH^- \rightleftharpoons Zn(OH)_4^{2-} + 2H_2O$$
(according to Brønsted–Lowry theory) $\Updownarrow$
$$ZnO_2^{2-} + 2H_2O$$

There are many compounds like zinc hydroxide which are considered to be amphoteric according to both the Brønsted–Lowry theory and the earlier theories; metal oxides and hydroxides form the most important group. In solution, metal oxides can be treated as hydroxides because most of the reactions first involve hydration of the oxide to hydroxide. Amphoteric properties are exhibited by the hydroxides of those elements which, in relation to the oxygen atom of the OH group, are neither very strong nor very weak acceptors of an electron pair. If an atom of an element is much more electronegative than an oxygen atom, it polarizes its electron-pair bond with oxygen, which results in polarization of the oxygen–hydrogen bond, and this will facilitate loss of the proton. An example is hypochlorous acid ClOH

$$Cl\overset{\frown}{-}O\overset{\frown}{-}H + H_2O \rightleftharpoons ClO^- + H_3O^+$$

Thus, if hypochlorous acid is present in a medium containing molecules with strong acceptor properties, the proton dissociates from the HClO molecule.

If the atom bound to the oxygen atom of the OH^- group is much less electronegative than oxygen, the oxygen atom polarizes the electron-pair bond in the opposite direction to that just discussed, and basic dissociation is favoured, particularly in the presence of proton donors, i.e. in an acid medium. The most strongly basic hydroxides are the hydroxides of the alkali metals, especially CsOH.

Many elements exhibit intermediate electronegativity; neither of the two opposite tendencies then predominates. In the presence of strong proton donors the compounds undergo basic dissociation; in the presence of strong acceptors they undergo acid dissociation. Well-known examples of such compounds include zinc hydroxide, which has been discussed already, and also some other hydroxides, such as those of aluminium, chromium(III), arsenic(III), and antimony(III). However, nearly all hydroxides of heavy metals will form anions to some extent. Should they all be regarded as amphoteric? How should we regard such compounds as bismuth hydroxide $Bi(OH)_3$, iron(III) hydroxide $Fe(OH)_3$ and cadmium hydroxide $Cd(OH)_2$? The decisive factor here is the value of the acid dissociation constant of the metal hydroxide.

For example, for the reaction

$$Bi(OH)_3 + H_2O \rightleftharpoons H_3O^+ + H_2BiO_3^-$$

the value of the dissociation constant is

$$K_{a1} = [H_3O^+][H_2BiO_3^-] = 6 \times 10^{-19} \tag{2.21}$$

and it is easy to calculate that the concentration of $H_2BiO_3^-$ anions in a strongly alkaline solution ($[H_3O^+] = 10^{-14}$) is

$$[H_2BiO_3^-] = \frac{6 \times 10^{-19}}{[H_3O^+]} = 6 \times 10^{-5} \, M \tag{2.22}$$

The concentration is thus very small, so in analytical practice, which is usually concerned with 0.1–$0.01 M$ solutions, it can be assumed that bismuth hydroxide does not show amphoteric properties.

However, for the reaction

$$Pb(OH)_2 + H_2O \rightleftharpoons H_3O^+ + HPbO_2^-$$

the acid dissociation constant is $K_a = 1 \times 10^{-15}$, and the solubility of lead hydroxide in a solution with $[H_3O^+] = 10^{-14}$ is equal to the $HPbO_2^-$ concentration, i.e.

$$[HPbO_2^-] = \frac{1 \times 10^{-15}}{10^{-14}} = 0.1 M \tag{2.23}$$

This value is fairly large, so lead hydroxide is usually regarded as an amphoteric compound (and experiment confirms this).

These examples indicate that a decision to regard a hydroxide as an amphoteric substance depends on two criteria. The first is objective—the value of the acid dissociation constant of the hydroxide. The larger it is, the more easily the hydroxide dissolves as the H_3O^+ concentration decreases. The second criterion is more subjective, and depends on the concentrations involved and on the scale of working. In very dilute systems (e.g. 10^{-5}–$10^{-4}M$) more substances can be regarded as amphoteric.

Let us consider, for example, a solution of silver hydroxide, AgOH. Its actual solubility in $0.1M$ sodium hydroxide is approximately $3 \times 10^{-5}M$. Thus, at the $\sim 10^{-5}M$ level, we may regard silver hydroxide as soluble in an alkaline medium, i.e. as being amphoteric. However, if we wanted to precipitate the hydroxide from a concentrated solution of a silver salt (e.g. $0.1M$ AgNO$_3$) and then to dissolve it by raising the pH to 13 ([OH$^-$] = $0.1M$). this would prove impossible because the maximum silver concentration in solution would still be only $3 \times 10^{-5}M$. In this case it would be logical to state that silver hydroxide is not amphoteric. Thus the division into amphoteric and non-amphoteric compounds is not very sharp.

Table 2.3

Amphoteric and non-amphoteric hydroxides of metals

	Reaction of acid dissociation	K_a	pK_a
Usually treated as amphoteric	$As(OH)_3 \rightleftharpoons H_3O^+ + AsO_2^-$	5×10^{-10}	9.3
	$Ga(OH)_3 + H_2O \rightleftharpoons H_3O^+ + H_2GaO_3^-$	2×10^{-14}	13.7
	$Al(OH)_3 \rightleftharpoons H_3O^+ + AlO_2^-$	4×10^{-15}	14.4
	$Sn(OH)_2 + H_2O \rightleftharpoons H_3O^+ + HSnO_2^-$	2.5×10^{-15}	14.6
	$Pb(OH)_2 + H_2O \rightleftharpoons H_3O^+ + HPbO_2^-$	1×10^{-15}	15.0
	$Cr(OH)_3 \rightleftharpoons H_3O^+ + CrO_2^-$	8×10^{-16}	15.1
	$Zn(OH)_2 + H_2O \rightleftharpoons H_3O^+ + HZnO_2^-$	5×10^{-17}	16.3
	$Ni(OH)_2 + H_2O \rightleftharpoons H_3O^+ + HNiO_2^-$	2×10^{-17}	16.7
	$AgOH + H_2O \rightleftharpoons H_3O^+ + AgO^-$	3×10^{-18}	17.5
Usually treated as non-amphoteric	$Fe(OH)_2 + H_2O \rightleftharpoons H_3O^+ + HFeO_2^-$	1.3×10^{-18}	17.9
	$Bi(OH)_3 + H_2O \rightleftharpoons H_3O^+ + H_2BiO_3^-$	6×10^{-19}	18.2
	$Fe(OH)_3 + H_2O \rightleftharpoons H_3O^+ + H_2FeO_3^-$	4×10^{-19}	18.4
	$Co(OH)_2 + H_2O \rightleftharpoons H_3O^+ + HCoO_2^-$	2×10^{-19}	18.7
	$Cu(OH)_2 + H_2O \rightleftharpoons H_3O^+ + HCuO_2^-$	1.3×10^{-19}	18.9
	$Mn(OH)_2 + H_2O \rightleftharpoons H_3O^+ + HMnO_2^-$	4×10^{-20}	19.4
	$Cd(OH)_2 + H_2O \rightleftharpoons H_3O^+ + HCdO_2^-$	8×10^{-21}	20.1

Table 2.3 lists values for the acid dissociation constants of some metal hydroxides, in order of decreasing acid strength. As a working rule, it can be assumed that the limiting value of pK_a for a hydroxide to be amphoteric is 16.

However, this criterion is a considerable simplification because it takes into account only the first acid-dissociation step of a hydroxide. The situation is a little different when there is a two-step dissociation, such as occurs with zinc hydroxide:

$$Zn(H_2O)_2(OH)_2 \rightleftharpoons Zn(H_2O)(OH)_3^- \rightleftharpoons Zn(OH)_4^{2-}$$

We find that despite its comparatively large pK_a value of 16.3, zinc hydroxide easily dissolves in strong bases and is a typical example of an amphoteric compound. By contrast, nickel hydroxide, which has a very similar pK_a value dissolves in alkalis only to a slight extent.

The amphoteric character of hydroxides can be assessed from logarithmic concentration diagrams. Any amphoteric behaviour will result in increased solubility of hydroxides in alkaline solutions, and this will be shown by the right-hand part of the curve in the area corresponding to

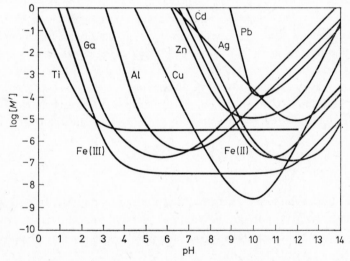

Fig. 2.3. Dependence of the logarithm of the total concentration of metal ions in equilibrium with hydroxide precipitate on the pH of the solution.

high pH values (Fig. 2.3). The larger the values of the logarithm of the relevant concentration, the more amphoteric is the metal hydroxide. The factors which affect the position of the right-hand part of the curve are the value of the concentration minimum, its position on the pH axis, and the slope of the right-hand part of the curve. The value of the minimum depends on the solubility product of the precipitate $M(OH)_n$, and its position on the pH axis is connected both with the solubility product and the basic dissociation constants of the hydroxide (the left-hand part of the curve) and with the acid dissociation constant of the hydroxide (the right-hand part of the curve). The slopes of these segments increase with the number of protons involved in the equilibria. A more detailed quantitative discussion of the reactions and equilibria involved will be given later (Section 3.10).

Amphoteric compounds are weak electrolytes, which follows from the fact that amphoterism expresses two opposing tendencies. If in some particularly favourable conditions a molecule may demonstrate both tendencies, then neither of them will be expressed strongly; the compound will be neither a strong acid (which would prevent the appearance of base properties) nor a strong base (which would prevent reaction as an acid). Also, the ionic product of water would be exceeded if an amphoteric hydroxide simultaneously dissociated to a high degree both as an acid and as a base, and that is clearly impossible.

Strictly speaking amphoteric behaviour is a property only of oxides and hydroxides, but we often speak about acid, amphoteric and basic sulphides of metals. Acid sulphides, analogously to acid oxides, are sulphides which dissolve in basic solutions, producing thioanions. Basic sulphides dissolve in acids, producing metal ions and hydrogen sulphide. Amphoteric sulphides may react in both ways, depending on the reaction medium.

The antimony(III) compounds are an example of such behaviour. The reactions which show the amphoteric properties of antimony(III) oxide are:

$$Sb_2O_3 + 6H_3O^+ \rightleftharpoons 2Sb^{3+} + 9H_2O$$

$$Sb_2O_3 + 2OH^- \rightleftharpoons 2SbO_2^- + H_2O$$

It should be remembered, of course, that both antimony(III) oxide and the ions containing antimony are in reality always hydrated, so other formulae can be written. The reactions for antimony(III) sulphide can be presented similarly:

$$Sb_2S_3 + 6H^+ \rightleftharpoons 2Sb^{3+} + 3H_2S$$

$$Sb_2S_3 + 2SH^- \rightleftharpoons 2SbS_2^- + H_2S$$

These reactions would proceed quite analogously to those for Sb_2O_3 if they took place not in water but in liquid hydrogen sulphide, where the solvated proton could be denoted by the symbol H_3S^+. In aqueous solutions, however, there exist both oxygen compounds (derivatives of water) and sulphur compounds (derivatives of hydrogen sulphide). The presence of various competing ions and molecules obscures and complicates the reaction; however, the amphoteric properties of antimony(III) sulphide can also be explained in this manner. The processes taking place in aqueous solutions are therefore

$$Sb_2S_3 + 6H_3O^+ \rightleftharpoons 2Sb^{3+} + 3H_2S + 6H_2O$$

$$Sb_2S_3 + HS^- + OH^- \rightleftharpoons 2SbS_2^- + H_2O$$

or

$$2Sb_2S_3 + 4OH^- \rightleftharpoons 3SbS_2^- + SbO_2^- + 2H_2O$$

Similar reactions can be written for the arsenic and tin compounds.

The anions which are produced when a base acts on an amphoteric hydroxide can usually be written to represent the ion with differing degrees of solvation. For zinc, for example, we could write either ZnO_2^{2-} or $Zn(OH)_4^{2-}$. These formulae differ only in the number of solvating water molecules. In a solution it is usually impossible to ascertain this number, and the solid-state composition of a substance does not always corresponds to its composition in solution. However, X-ray structural studies of a large number of solid hydrates, e.g. $Na_2SnO_3 \cdot 3H_2O$, $NaAuO_2 \cdot 2H_2O$, $NaSbO_3 \cdot 3H_2O$ and $Na_2PtO_3 \cdot 3H_2O$, have shown that the structures of these compounds are better described by the formulae $Na_2Sn(OH)_6$, $NaAu(OH)_4$, $NaSb(OH)_6$ and $Na_2Pt(OH)_6$. Hence it may be supposed that in solution too, the structures of the anions are better respresented by formulae corresponding to maximum hydration. The formulae would then be analogous to those of the chloride complexes of those metals, i.e. $SnCl_6^{2-}$, $AuCl_4^-$, $SbCl_6^-$ and $PtCl_6^{2-}$.

The stability of a hydroxo complex can be characterized by the value of the *stability constant*. If the acid properties of aluminium hydroxide are represented by means of the equation

$$Al(OH)_3 + H_2O \rightleftharpoons H_3O^+ + H_2AlO_3^-$$

then the equilibrium constant of the reaction is the acid dissociation constant of $Al(OH)_3$

$$K_a = [H_3O^+][H_2AlO_3^-] \tag{2.24}$$

On the other hand, if the reaction is presented in the form

$$Al(OH)_3 + OH^- \rightleftharpoons Al(OH)_4^-$$

then the equilibrium constant is the stepwise stability constant of the $Al(OH)_4^-$ complex

$$K_4 = \frac{[Al(OH)_4^-]}{[OH^-]} \tag{2.25}$$

The reciprocal of the stability constant of the complex (the instability constant) may be regarded as the basic dissociation constant of $Al(OH)_4^-$, i.e. it corresponds to the reaction

$$Al(OH)_4^- \rightleftharpoons H_2AlO_3^- + H_2O \rightleftharpoons Al(OH)_3 + OH^-$$

$$K_b = \frac{1}{K_4} = \frac{[OH^-]}{[H_2AlO_3^-]} \tag{2.26}$$

This implies that division of the dissociation constant K_a by the stability constant K_4 yields the ionic product of water. Thus, small values of K_a, correspond to small values of the stability constant of the hydroxo complex.

We must emphasize once more that the concept of amphoteric substances or ampholytes refers to substances with acid and base properties in the sense of the Brønsted–Lowry theory. The term amphoteric substance should be reserved, in the sense of the classical acid–base theories, for chemical compounds (neutral molecules). We then avoid such unfortunate expressions as "a hydrogencarbonate ion is an amphoteric substance". Moreover, we regard it as not very appropriate to use the concept of amphoteric behaviour in any wider sense; e.g. to describe the tendency of a substance to behave either as an oxidant or a reductant.

2.11 ACIDS AND BASES IN NON-AQUEOUS SOLVENTS

The Brønsted–Lowry theory can also be applied to other amphiprotic solvents. The dissociation range of a solvent depends on its donor and acceptor properties and on its relative permittivity. These factors, which are related to each other, are the main influences on acid and base dissociation in these solvents. If the relative permittivity (dielectric constant) is large, the work needed to separate the ions of an acid, a base or a salt (which is inversely proportional to the relative permittivity) is small and the interactions between the ions are weak, whereas, if the relative permittivity is small the attraction between the ions is strong and the ions tend to form ion pairs.

The dissociation constant of an electrolyte AB which produces the positive ion A^+ and the negative ion B^- is described approximately by the equation

$$pK = a + \frac{b}{\varepsilon} \tag{2.27}$$

where a and b are constants and ε is the relative permittivity.

From the dissociation of the amphiprotic solvent HR we obtain the positive ion H_2R^+, which is composed of a proton and a molecule of the solvent (i.e. the solvated proton) and the negative ion R^-, produced by the removal of a proton from a molecule of the solvent. An excess of H_2R^+ ions (*lyonium ions*) in a solution in relation to R^- ions (*lyate ions*) indicates that the dissolved substance is acidic, whereas an excess of lyate ions indicates that it is basic.

Just as in water the ion H_3O^+ is the strongest proton donor, in any other solvent HR the ion H_2R^+ is the strongest acid existing in the solution, and the ion R^- (analogously to the OH^- ion in water) is the strongest acceptor. If, for example, perchloric acid is dissolved in ethanol, the reaction will proceed in the following way:

$$HClO_4 + C_2H_5OH \rightarrow C_2H_5OH_2^+ + ClO_4^-$$

Since this reaction goes entirely to the right, we can state that $HClO_4$ is a strong acid in ethanol as well as in water. The dissolution in ethanol of an ionic compound such as sodium ethoxide, C_2H_5ONa, produces in the solution a large concentration of $C_2H_5O^-$ ions, i.e. sodium ethoxide is a strong base. However, if we dissolve HCl in ethanol, the proton transfer is no longer complete:

$$HCl + C_2H_5OH \rightleftharpoons C_2H_5OH_2^+ + Cl^-$$

Hydrogen chloride in this solvent is an acid of at most medium strength, only partially dissociated. The equilibrium constant of the reaction, i.e. the dissociation constant of hydrogen chloride will have a practical meaning, since at a given temperature and given total concentration there will exist in the solution a strictly defined concentration of undissociated molecules of HCl. Thus, ethanol is a differentiating solvent towards the acids that are strong in water. In ethanol solutions, just as for aqueous solutions, the dissociation constant is the equilibrium constant of the reaction of the acid with the solvent:

$$K = \frac{[C_2H_5OH_2^+][Cl^-]}{[HCl]} \tag{2.28}$$

The differentiating ability of a solvent towards acids depends on the strength of the acceptor properties of the solvent. The stronger the acceptor properties of the solvent, the more easily will it remove protons from acids, to and thus the less differentiating and more levelling will it be in relation to acids. Therefore, in solvents which are more basic than water, such as liquid ammonia, many more acids will be dissociated completely. The opposite is true for solvents which are more acidic than water. For example, anhydrous acetic acid is a weak acceptor, so there are numerous acids which are completely dissociated in water but only partially dissociated in acetic acid.

The donor and acceptor properties of solvents are not always related in the same way, since they are also connected with the relative permittivity. This in turn may influence in various ways the dissociation of different types of solutes. It is impossible, therefore, to give an exact and unequivocal sequence of solvents in which there is a gradual change in the donor and the acceptor character. However, an approximate sequence is sulphuric acid, trichloroacetic acid, acetic acid, water, ethanol, ammonia, pyridine. The solvents at the beginning of the series are strong donors and those at the end are strong acceptors.

The weaker (less acid) the donor properties of a solvent, the more it will differentiate bases. The stronger a donor it is, the more easily will it give a proton to bases, even not very strong ones, and it will have a levelling effect.

If we want to consider the properties of a substance in various solvents, we cannot say that a given substance has donor, acceptor or amphiprotic properties in an absolute sense, because the properties depend on the reaction medium and on the other reactant.

As an example, we mention the properties of acetone, which in aqueous solutions behaves as a neutral species, whereas in concentrated hydrochloric acid or sulphuric acid medium it behaves as a base which accepts a proton according to the reaction:

$$
\underset{\substack{\| \\ CH_3-C-CH_3}}{O} + H^+ \rightleftharpoons \underset{\substack{\| \\ CH_3-C-CH_3}}{OH^+}
$$

Acetone reacts differently again in dimethylsulphoxide solution. The addition of a substance with strong basic properties, e.g. the methoxide ion, reveals the acidic properties of acetone

$$
\underset{\substack{\| \\ CH_3-C-CH_3}}{O} + CH_3O^- \rightleftharpoons \underset{\substack{| \\ CH_3-C=CH_2}}{O^-} + CH_3OH
$$

Values may be determined for the dissociation constants of strong acids in differentiating solvents. Then, because for two different solvents the ratio of the dissociation constants of acids of the same type (molecular acids, cationic acids or anionic acids) is approximately constant, it is possible to calculate values for the dissociation constants of the strong acids in water. Although these data for aqueous solutions have neither physical nor practical meaning, they are mentioned in textbooks and tables of equilibrium constants in order to convey some idea of the strength of the proton-donor properties of a given molecule. The data given for hydrochloric acid—

$10^{3.7}$, 10^7, 10^6, 10^3 and $10^{2.6}$—indicate how divergent these numbers can be, because of differing measurement conditions and methods of calculation. Nevertheless, these data can be useful, for example for putting acids into a series in order of decreasing proton-donor properties:

$$HClO_4 > HI > HBr > HCl > H_2SO_4 > HNO_3 > H_3O^+$$

It should be remembered, however, that this series reflects the donor character of molecules, but in any given solvent the strongest acid is always the solvated proton (the lyonium ion), and the strongest base is the lyate ion. In the neutralization reaction of strong acids by strong bases the reaction will be between those two ions. Thus, in liquid ammonia the neutralization of a strong acid, e.g. ammonium chloride NH_4Cl, by a strong base, e.g. sodium amide NH_2Na, proceeds according to the equation

$$NH_4^+ + NH_2^- \rightleftharpoons 2NH_3$$

2.12 A SCALE OF ACIDITY AND BASICITY—pH

The characteristic properties of acids and bases depend on the donor or acceptor character of molecules. The quantitative measure of this character, in relation to the solvent, is the value of the dissociation constant. Thus, the dissociation of acids is the source of hydrogen ions in a solution, and the dissociation of bases is the source of hydroxide ions. An analytical chemist is interested in the acidity or basicity of a solution, and this depends on the kinds of acids and bases dissolved and on their concentrations. Solutions of strong acids, which are completely dissociated, are more acidic than solutions of weak acids, which are dissociated to only a small extent.

The best objective measure of the acidity of an aqueous solution is the value of the concentration of hydrated hydrogen ions actually present in the solution. The measure of basicity, on the other hand, is the concentration of hydroxide ions. In aqueous solutions, the two concentrations are related by the ion product of water, $K_w = [H_3O^+][OH^-] = 10^{-14}$, which is constant at a given temperature. Thus a decrease in the concentration of H_3O^+ ions involves an increase in the concentration of OH^- ions. Consequently, the knowledge of one of the concentrations is sufficient to characterize a given system. Usually it is the concentration of hydrogen ions that is calculated to find out whether a solution is acidic or basic. For example, if the concentration of OH^- ions is known to be $0.005M$, it is easy to calculate that

$$[H_3O^+] = \frac{K_w}{[OH^-]} = \frac{10^{-14}}{5 \times 10^{-3}} = 2 \times 10^{-12} M \qquad (2.29)$$

In very pure water the concentration of H_3O^+ ions is equal to the concentration of OH^- ions, and this equality is regarded as a characteristic feature of a neutral solution. From the value of the ion product of water it follows that at a temperature of 25°C in a neutral solution

$$[H_3O^+] = [OH^-] = \sqrt{K_w} = 10^{-7} M \qquad (2.30)$$

At higher temperatures the higher value of K_w will make the concentration of H_3O^+ ions in a neutral solution greater than $10^{-7} M$. A neutral solution is always taken to be one in which the concentrations of H_3O^+ ions and OH^- ions are equal.

If the concentration of H_3O^+ ions exceeds the concentration of OH^- ions, for example as a result of the presence of an acid in the solution, the solution is said to be *acidic*. When the concentration of OH^- ions exceeds the concentration of H_3O^+ ions, the solution is said to be *basic* or *alkaline*.

Because the concentration of H_3O^+ may change by ten or more orders of magnitude, it is advantageous to express it in logarithmic units. In 1909 Sørensen suggested that, because in a potentiometric measurement the potential is proportional to the logarithm of the concentration of H_3O^+ ions, a useful measure of the acidity of a solution would be the negative logarithm of the hydrogen ion concentration; he called this pH

$$pH = -\log[H^+] \qquad (2.31)$$

The symbol H^+ denotes of course the hydrated hydrogen ion. Thus, in a neutral solution at 25°C, pH = 7; for acid solutions the pH is < 7, and in basic solutions > 7.

The fundamental scale of pH in dilute aqueous solutions consists of 14 units from 0 to 14. In a solution with pH = 0, the concentration of H_3O^+ ions equals $10^0 = 1M$. In a solution with pH = 14, the concentration of H_3O^+ ions is $10^{-14} M$, so the concentration of OH^- ions in the solution is $10^0 = 1M$. In principle, there can exist concentrated solutions of strong (completely dissociated) acids where $[H_3O^+] > 1M$, i.e. the pH has a negative value. Similarly, we can imagine pH > 14 in a solution of a concentrated base in which $[OH^-] > 1M$, i.e. $[H_3O^+] < 10^{-14} M$. It should be remembered that the interactions between ions in very concentrated solutions are so strong that they distort this simple picture.

Just as we use pH to denote the negative logarithm of $[H_3O^+]$, we can use the negative logarithm of the concentration of OH^- ions, denoted by

Table 2.4

Properties that decide the acidity and basicity of aqueous solutions

Acidic solution	Basic solution	Remarks
$[H_3O^+] > [OH^-]$	$[OH^-] > [H_3O^+]$	from the definition
$[H_3O^+] > 10^{-7}$	$[H_3O^+] < 10^{-7}$	
$[OH^-] < 10^{-7}$	$[OH^-] > 10^{-7}$	exactly true only at 25°C
pH < 7	pH > 7	
pOH > 7	pOH < 7	

pOH. Table 2.4 lists the criteria by which the acidity and basicity of aqueous solutions can be assessed; these criteria are based on the concentrations of H_3O^+ and OH^-.

Converting $[H_3O^+]$ into pH and vice versa, is a simple calculation.

Example 2.1. Calculate the pH of a solution in which the concentration of hydrogen ions is $5.6 \times 10^{-4}M$.

$$pH = -\log[H_3O^+] = -\log(5.6 \times 10^{-4})$$
$$= -\log 5.6 - \log 10^{-4} = -0.75 + 4 = 3.25$$

Example 2.2 Calculate $[OH^-]$ for a solution of pH 2.20.

$$pOH = -\log K_w - pH = 14 - 2.20 = 11.80$$
$$pOH = -\log[OH^-] = 11.80$$
$$\log[OH^-] = -11.80 = -12.00 + 0.20$$
$$[OH^-] = (\text{antilog } 0.20) \times 10^{-12} = 1.6 \times 10^{-12}M$$

In calculations of this kind, it must be remembered that an increase in pH of one unit corresponds to a tenfold decrease is the concentration of H_3O^+ ions. Conversely, a 1-unit decrease in pH corresponds to a tenfold increase in the concentration of H_3O^+ ions.

2.13 pH AND THE ACTIVITY OF HYDROGEN IONS

The original definition of pH given by Sørensen was based on the *concentration* of hydrogen ions calculated from the stoichiometric concentration of the acid, taking into account the apparent degree of dissociation of a strong acid. However, it was found necessary to replace concentration by the *activity* of hydrogen ions. Accordingly, the current definition of pH is

$$pH = -\log a_{H^+} = -\log[H^+]f_{H^+} \tag{2.32}$$

where f_{H^+} denotes the activity coefficient of the hydrated hydrogen ion and a_{H^+} denotes its activity.

A significant difficulty in the realization of the pH scale based on hydrogen-ion activities is the need to measure or calculate activity coefficients. This cannot be done accurately, either empirically or theoretically, from purely thermodynamic assumptions. Nevertheless, for an electrolyte it is possible to measure the mean activity coefficient, which is the geometrical mean of the activity coefficients of the ions (in the case of an electrolyte which dissociates into two ions). From this principle and from the Debye–Hückel theory, we can calculate ionic activity coefficients which agree fairly well with theory.

However, in analytical practice, such a procedure is not generally used. Instead, pH is defined by means of standard solutions for which the pH value, i.e. the activity of hydrogen ions, is known, sometimes very exactly: pH values may then be determined by comparison with these standard solutions.

The difference between the thermodynamic (activity) value and the concentration value of pH is equal to the value of $-\log f_{H^+}$; that is

$$-\log a_{H^+} = -\log[H^+] - \log f_{H^+} \tag{2.33}$$

Since f_{H^+} is less than one, the thermodynamic pH value is somewhat larger than the concentration value. The difference is not large, since for ionic strengths of 0–0.5, the value of $\log f_{H^+}$ is not less than -0.10.

The acidity scale $pa_H = -\log(c_H f_H)$ might seem correct and well justified thermodynamically, but in practice its determination is not simple. The difficulties follow from the need to estimate an individual activity coefficient for hydrogen ion, which, when accurate measurements are required, cannot be replaced by the mean activity coefficient of an electrolyte dissociating into two ions (1–1). Thus, the scale $p(a_H f_{Cl})$ has been proposed, for which

$$p(a_H f_{Cl}) = -\log(c_H f_H f_{Cl}) \tag{2.34}$$

which has an obvious chemical sense, as it does not require the use of an individual activity coefficient but only of the product of the coefficients for two ions. Empirically, this scale can be established on the basis of measurement of the emf values for cells (Section 2.15) without liquid junction, in which a hydrogen electrode and a silver–silver chloride electrode are immersed in the same solution. This allows determination of values of $p(a_H f_{Cl})$, for specified concentrations of chloride. The various scales of pH are compared in Fig. 2.4, which shows the acidity of an equimolar acetate buffer at 25°C, as a function of concentration. All the scales, pc_H

(concentration), pa_H (activity), and $p(a_H f_{Cl})$, tend to a common value as the concentration decreases.

In most cases encountered in analytical chemistry, calculations are based on the concentration scale of pH. Their accuracy is usually sufficient, and the differences between the activity scale and the concentration scale

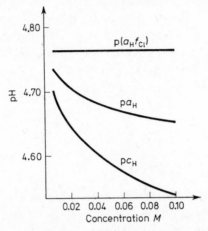

Fig. 2.4. Dependence of the acidity of an acetate buffer on its molar concentration for equimolar concentrations of the acid and conjugate base, expressed by means of different pH functions.

are not large. However, if it is necessary and practically possible to use more exact measurements and calculations, then we should take into account the values of the activity coefficients and use the activity scale pa_H.

In everyday laboratory practice the precision of measurement of pH is about ± 0.02. The imprecision results from variability of the activity coefficients, fluctuations of temperature, the effect of the *liquid-junction potential* in potentiometric measurements, and other experimental errors. It corresponds in concentration terms to about 5% error. Indeed, if the pH of two solutions differs by 0.02, then

$$pH - pH' = 0.02 \quad \text{or} \quad \log[H^+]' - \log[H^+] = 0.02$$

so

$$\log \frac{[H^+]'}{[H^+]} = 0.02$$

and therefore

$$\frac{[H^+]'}{[H^+]} = 1.05$$

Thus in calculations of ionic equilibria, neglect of the terms that have values of less than 5% of the largest term may at most introduce a systematic error within the normal experimental precision range. Also, quoting pH values of solutions with a precision greater than 0.01, except in particularly precise and exact physicochemical measurements, shows a lack of understanding of the theory and technique of measurement.

If the activity coefficients of ions are taken into account, the thermodynamic values of the equilibrium constants (dissociation constants, ionic products, stability constants of complex compounds) differ somewhat from the concentration constants. Similarly, the ion product of water, calculated on the basis of activity, is slightly less than 10^{-14}, so pa_H in neutral solutions is a little less than 7.

2.14 ACID–BASE INDICATORS

The simplest method of pH determination is the use of acid–base indicators. The accuracy of this method is usually about 1 pH unit, and only in exceptional cases it is as good as 0.2 pH unit.

Most *indicators* are organic compounds which change colour under the influence of an acid or base. The chemical structure of their molecules is very varied; the common feature is that they undergo acid or base dissociation in solution and are weak electrolytes. The general equation for the dissociation of an indicator is

$$HIn + H_2O \rightleftharpoons H_3O^+ + In^- \qquad (2.35)$$

where HIn and In$^-$ denote the acid and basic forms of the indicator, i.e. they are a conjugate acid–base pair. For a compound to function as an indicator, the two forms must differ in colour. The dissociation process itself is seldom accompanied by a distinct change of colour, but in a sufficiently complicated organic molecule, the dissociation will result in structural changes caused by shift of electrons. Let us consider the structural changes of two of the most common acid–base indicators, Methyl Orange and phenolphthalein.

Methyl Orange, which is a two-colour indicator, exists in neutral and basic solution as the yellow anion

In an acid medium a proton is attached to give the red product

or

This molecule is neutral overall, but the electrical charges are not evenly distributed.

The equilibrium

$$HIn + H_2O \rightleftharpoons H_3O^+ + In^- \tag{2.36}$$

can be shifted to the left by raising the concentration of H_3O^+ ions or to the right by decreasing their concentration.

A quantitative measure of the equilibrium is the dissociation constant of Methyl Orange

$$K_{In} = \frac{[H_3O^+][In^-]}{[HIn]} = 1.6 \times 10^{-4} \tag{2.37}$$

The colour of the solution depends on the ratio of the concentrations of the two forms of the indicator, i.e.

$$\frac{[In^-]}{[HIn]} = \frac{K_{In}}{[H_3O^+]} \tag{2.38}$$

It is usually assumed that because of the inability of the human eye to distinguish small changes in hue, no colour change in the acid form of the indicator is perceived until the $[In^-]:[HIn]$ ratio is at least $1:10$. Analogously, when the ratio $[In^-]:[HIn]$ exceeds $10:1$ no further colour change is seen, and the colour is that of the basic form. Simple calculations show that the change of concentration ratio from $1:10$ to $10:1$ corresponds to a hundredfold change in the H_3O^+ concentration, i.e. a 2-unit change of pH. The range of colour change actually observed for different indicators, i.e. the range where the "mixed" colour appears, is usually rather smaller; for Methyl Orange it is from pH 3.2 to 4.5, i.e. 1.3 pH units. Thus, the original assumption is not strictly correct, and is based on an over-optimistic estimate of the sensitivity of the eye to colour changes.

When the concentrations of the basic and acid forms are equal,

$$K_{In} = \frac{[H_3O^+][In^-]}{[HIn]} = [H_3O^+] \tag{2.39}$$

That is, the dissociation constant of the indicator is numerically equal to $[H_3O^+]$. The equation (2.37) can be presented in the logarithmic form as

$$pH = pK_{In} + \log \frac{[In^-]}{[HIn]} \tag{2.40}$$

This expression shows that the colour-change interval of a two-colour indicator depends only on the ratio of the concentrations of the two forms and does not depend on the total quantity of indicator.

Phenolphthalein, which is colourless in an acid solution and red in a basic solution, is an example of a one-colour indicator. In concentrated alkaline solutions, a hydroxide ion can be attached to yield another colourless form. The reactions are as follows:

Only the middle one of these forms, characterized by the presence of a quinonoid ring instead of a benzene ring, is intensely coloured. The decolourization of phenolphthalein in strongly basic medium is not utilized for pH determination.

Unlike two-colour indicators, one-colour indicators change colour over a range that does depend on the total concentration of indicator. If the acid form HIn is colourless and the basic form In^- is coloured, then, as the pH is increased, a colour will become visible when a definite concentration of the In^- form is present. This concentration is relatively low in comparison with the total concentration of the indicator C_{In}. The basic equation is, as before

$$pH = pK_{In} + \log \frac{[In^-]}{[HIn]} = pK_{In} + \log \frac{[In^-]}{C_{In} - [In^-]} \tag{2.41}$$

If we assume then, that

$$C_{In} - [In^-] \cong C_{In} \tag{2.42}$$

we may write

$$pH = pK_{In} + \log \frac{[In^-]}{C_{In}} \tag{2.43}$$

or

$$pH = pK_{In} + \log [In^-] - \log C_{In} \tag{2.44}$$

If we are interested in the pH value at which colour begins to appear, then $[In^-]$ is the minimum concentration of the basic form that causes a visible colour in the solution. The equation shows that this pH value depends on the total concentration of the indicator; a tenfold increase of concentration makes the colour appear at a pH lower by a unit. Therefore, if one-colour indicators are used, it is necessary to control the indicator concentration.

Some indicators have several coloured forms, and these allow pH to be measured in different ranges of values. An example is Thymol Blue, which changes colour at pH 1.2–2.8 from red to yellow, and at pH 8.0–9.6 from yellow to blue. To each of these changes can be assigned a separate value of pK_{In}: 1.65 to the first, and 9.20 to the second. The protolytic reactions are expressed by the following equations:

Red   Yellow   Blue

Usually, for measurement of pH, only a small quantity of indicator is needed, because of the intensity of the colours. It should be remembered, however, that the acid–base properties of an indicator may have a significant effect in very dilute solutions of acids and bases. If one drop of a very dilute solution of an acid or a base is added to a few drops of the indicator, then the pH of the mixture of the solution and the indicator will depend mainly on the dissociation constant of the indicator. Thus, the original solution equilibrium will be significantly disturbed, and the indicator equilibrium will not reflect its original value: hence the result of the measurement will be wrong.

Indicators normally yield correct pH values only in dilute aqueous solutions. If a solution contains substances which can influence the activity coefficients of the various forms of the indicator, the measurement is invalidated. For example, large concentrations of neutral salts, causing a salt effect, have a disturbing influence to an extent that depends on the type of acid dissociation of the indicator. The main types of dissociation are:

$$HIn^+ + H_2O \rightleftharpoons H_3O^+ + In$$
$$HIn \; + H_2O \rightleftharpoons H_3O^+ + In^-$$

$$HIn^- + H_2O \rightleftharpoons H_3O^+ + In^{2-}$$

The greatest disturbances occur with the third type. The presence of organic solvents (e.g. ethanol), colloidal substances (e.g. proteins) or even some inorganic ions reacting with one of the forms of the indicator, can also cause loss of accuracy of the pH measurement. It should also be remembered that the effect of temperature on the dissociation constant is considerable, so determinations should not be done at greatly different temperatures from those used for determination of the indicator properties.

For measurement of pH by means of indicators we can use solutions of indicators or indicator papers, i.e. strips of filter paper saturated with a solution of the indicator and then dried. A drop of the solution to be tested is placed on the indicator paper. Indicator papers are usually supplied together with a colour scale. They may contain a single indicator, for estimation of pH over a small range, with an accuracy of perhaps 0.2–0.4 pH unit, or a mixture of different indicators which exhibit several changes of colour over a wide range of pH. For example, over the pH range 1–10, a transition from red, through orange, yellow, and green, to green-blue may be obtained. Such 'universal indicator papers' are particularly useful in the laboratory for quick though approximate estimation of the pH of an unknown solution. Better discrimination can be obtained by use of a paper carrying a band of indicator and several bands of the same indicator plus various buffer mixtures of known pH. The indicator band can then be matched with the buffer band closest to it in colour.

Table 2.5

Properties of some acid–base indicators

Name of indicator	Range of pH of the colour change	Colour of acidic form	Colour of basic form
Methyl Violet	0.1–0.5	yellow	green
	1.0–1.5	green	blue
	2.0–3.0	blue	violet
Thymol Blue	1.2–2.8	red	yellow
	8.0–9.6	yellow	blue
Bromophenol Blue	3.0–4.6	yellow	violet
Methyl Orange	3.2–4.5	red	yellow
Methyl Red	4.4–6.2	red	yellow
Bromothymol Blue	6.0–7.6	yellow	blue
Phenol Red	6.8–8.0	yellow	red
Phenolphthalein	8.2–10.0	colourless	red
Thymolphthalein	9.4–10.6	colourless	blue

Table 2.5 shows the colour changes and pH ranges of some common indicators. More comprehensive lists of indicators can be found in chemical handbooks and in suppliers' catalogues.

In opaque or coloured solutions, pH measurement by means of indicators can be difficult. In such solutions, we can use fluorescent indicators, which act on the same principle as coloured indicators, but the displacement of the protolytic equilibrium causes the appearance, disappearance or change of tint of the fluorescence, which usually occurs under ultraviolet radiation. As an example we can mention 4-ethoxyacridone

When the pH is increased from 1.4 to 3.2, its fluorescence changes from green to blue.

The colour change of acid–base indicators occurs over a range of approximately two units of pH. A sharper colour change can sometimes be obtained by adding to the indicator a substance with a colour complementary to that of one of the indicator forms. The overlapping of these colours makes the solution appear grey. For example, if Bromocresol Blue is mixed with Methyl Red, the resultant colour will be reddish-orange in acid solutions (pH < 3.8) and green at pH > 6.2. If the two indicators are present in the correct proportions (3:2 by weight), an intermediate grey colour appears at pH 5.1, as a result of the summing of red (Methyl Red) and green (the transition colour for Bromocresol Blue).

The sharp colour change for mixed indicators can occur over as little as 0.2 pH units. These indicators are particularly useful in acid–base titrations in which the pH change at the end-point is small (cf. Chapter 4).

2.15 METHODS FOR MEASUREMENT OF pH IN SOLUTIONS

Despite the limited accuracy they offer for pH measurement, acid–base indicators are very useful in the laboratory, as discussed in the previous section.

However, if the pH of a solution is to be determined with an accuracy of better than 1–2 pH units, more objective methods of measurement must be used. In principle, two methods may be used: the colorimetric (spectrophotometric) method and the potentiometric method. Both require some laboratory equipment, skill in using it and knowledge of the physico-chemical basis of the method used.

In this work it is not possible to treat the theory comprehensively; instead the principle of measurement will be discussed briefly in a manner that is descriptive rather than exact.

The *colorimetric method* involves instrumental measurement of an indi-cator equilibrium. To a solution of known pH is added a definite quantity of indicator; the colour obtained is compared with that produced in a test solution, the experimental conditions being the same. According to Beer's law, the absorbance is proportional to the concentration of the absorbing species, so measurement of the absorbance of a solution by means of an absorptiometer or spectrophotometer, at a wavelength at which only one form of the indicator absorbs radiation, allows determination of the con-centration of that particular form of the indicator. Equations (2.40) or (2.44) can then be used to calculate fairly accurately the pH of the test solution.

The second (and much more frequently used) method of pH measure-ment, is *potentiometry*. The instrument used, called a 'pH meter', is a special form of potentiometer designed to measure the potential of a cell that consists of two electrodes immersed in the solution tested. One of these electrodes functions as a reference electrode, i.e. an electrode with a constant and known potential. The second electrode, the 'indicator electrode', has the property that its potential is linearly dependent on pH. Several types of electrode have this property.

For ultimate reference purposes, the *hydrogen electrode* is used as indicator electrode. This consists of a platinum foil covered with platinum black, which is immersed in the solution and has a stream of hydrogen at exactly specified pressure bubbled over it. The redox reaction for the hy-drogen electrode is $2H^+ + 2e \rightleftharpoons H_2$, so the potential of the electrode is expressed by the equation

$$E = E^0 + \frac{RT}{2F} \ln \frac{(a_{H^+})^2}{a_{H_2}} \tag{2.45}$$

At constant pressure of gaseous hydrogen equal to 101325 Pa we obtain

$$E = E^0 + \frac{RT}{F} \ln a_{H^+} = E^0 - \frac{2.303RT}{F} \text{pH} \tag{2.46}$$

In this expression, it is convention to define E^0 as equal to 0, independent of temperature. The equation shows that the potential of the hydrogen electrode depends linearly on pH. At 25°C, this expression becomes

$$E = E^0 - 0.05916 \text{ pH} \tag{2.47}$$

That is, if $[H_3O^+]$ changes by a factor of 10, i.e. the pH changes by 1 unit, the electrode potential changes by 59.16 mV, at 25°C.

Almost all practical pH measurements are made with *glass electrode* (which responds to changes in hydrogen ion activity) invented by Haber and Klemensiewicz in 1909. The electrode is a glass tube with a small bulb blown in the end to form a thin membrane. The glass used has a strictly defined composition. Inside the tube is a solution of constant composition in which a silver wire coated with silver chloride is immersed. Since the internal solution has a constant chloride concentration (e.g. $0.1M$ HCl), the potential of this inner silver–silver chloride electrode is constant. The bulb is immersed in the sample solution.

The surface layer of the glass membrane becomes hydrated in contact with an aqueous solution, and ion-exchange processes occur in it between the glass and the solution. Also, hydrogen ions diffuse within the hydrated (gel) layer of the glass. As a result, a potential develops across the membrane that is linearly dependent on pH. That is, the interfacial potential of the glass electrode has the same dependence on pH as the redox Pt/H_2 electrode does. The glass electrode for pH measurement is an example of a membrane electrode. It should be noted that only a glass electrode designed to have a hydrogen function will behave as described; it is also possible, by changing the glass composition, to prepare glass electrodes with a sodium function or a potassium function. Only a very small number of substances interfere with pH measurements made with glass electrodes.

In the past, two other electrodes were used for pH measurement. The *quinhydrone electrode*, made from the very sparingly soluble equimolar mixture of quinone and hydroquinone that is called quinhydrone, has its potential given by the equation

$$E = E^0 + \frac{RT}{2F} \ln \frac{a_{(quinone)}(a_{H^+})^2}{a_{(hydroquinone)}} \tag{2.48}$$

Because the activities of quinone and hydroquinone are equal, the potential of the electrode depends, within certain limits, only on pH.

The *antimony electrode*, which is still used in some circumstances, consists of a surface-oxidized antimony rod. Its potential is that of the reaction

$$Sb_2O_3 + 6H^+ + 6e^- \rightleftharpoons 2Sb + 3H_2O$$

In this reaction the only reactant which may change its activity is the hydrogen ion, thus the potential of the antimony electrode depends in principle only on pH. Table 2.6 presents a summary of the main characteristics of pH electrodes.

The potential of a pH electrode is determined indirectly from the emf of the cell:

| pH electrode | test solution | reference electrode |

or

| pH electrode | test solution | salt bridge | reference electrode |

In the first case both electrodes are reversible to the ions present in the solution, and the measured emf of the cell without liquid junction is the algebraic sum of the potentials of the two electrodes. Such measurements are rare in practice, because of the difficulty of finding a suitable reference electrode despite the greater accuracy of measurement. In the second case the choice of reference electrode is easy; usually a calomel electrode is used. However, the measured emf includes the liquid-junction potential arising at the interface between the test solution and the salt bridge. This potential, because of its variability in strongly acid and basic solutions, decreases the accuracy of measurement.

Practical potentiometric measurement consists in measuring the cell emf for the test solution (E_x) and for a standard buffer solution with known $pH_{st}(E_{st})$. The pH of the unknown solution can then be calculated from the simple relation

$$pH_x = pH_{st} + \frac{(E_x - E_{st})F}{2.303RT} \tag{2.49}$$

In practice, such calculations are rare. Instead, the buffer solution is used to calibrate the pH meter, which can then be used to give direct pH readings.

For accurate pH determinations, the buffer used for calibration should have a pH close to that of the test solution. When great accuracy is not required, the difference may be a few pH units; in measurements of high accuracy it must not be greater than 1–1.5 pH units.

Potentiometric methods of pH measurement are more accurate than the other methods available. With good instruments, it is possible to obtain an accuracy of 0.02 pH unit, and the best instruments permit pH measurement with an accuracy of 0.001 pH unit. However, in analytical practice, such accuracy is unnecessary; if the measurements are to have a physical meaning, numerous additional factors must be taken into account, e.g. the activity coefficients of the ions and the liquid-junction potential.

Table 2.6

Some characteristic data for pH electrodes

	Hydrogen electrode	Quinhydrone electrode	Antimony electrode	Glass electrode
Range of pH	unlimited	0–8	1–11	0–13
Accuracy of measurements	±0.001	±0.002	±0.1	±0.005
Theoretical value of the potential for pH 0 (vs. standard hydrogen electrode)	0.000	+0.6996	~ +0.15; depends on the preparation of electrode	depends on the internal solution
Time of electrode response (min)	30–60	5	3	1
Ease of operation	difficult	moderate	easy	easy
Electrical resistance	small	small	small	large
Interfering substances	strong oxidants, reductants, substances which poison platinum black, ions of noble metals	large concentrations of salts, proteins, some amines, oxidants and reductants	oxidants, substances complexing antimony (e.g. tartrates)	some colloids, large concentrations of sodium salts, hydrogen fluoride

The value of pH measured potentiometrically corresponds to the activity of hydrogen ions and not to their concentration. Sometimes, if we want to emphasize this, we use the symbol p_aH.

For example, the potentiometric measurement of pH in a $0.02M$ solution of $(NH_4)_2SO_4$ gives pH $= 5.30$; that is $p_aH = 5.30$. In order to obtain the hydrogen ion concentration, we must calculate, on the basis of the ionic strength of the solution, the activity coefficient of the hydrogen ion. In the solution in question the ionic strength is

$$I = \tfrac{1}{2} \sum C_i z_i^2 = \tfrac{1}{2}(0.04 \times 1 + 0.02 \times 4) = 0.06$$

Thus $f_{H^+} = 0.85$, and $\log f_{H^+} = -0.07$. Since

$$C_{H^+} = \frac{a_{H^+}}{f_{H^+}}$$

we have

$$\log C_{H^+} = \log a_{H^+} - \log f_{H^+}$$

whence

$$-\log C_{H^+} = p_aH + \log f_{H^+} = 5.30 - 0.07 = 5.23$$

Since even in solutions with ionic strength 0.5, the activity coefficient of the hydrogen ion is not less than 0.8 ($\log f_{H^+} > -0.10$), the difference between pH and p_aH does not exceed 0.1 unit. Therefore, in most calculations used in analytical chemistry we may neglect the value of the activity coefficient and use the value of pH whether it refers to the concentration or the activity of the hydrogen ion.

2.16 ACIDITY SCALES IN OTHER SOLVENTS

One of the merits of the Brønsted–Lowry acid–base theory is the possibility of uniform treatment of different acid–base systems in a variety of solvents. In order to develop a suitable treatment, we must first consider the factors that influence the dissociation process, and their connection with solvent properties, the most important of which are the donor–acceptor properties and the relative permittivity. A clear picture of the behaviour of different solvents is complicated by the fact that these two properties change in different ways from one solvent to another, and the effect observed is a resultant of the two. Moreover, some solvents may influence acid–base systems in an individual specific fashion.

More and more solvents are finding application in chemistry today. In principle, every substance can be used as a solvent at any temperature between its melting point and its boiling point. Substances which are gaseous at room temperature, e.g. ammonia, can be used as solvents at low enough temperatures, and solid substances, e.g. salts, when molten, i.e. at temperatures up to 1000°C. In all these systems acid–base reactions can proceed.

In the classification of solvents, the most general division is between molecular solvents and ionic solvents. *Molecular solvents* are characterized by negligible conductivity when pure; the molecules of such solvents are often associated. They are generally used at room temperature, but some need low temperatures (e.g. NH_3, SO_2), and some can be used at higher temperatures (e.g. acetamide, m.p. 82°C). Some compounds which are traditionally thought of as salts but are undissociated when pure are really molecular species, e.g. $SbCl_3$ and $HgCl_2$.

Ionic solvents are characterized by good electrical conductivity, resulting from the mobility of the ions of which they consist. They are mostly salts with high melting points (e.g. NaCl, 808°C); hence their investigation and use is possible only at high temperature. To lower the temperature needed, eutectic mixtures of salts are sometimes used. A mixture of the nitrates of potassium, lithium and sodium in 53:30:17 mole ratio melts at 120°C, whereas lithium nitrate, which has the lowest melting point of the nitrates mentioned, has a melting point of 254°C.

Sometimes liquid metals are regarded as a separate class of solvents; they are not yet well studied.

Acid–base processes can occur in both molecular solvents and ionic solvents. However, since the Brønsted–Lowry theory refers to protic systems, which mainly occur in molecular solvents, we shall consider these solvents in greatest detail.

With reference to donor–acceptor properties, solvents may be inert or active. *Inert solvents* (i.e. *aprotic solvents*) do not react with acids or bases dissolved in them and thus do not show acid–base properties themselves. This group of solvents includes aliphatic and aromatic hydrocarbons, chloroform, carbon tetrachloride and others. *Active solvents* are divided into *basic solvents*, which are proton acceptors and can react with acids (such as pyridine, ethers, ketones and some amides, e.g. N, N-dimethylformamide) and *amphiprotic solvents* (such as water, organic acids, alcohols, ammonia and many others). It is interesting to note that no solvents are known that exhibit only acidic and no basic properties.

Amphiprotic solvents have greatly varied proton–donor and proton–acceptor properties. Therefore, for analytical practice, it is convenient to divide them into three groups. The first consists of *solvents with water-*

like acid–base properties; it includes water and aliphatic alcohols. The second group consists of *protogenic solvents*; in comparison with water these are stronger acids but weaker bases. These properties are found in organic acids, e.g. acetic acid or formic acid, and inorganic acids, e.g. sulphuric acid. The third group, *protophilic solvents*, consists of solvents which are stronger bases but weaker acids than water. Examples are ammonia and ethylenediamine.

Let us now consider the behaviour of acids of different strengths in a solvent (HR) which shows basic properties. There are two fundamental cases.

The first refers to acids (denoted by HA_I) which are stronger donors than the solvated proton (H_2R^+), so the reaction

$$HA_I + HR \rightarrow RH_2^+A_I^- \rightarrow H_2R^+ + A_I^- \tag{2.50}$$

which proceeds through an intermediate stage of solvolysis, results similarly to the case for water, in the formation of a solvated proton and an anion A_I^-. If the electrostatic attraction between the ions is strong, i.e. the solvent has a low relative permittivity, the reaction can proceed only up to the intermediate stage. We then speak of ionization leading to the formation of an ion pair $RH_2^+A_I^-$.

The second case concerns acids (denoted by HA_{II}), that are weaker donors than the solvated proton. The conjugate base A_{II}^-, is therefore a weaker proton acceptor than the solvent molecule. The reaction with the solvent then occurs to a small degree and the equilibrium

$$HA_{II} + HR \rightleftharpoons H_2R^+ + A_{II}^- \tag{2.51}$$

usually lies to the left.

To compare the strengths of different acids we can consider the equilibrium constant of their reactions with a reference base B. To do this, for the reaction

$$HA + B \rightleftharpoons BHA \tag{2.52}$$

we write the reciprocal of the equilibrium constant, which is thus the dissociation constant of the BHA ion pair

$$K = \frac{[B][HA]}{[BHA]} \tag{2.53}$$

We can then define a logarithmic (Fig. 2.5a), acidity scale, with zero corresponding to a $1M$ strong base solution ($\log 1 = 0$) and the different pK values for the equilibria of various acids with that base placed in order of magnitude. If reaction (2.52) proceeds from left to right the pK values are positive, but if an acid (we will call it HA_{III}) is too weak to react with base B, reaction (2.52) will scarcely occur, and the pK value would be negative, thus below zero on the acidity scale.

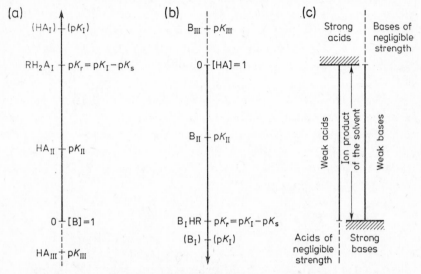

Fig. 2.5. Acidity scale limited by the basic properties of the solvent (a); basicity scale limited by the acidic properties of the solvent (b); scale limited by the acidic and basic properies of the solvent (c).

The acids that we have defined as strong (HA_I) react completely with base B, and are high on the acidity scale. However, it must be remembered that if the reaction between B and HA_I takes place in solvent HR, then reaction (2.50) occurs, and then we have a reaction between the base B and RH_2A_I or H_2R^+ (after dissociation)

$$RH_2A_I + B \rightleftharpoons BHA_I + HR$$

or

$$H_2R^+ + B \rightleftharpoons BH^+ + HR \qquad (2.54)$$

These reactions proceed in the same way for all strong acids, so the position of the point pK_r corresponding to the acid RH_2A or H_2R^+ depends, for a given reference base, on the properties of the solvent; since pK_r is also the pK of reaction (2.54), the difference between pK_I and pK_r corresponds to the value of pK_s for reaction (2.50), i.e. the solvolysis of acid HA_I by a solvent.

For amphiprotic solvents, such as water or ammonia, we generally use as the reference base the lyate ion of the solvent, i.e. OH^- or NH_2^- respectively. Then the value of K_r corresponds to the ion product of water, which on the pK scale is equal to 14, or to the ion product of ammonia, equal to 32 (at $-60°C$). The limitation of the acidity scale by the basic properties of the solvent causes the levelling effect of the solvent towards strong acids. The values of the equilibrium constants K correspond in these cases to the values of the dissociation constants of the conjugate bases $K_b = K_w/K_a$. In the vicinity of $pK = 0$ are the acid dissociation constants with values close to the ion product of the solvent, and in the vicinity of pK_r are the dissociation constants with values close to one.

In a similar way, we can consider the strengths of different bases in an acidic solvent and plot them on a basicity scale (Fig. 2.5b). The zero point of the scale is for a $1M$ solution of a strong acid taken as the reference acid ($[HA] = 1$). At negative pK values we have the bases (B_{III}) which are too weak to react with acid HA. Then weak bases (B_{II}) which react according to the equation

$$B_{II} + HA \rightleftharpoons B_{II}HA, \qquad K_{II} = \frac{[B_{II}][HA]}{[B_{II}HA]} \qquad (2.55)$$

will have small pK_{II} values and strong bases (B_I) will have very high pK_I values for the equilibrium with acid HA. Again in reality the solvent HR reacts with bases according to the equation

$$B_I + HR \rightarrow B_IHR \rightarrow B_IH^+ + R^- \qquad (2.56)$$

and in this reaction, just as in reaction (2.50), we may obtain the solvolysis product B_IHR or this may dissociate to form B_IH^+ and R^- ions. This depends, as in the case of acid dissociation, on the relative permittivity of the solvent. Thus, for strong bases in a solvent with acid properties the reaction is

$$B_IHR + HA \rightleftharpoons B_IHA + HR$$

or

$$R^- + HA \rightarrow A^- + HR \qquad (2.57)$$

irrespective of the nature of the strong base and dependent only on the reference acid HA and on the acid properties of the solvent. If in an amphiprotic solvent, i.e. in a solvent which has both acid and base properties, we use as the reference acid the lyonium ion (in water the H_3O^+ ion and in ammonia the NH_4^+ ion) then the value of the equilibrium constants of reaction (2.57) corresponds to the ion product of the solvent. The value of the equilibrium constant corresponds to the value of the acid-dissociation constant of the conjugate acid; the weaker the base, the stronger the conjugate acid and the higher the value of pK.

If the solvent is only a donor or only an acceptor of protons then the scale is bounded only on one side, as shown in Fig. 2.5a, b. However, a solvent which has both these properties will have a scale bounded on both sides (Fig. 2.5c); this diagram is applicable to an amphiprotic solvent. It should be noted that when we were considering solvents with only basic or only acid properties we gave as examples water or ammonia. This was of course an oversimplification, in which we intentionally omitted to mention the opposite property of the solvent.

The acidity (or basicity) scales for a given solvent may be considered to be pH scales, if we define pH as the negative logarithm of the concentration of the solvated proton. By starting from the values of the dissociation constant of acids and assuming that the concentrations of an acid and its conjugate base are equal, we can obtain a scale on which the values of pK correspond to the values of pH for a given solvent. For an aprotic solvent this scale is not bounded on either side; when donor or acceptor properties appear, the scale is bounded on one side; for amphiprotic solvents the scale is bounded on both sides and its length corresponds to the negative logarithm of the ion product of the solvent.

We have already mentioned several times the dependence of the dissociation of acids and bases on the relative permittivity of the solvent. Solvents may be classified, according to the value of the relative permittivity as dissociating ($\varepsilon > 40$), intermediate ($15 < \varepsilon < 40$) and weakly dissociating ($\varepsilon < 15$). The first group contains solvents in which, analogously to water, partial or sometimes complete dissociation occurs. This group includes water, formic acid, sulphuric acid and amides. The second, intermediate group consists of methanol, ethanol, ammonia, acetonitrile, N, N-dimethylformamide, nitrobenzene and other solvents. The third group, in which ionic dissociation can virtually be neglected, includes hydrocarbons and halogenated hydrocarbons, dioxan, acetic acid, pyridine and other solvents. In these solvents reactions of an acid with a base give an undissociated product

$$HA + B \rightleftharpoons HB^+A^- \tag{2.58}$$

Such a product is formed, for example, in the reaction of pyridine with perchloric acid in the form of the ion pair $C_5H_5NH^+ClO_4^-$. This reaction takes place both in an inert solvent, and also in pyridine itself as a solvent.

For systems in which ionic dissociation does not take place acidity or basicity scales can be drawn, but because of ion pair formation the individual properties of the acids or the bases play a much more important role.

The scales retain their validity only for systems which have one reagent (e.g. an acid) in common. When the nature of the reagent is changed, the scale shifts, but the relative position of the second partner often remains constant.

The acid dissociation constant K_a in a given solvent is a function of two constants: the solvolysis constant

$$K_s = \frac{[HA]}{[RH_2^+AR^-]}$$

for the reaction $HA + 2HR \rightleftharpoons RH_2^+AR^-$ (2.59)

and the ionic dissociation constant

$$K_i = \frac{[RH_2^+][AR^-]}{[RH_2^+AR^-]}$$

for the reaction $RH_2^+AR^- \rightleftharpoons RH_2^+ + AR^-$ (2.60)

The total equilibrium constant characterizes the equilibrium between the ions produced in the reaction and the total concentration of solvated and unsolvated forms of the acid, i.e. it is determined by the equation

$$K_a = \frac{[RH_2^+][AR^-]}{[HA] + [RH_2^+AR^-]} = \frac{K_i}{1 + K_s}$$ (2.61)

If as a result of a high relative permittivity there is complete dissociation of the ion pair formed, then $K_s \ll 1$, so $K_a = K_i$. In solvents with a small relative permittivity $K_s \gg 1$, so $K_a = K_i/K_s$.

The effect of changes in the relative permittivity is to change the force of electrostatic attraction between ions, so the influence of ε depends on the kind of dissociation reaction.

If the reaction proceeds according to

$$BH^+ + HR \rightleftharpoons H_2R^+ + B$$ (2.62)

or

$$A^- + HR \rightleftharpoons HA + R^-$$ (2.63)

(where, for example, the NH_4^+ ion might be the acid or acetate ion the base) the values of the dissociation constant do not depend on the relative permittivity of the solvent, because in the dissociation process there is no separation of ions with opposite signs.

If a neutral molecule of an acid (e.g. acetic acid) dissociates according to the equation

$$HA + HR \rightleftharpoons H_2R^+ + A^- \tag{2.64}$$

or a neutral molecule of a base (e.g. pyridine) dissociates according to the equation

$$B + HR \rightleftharpoons BH^+ + R^- \tag{2.65}$$

then, because of the separation of charges, the dissociation constant does depend on the values of the relative permittivity of the solvent. An approximate expression for this dependence (already mentioned in Section 2.11) is

$$pK = a + \frac{b}{\varepsilon} \tag{2.66}$$

i.e. pK is a linear function of the reciprocal of the relative permittivity.

Similarly, consideration of systems resulting in higher electrical charges, e.g. the dissociation of an anionic acid

$$HCO_3^- + H_2O \rightleftharpoons H_3O^+ + CO_3^{2-} \tag{2.67}$$

or a cationic base

$$\tag{2.68}$$

$$NH_2CH_2CH_2NH_3^+ + H_2O \rightleftharpoons NH_3^+CH_2CH_2NH_3^+ + OH^-$$

shows that the electrostatic interactions are doubled

$$pK = a + \frac{2b}{\varepsilon} \tag{2.69}$$

These expressions are illustrated in Fig. 2.6.

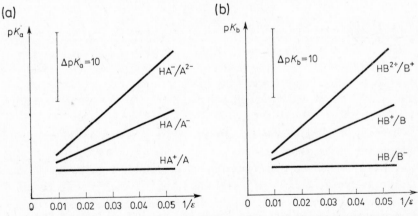

Fig. 2.6. Effect of the relative permittivity on the acid dissociation constants (a) and on the basic dissociation constant (b) for various acid–base systems.

If we take into account the joint influence of the acid–base properties and of the relative permittivity, we can construct pH scales relating to different solvents (Fig. 2.7). For each solvent a pH scale is made, on which the pK values of individual acid–base systems are marked. These scales are shifted with respect to each other along the x-axis according to the value of the reciprocal of the relative permittivity of the solvent, this value being marked on the axis. The positions of different scales along the y-axis are determined on the assumption that the dissociation constants of acids of the type BH^+ or bases of type A^- are independent of the relative permittivity of the solvent. Such a diagram (Tab. 2.7) is only approximately valid.

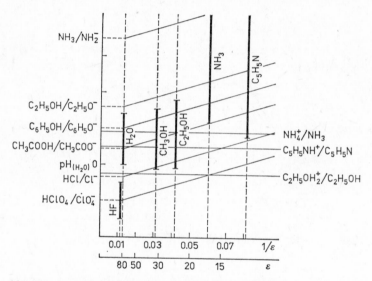

Fig. 2.7. Approximate acidity scales in various solvents.

The approximation arises from our neglect of the specific interaction of protons with the molecules of the solvent; in many cases this interaction can greatly change the expressions given above. These were derived from consideration of only the 'classical' electrostatic interactions.

If we now consider also the Brønsted–Lowry theory, i.e. if we regard the proton as the particle which is responsible for acidity, we can readily conclude that the more strongly the solvent molecule binds the proton, the weaker are its acid properties in the solvent. Thus, when the proton is strongly solvated the acidity will be smaller, although both the concentration and the activity of the solvated proton are comparable or even equal. Thus, to be able to compare in an absolute way the acidity of solutions

Table 2.7

Characteristics of some amphiprotic solvents

Solvent HR	Cation H_2R^+	Anion R^-	Ion product pK_s, 25°C	Relative electric permittivity
Acetamide	$CH_3CONH_3^+$	CH_3CONH^-	14.6_d^a	59.2^a
Acetic acid	$CH_3COOH_2^+$	CH_3COO^-	15.2	6.2
Acetonitrile	CH_3CNH^+	CH_2CN^{-e}	19.5	36.2
Ammonia	NH_4^+	NH_2^-	32^b	22.4^d
Dimethylsulphoxide	$C_2H_6SOH^+$	$C_2H_5SO^-$	33.3	48.9
Ethanol	$C_2H_5OH_2^+$	$C_2H_5O^-$	18.9	24.3
Hydrazine	$N_2H_5^+$	$N_2H_3^-$	13	52.9
Hydrofluoric acid	H_2F^+	F^-	10.7^c	83.6^c
Methanol	$CH_3OH_2^+$	CH_3O^-	16.7	32.6
Sulphuric acid	$H_3SO_4^+$	HSO_4^-	2.9	110
Water	H_3O^+	OH^-	14.0	81.0

[a] At 100°C; [b] at −60°C; [c] at 0°C; [d] at −33°C; [e] the CH_2CN^- ion is unstable, so it is often suggested that acetonitrile is a proton-acceptor solvent and not an amphoteric one.

in different solvents, we need a method of measurement which can estimate the strength of bonding between the proton and the solvent. However, our present knowledge does not provide a direct method of this kind, since it would require measurement of the absolute activity of a single ion in each of the solvents considered.

It might appear that the measurement of proton activity could be made in any solvent by use of a hydrogen electrode (see Section 2.15). However, this theoretically sound idea fails because every potentiometric measurement needs two electrodes. Here, the first electrode is the hydrogen electrode, and we wish to measure its potential in various solvents. However, we need a second reference electrode, which must have a constant potential, whatever the solvent. It is virtually impossible to find such an electrode.

For example, if we used a platinum electrode in the presence of ferric and ferrous ions $[Pt/(Fe^{3+}, Fe^{2+})]$, as the reference electrode, its potential would depend on the manner of solvation, in a given solvent, of the ions which control the potential of the electrode. A similar situation arises in measurements in aqueous solutions, where the presence of complexing agents, e.g. F^-, CN^-, that complex the two ions to different degrees, causes the potential of the electrode to change even though the analytical concentrations of Fe(III) and Fe(II) are constant.

To overcome this difficulty it has been suggested that a suitable reference electrode is one in which a platinum foil is placed in a solution of two species of different oxidation state, for which it can be assumed that the relative solvation is independent of the solvent.

It is not easy to find such species, and the choice is always arbitrary; however, fairly good results have been obtained by using, as the reduced form, a complex of iron(II) with cyclopentadiene anions, called ferrocene, and as the oxidized form, the ferricinium ion, containing iron(III).

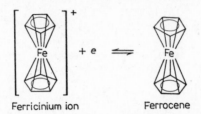

Ferricinium ion   Ferrocene

Both species belong to a group of complexes called *sandwich complexes*, in which the central ion is completely protected from the action of the solvent by ligands, and so it is presumably reasonably true that the relative solvation of the two substances is independent of the solvent.

Thus, if we make potentiometric measurements of proton activity by measuring the emf of the cell $Pt/H_2|H^+$, ferrocene, ferricinium ion|Pt in various solvents, we should obtain values for the actual proton activity, which may be an absolute measure of acidity.

Table 2.8

Values of the transfer activity coefficient of a proton $\gamma_{t(H^+)}$ for different solvents, in relation to water

Solvent	$\log \gamma_{t(H^+)}$
Acetamide	-4.1
Acetonitrile	$+6.1$
Ammonia	-18.0
Dimethylformamide	-5.0
Dimethylsulphoxide	$+5.8$
Ethanol	-1.5
Hydrazine	-13.2
Acetic acid	$+7.0$
Sulphuric acid	$+10.0$
Trifluoroacetic acid	$+16.3$
Methanol	-1.8
Pyridine	-10.0
Water	0

Such measurements made for two different solvents will permit us to determine the relative activity of a proton in one solvent in relation to that in the other. If we choose a certain solvent as a standard system, we may determine, in relation to the proton activity in that solvent, the activities of the proton in other solvents. Water seems to be an obvious reference solvent; it is possible to corelate the acidity scales in any protic solvent with the acidity scale in water.

Thus, if we denote by $(a_{H^+})_s$ the activity of the solvated proton in solvent S, and by $(a_{H^+})_w$ the activity of the hydrated proton (in water, of course), then

$$(a_{H^+})_w = (a_{H^+})_s \gamma_{t(H^+)} \tag{2.70}$$

where $\gamma_{t(H^+)}$ is called the *activity coefficient* for transfer of the H^+ ion from solvent S into water (at infinite dilution). Numerical values of $\log \gamma_{t(H^+)}$ are given in Table 2.8 for several solvents.

We may then write

$$pH_w = pH_s - \log \gamma_{t(H^+)} \tag{2.71}$$

which allows comparison of the two pH scales if the value of $\gamma_{t(H^+)}$ is known. Thus, experimental determination of this value allows us to obtain a nearly perfect system for correlation of various solvents, as shown in Fig. 2.8.

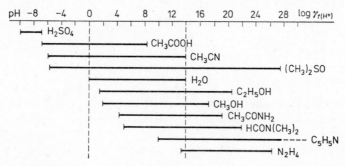

Fig. 2.8. Acidity scales in various solvents, based on the values of transfer activity coefficients $\gamma_{t(H^+)}$.

A system of this kind is said to be 'nearly perfect' because experimental determination of transfer coefficients gives somewhat divergent values, depending on the original assumptions made.

We have assumed here that the reduced and oxidized forms of ferrocene are solvated to the same extent in all the solvents tested. This is not necessarily quite correct; nevertheless, the establishment of the principle should lead to improvements in the practical implementation.

PROBLEMS

1. Calculate the basic dissociation constant of the conjugate base of the following acids:

    a. hydroxylammonium ion, $K_a = 10^{-6.1}$;

    b. hydrogen arsenate ion, $K_a = 10^{-11.4}$;

    c. selenious acid, $K_a = 10^{-2.6}$;

    d. hydrogen selenite ion, $K_a = 10^{-8.3}$;

    e. hydrazoic acid, $K_a = 2 \times 10^{-5}$;

    f. hydrocyanic acid, $K_a = 4.37 \times 10^{-10}$;

    g. hydrogen sulphide, $K_a = 6.3 \times 10^{-8}$;

    h. benzoic acid, $K_a = 6.3 \times 10^{-5}$.

2. Calculate the acid dissociation constant of the conjugate acid of the following bases:

    a. pyridine, $K_b = 2.0 \times 10^{-9}$;

    b. aniline, $K_b = 4.0 \times 10^{-10}$;

    c. trimethylamine, $K_b = 6.3 \times 10^{-5}$;

    d. ethylenediamine, $K_b = 3.2 \times 10^{-7}$;

    e. hydrogen ethylenediammonium ion, $K_b = 4.0 \times 10^{-4}$;

    f. bipyridyl, $K_b = 10^{-9.6}$.

3. Calculate the pH of solutions containing the following concentrations of hydrogen ions:

    a. $2.1 \times 10^{-2} M$;

    b. $0.2 \times 10^{-5} M$;

    c. $7.0 \times 10^{-7} M$;

    d. $1.0 \times 10^{-9} M$.

4. Calculate the hydrogen ion concentrations in solutions with following pH values:

    a. pH 2.0;

    b. pH 5.15;

    c. pH 7.0;

    d. pH 14.0.

5. Calculate the hydroxide ion concentrations in the following solutions. Assume that the ion product of water is 1.0×10^{-14}.

    a. pH 0.3;

    b. pH 7.0;

    c. pH 12.3;

    d. pH 11.7;

    e. pH 14.5;

    f. pH −0.1.

6. Calculate the pH value of a neutral solution of the following pure solvents:

    a. water: ion product $= 1.0 \times 10^{-14}$;

    b. ethanol: ion product $= 1.26 \times 10^{-19}$;

    c. methanol: ion product $= 2.0 \times 10^{-17}$;

    d. sulPhuric acid: ion product $= 1.26 \times 10^{-3}$.

CHAPTER 3

Calculation of the pH in solutions of acids and bases

3.1 SOLUTIONS OF STRONG ACIDS AND STRONG BASES

In aqueous solutions the dissociation of a strong acid is complete if either the molecule originally contained a hydrogen ion, which simply became hydrated or a covalently bound hydrogen atom formed a hydrated ion in solution on reaction with water. Because in a solution, one hydrogen ion is produced from every molecule of a monoprotic strong acid, the concentration of H_3O^+ ions is numerically equal to the total concentration of the acid

$$[H_3O^+] = C_{HA} \qquad (3.1)$$

i.e.

$$pH = -\log C_{HA} \qquad (3.2)$$

No strong di- and polyprotic acids are known. Even sulphuric acid, which in its first dissociation step is a strong acid, is not completely dissociated in the second step (at concentrations $\gtrsim 10^{-5}M$) which indicates that the HSO_4^- ion is at most an acid of medium strength. The use of Eq. (3.1) is illustrated by the following example.

Example 3.1. Calculate the pH of a $0.050M$ solution of HCl.

The concentration of H_3O^+ ions in the solution equals the total concentration of the acid, i.e. $[H_3O^+] = 0.050M$. Hence

$$pH = -\log 0.050 = 1.30$$

In solutions of strong (completely dissociated) bases the concentration of OH^- ions depends on the concentration of the dissolved base and on the number of moles of OH^- ions produced from one mole of the base on dissolution. Many strong bases which are not ionic hydroxides, undergo a conjugate acid–base reaction with water, with release of an equivalent amount of hydroxide ions (see levelling effect, p. 50). Such a base is sodium ethoxide C_2H_5ONa, the anion of which is a strong base in water, and from one mole of C_2H_5ONa one mole of OH^- ions is produced. For ionic hydroxides, of course, (e.g. NaOH or $Ba(OH)_2$) in cleavage of the crystal lattice in the dissolution process the hydroxide ions undergo hydration and their concentration follows from the number of moles of base dissolved and the stoichiometric coefficient. Hence

$$[OH^-] = nC_B \qquad (3.3)$$

where C_B is the concentration of the strong base $M(OH)_n$ and n is the stoichiometric coefficient, (in practice, this takes values of only 1 or 2). If we know the value of pOH, we can calculate the pH, because

$$pOH = -\log(nC_B) \qquad (3.4)$$

and

$$pH = pK_w - pOH = pK_w + \log(nC_B) \qquad (3.5)$$

Example 3.2. Calculate pH and pOH in a $Ba(OH)_2$ solution of concentration $2 \times 10^{-4}M$.

The concentration of OH^- ions is

$$[OH^-] = 2 \times 2 \times 10^{-4}M = 4 \times 10^{-4}M$$

i.e.

$$pOH = -\log 4 \times 10^{-4} = 3.40 \quad \text{and} \quad pH = 14 - 3.40 = 10.60$$

3.2 CALCULATION OF THE pH OF STRONG ACIDS AND STRONG BASES: GENERAL CASE

The method for calculating the pH of a strong acid given in Section 3.1 is based on the assumption that the only source of H_3O^+ ions is the dissociation of an acid or its complete reaction with water. In our reasoning we omitted to mention that H_3O^+ ions can also originate from the autodissociation of water, where equal quantities of H_3O^+ and OH^- ions are formed. The quantities of ions formed in this way are very small; moreover, in solutions containing strong acids or bases the water dissociation reaction is suppressed by the common ion effect. Therefore in most practical cases for moderately dilute solutions, the H_3O^+ ions originating from the dissociation of water can be neglected.

The situation is quite different for very dilute solutions in which the analytical concentration of the acid is less than $10^{-6}M*$. In such cases

---

* Such solutions are not used in analytical practice, but it is interesting to consider them from the theoretical point of view.

there are two significant sources of H_3O^+ ions, *viz.* the dissolved acid and the dissociating solvent,

$$[H_3O^+]_{total} = [H_3O^+]_{acid\ dissn.} + [H_3O^+]_{H_2O\ dissn.} \tag{3.6}$$

The concentration of H_3O^+ ions derived from the acid is equal to the concentration of the acid, and the concentration of H_3O^+ ions from the dissociation of water has to be equal to the concentration of OH^- ions which, as we know, is given by the ratio of the ion product of water to the total concentration of hydrogen ions. Thus

$$[H_3O^+] = C_{HA} + [OH^-] = C_{HA} + \frac{K_w}{[H_3O^+]} \tag{3.7}$$

This expression is a quadratic equation with respect to $[H_3O^+]$, which may be rearranged to give

$$[H_3O^+]^2 - C_{HA}[H_3O^+] - K_w = 0 \tag{3.8}$$

In the general case, this allows the concentration of H_3O^+ ions to be calculated for any concentration of the acid. If the acid concentration is not too low, we can neglect the term $-K_w$, and Eq. (3.8) reduces to $[H_3O^+] = C_{HA}$, which is the simplified relation that we used earlier.

Table 3.1

Concentration of H_3O^+ ions and pH in solutions of strong acids

C_{HA}	$-\log C_{HA}$	$K_w/[H_3O^+]$	$[H_3O^+]$	pH
1	0	10^{-14}	1	0
1×10^{-2}	2	10^{-12}	1×10^{-2}	2.00
1×10^{-4}	4	10^{-10}	1×10^{-4}	4.00
1×10^{-6}	6	10^{-8}	1×10^{-6}	6.00
5×10^{-7}	6.3	7×10^{-8}	5.07×10^{-7}	6.29
2×10^{-7}	6.7	4.2×10^{-8}	2.42×10^{-7}	6.62
1×10^{-7}	7.0	6.2×10^{-8}	1.62×10^{-7}	6.78
5×10^{-8}	7.3	8.1×10^{-8}	1.31×10^{-7}	6.88
1×10^{-8}	8.0	9.5×10^{-8}	1.05×10^{-7}	6.98
1×10^{-9}	9.0	9.95×10^{-8}	1.005×10^{-7}	6.998
0	—	1×10^{-7}	1×10^{-7}	7.00

Table 3.1 lists values of $[H_3O^+]$ and pH calculated for different concentrations of a strong acid. A comparison of the data in the columns headed $-\log C_{HA}$ and pH clearly shows the range of acid concentrations for which the simplified formula can be used. For concentrations less than $10^{-7}M$, the simplified formula leads to the obviously erroneous conclusion that

very dilute solutions of acids are alkaline. It is also apparent that for concentrations of the order of $10^{-8}M$ or less the only factor of importance is the dissociation of water.

It should be remembered that in practice we do not use such highly diluted solutions of strong acids; for one thing, the minute quantities of contaminants (e.g. CO_2) which are almost always present in pure water would affect the pH of the solution more than the acid dissolved in it.

Example 3.3. Calculate the pH of a $1.5 \times 10^{-7}M$ solution of $HClO_4$.

Since the concentration of the acid is less than $10^{-6}M$, it is necessary to use the general equation. Thus, the concentration of hydrogen ions is given by

$$[H_3O^+] = 1.5 \times 10^{-7} + \frac{10^{-14}}{[H_3O^+]}$$

which is the same as

$$[H_3O^+]^2 - 1.5 \times 10^{-7}[H_3O^+] - 10^{-14} = 0$$

which gives

$$[H_3O^+] = \frac{1.5 \times 10^{-7} \pm \sqrt{2.25 \times 10^{-14} + 4 \times 10^{-14}}}{2}$$

The negative value of $[H_3O^+] = -0.5 \times 10^{-7}M$ has, of course, no physical meaning, so the only solution is the positive root of the equation, i.e. $[H_3O^+] = 2.0 \times 10^{-7}M$. Thus the pH is 6.70.

For verification we can also calculate that

$$pOH = 14 - 6.70 = 7.30, \quad \text{so} \quad [OH^-] = 5.0 \times 10^{-8}M$$

whence

$$[H_3O^+] = C_{HA} + [OH^-] = 1.5 \times 10^{-7} + 5.0 \times 10^{-8} = 2.0 \times 10^{-7}M$$

which confirms the correctness of the calculations.

In a similar way, we can calculate the pOH, and hence the pH for dilute solutions of strong bases. In these cases

$$[OH^-]_{total} = [OH^-]_{base\ dissn.} + [OH^-]_{H_2O\ dissn.} \tag{3.9}$$

and on substitition of the total concentration of base C_B and $[OH^-]$ from the dissociation of water, which equals $[H_3O^+]$, we obtain

$$[OH^-] = C_B + [H_3O^+] \tag{3.10}$$

and thus

$$[OH^-] = C_B + \frac{K_w}{[OH^-]} \tag{3.11}$$

Hence the quadratic equation in $[OH^-]$ has the form

$$[OH^-]^2 - C_B[OH^-] - K_w = 0 \tag{3.12}$$

From this we can calculate $[OH^-]$ and the pH of the solution.

Example 3.4. Calculate the pH of a $2 \times 10^{-7} M$ solution of $Ba(OH)_2$.

In this case, the concentration of OH^- ions from the dissolution of the base is twice the molar concentration of the base. Thus, in the equation, we must write $2C_B$ instead of C_B. Hence

$$[OH^-] = 2 \times 2 \times 10^{-7} + 10^{-14}/[OH^-]$$

In this case the quadratic equation has the form

$$[OH^-]^2 - 4 \times 10^{-7}[OH^-] - 10^{-14} = 0$$

Thus

$$[OH^-] = \frac{4 \times 10^{-7} \pm \sqrt{16 \times 10^{-14} + 4 \times 10^{-14}}}{2}$$

and the positive value for $[OH^-]$ is $4.25 \times 10^{-7} M$. Thus pOH = 6.37, and pH = $14 - 6.37 = 7.63$.

3.3 SOLUTIONS OF WEAK ACIDS

To calculate the pH of a solution of a weak acid we must know the dissociation constant as well as the total concentration in solution. The basic equation of the dissociation reaction of acid HA is

$$HA + H_2O \rightleftharpoons H_3O^+ + A^- \tag{3.13}$$

If the dissociation of HA molecules is the only source of H_3O^+ ions and A^- ions, the concentrations of the ions produced in the solution are equal, i.e.

$$[H_3O^+] = [A^-] \tag{3.14}$$

The concentration of the undissociated acid is equal to the total concentration of acid C_{HA} minus the concentration of the dissociated molecules, i.e. $[H_3O^+]$ or $[A^-]$

$$[HA] = C_{HA} - [H_3O^+] \tag{3.15}$$

Substitution of (3.14) and (3.15) in the equation for the acid dissociation constant gives

$$K_a = \frac{[H_3O^+][A^-]}{[HA]} = \frac{[H_3O^+]^2}{C_{HA} - [H_3O^+]} \tag{3.16}$$

which yields the quadratic equation

$$[H_3O^+]^2 + K_a[H_3O^+] - K_a C_{HA} = 0 \tag{3.17}$$

which can be solved for $[H_3O^+]$

$$[H_3O^+] = \frac{-K_a + \sqrt{K_a^2 + 4K_a C_{HA}}}{2} \tag{3.18}$$

This expression is valid for solutions of any weak acid, except very dilute solutions, with concentrations below $10^{-6}M$ (see Section 3.5).

In many cases, however, it is possible to simplify the formula, because calculations need not be made with an accuracy better than 5%. Thus, if an acid is dissociated to a relatively small extent so that the dissociated fraction is less than 5% of the total concentration, we can assume that

$$[HA] = C_{HA} \tag{3.19}$$

Thus, Eq. (3.16) simplifies to

$$K_a = \frac{[H_3O^+]^2}{C_{HA}} \tag{3.20}$$

so

$$[H_3O^+] = \sqrt{K_a C_{HA}} \tag{3.21}$$

or in logarithmic form

$$pH = \tfrac{1}{2}pK_a - \tfrac{1}{2}\log C_{HA} \tag{3.22}$$

Because of the limited accuracy of pH measurements (Section 2.13) and because of the uncertainty in the values of dissociation constants, more exact calculations are just wasted effort when the degree of dissociation is less than 5%. In all our calculations, we shall consistently neglect terms that are less than 5% of the largest term of the sum. On this basis we can derive from Eq. (3.20) another approximate criterion for the use of the simplified formula (3.21). If we substitute in Eq. (3.16)

$$[H_3O^+] = [A^-] \leqslant 0.05 C_{HA} \tag{3.23}$$

then

$$\frac{C_{HA}}{K_a} \geqslant 400 \tag{3.24}$$

i.e. if the total concentration of an acid is at least 400 times the dissociation constant, Eq. (3.21) can be used to calculate $[H_3O^+]$ or Eq. (3.22) to calculate the pH.

Example 3.5. Calculate the pH of a $0.50M$ solution of acetic acid (dissociation constant 1.8×10^{-5}).

In this case the ratio C_{HA}/K_a is approximately 3×10^4, so we can use the simplified formula (3.21). Thus,

$$[H_3O^+] = \sqrt{1.8 \times 10^{-5} \times 0.50} = \sqrt{9.0 \times 10^{-6}} = 3.0 \times 10^{-3}M$$

i.e. pH = 2.52.

For verification we compare the calculated value of $[H_3O^+]$ with the original concentration of the acid. The concentration of H_3O^+ is 0.6% of the total concentration so the use of the simplified formula is justified.

Example 3.6. Calculate the pH of a $0.10M$ solution of dichloroacetic acid ($K_a = 8.0 \times 10^{-2}$).

In this case the proportion of the concentration of the acid to its dissociation constant is 1.25, so the simplified relation cannot be used and for calculating $[H_3O^+]$ the more general formula (3.17) must be used. Thus, we obtain

$$[H_3O^+]^2 + 8.0 \times 10^{-2}[H_3O^+] - 8.0 \times 10^{-2} \times 1.00 \times 10^{-1} = 0$$

whence

$$[H_3O^+] = \frac{-8.0 \times 10^{-2} + \sqrt{64 \times 10^{-4} + 32 \times 10^{-3}}}{2}$$

$$= 5.85 \times 10^{-2}M$$

and so

pH = 1.23

If we used the simplified formula (3.21) to solve this question, we would obtain

$$[H_3O^+] = \sqrt{8.0 \times 10^{-2} \times 1.00 \times 10^{-1}} = 8.95 \times 10^{-2}M$$

pH = 1.05

This result differs greatly from the correct one.

The method above for performing calculations is quite general and can be used both for molecular acids and for cationic acids, including the ammonium ion or hydrated ions of heavy metals. Thus it is possible to calculate the pH of solutions of a salt containing an ion with acid properties. If a salt with general formula BH^+X^- has an anion which does not show acid–base properties and a cation which is a weak acid, then in a solution of the salt the acid BH^+ dissociates according to the equation

$$BH^+ + H_2O \rightleftharpoons H_3O^+ + B \tag{3.25}$$

(In the past this reaction would have been described as a hydrolysis reaction.) If we start with the expression defining the acid dissociation constant of BH^+

$$K_a = \frac{[B][H_3O^+]}{[BH^+]} \tag{3.26}$$

and make simplifying assumptions as before, we obtain as the simplest formula

$$[H_3O^+] = \sqrt{K_a C_{BHX}} \quad \text{or} \quad pH = \tfrac{1}{2}pK_a - \tfrac{1}{2}\log C_{BHX} \tag{3.27}$$

where C_{BHX} denotes the analytical (total) concentration of the salt BH^+X^-.

A salt with a weak–acid cation and an inactive anion is often described as a salt of a strong acid HX and a weak base B, which is the conjugate base of the weak acid BH^+. For this reason, it is sometimes advisable to transform Eq. (3.27) so that, instead of the acid dissociation constant K_a, it contains the basic dissociation constant K_b of the conjugate base of the acid. Since $K_a = K_w/K_b$, we obtain, on substitution in Eq. (3.27)

$$[H_3O^+] = \sqrt{\frac{K_w C_{BHX}}{K_b}} \tag{3.28}$$

or

$$pH = \tfrac{1}{2}pK_w - \tfrac{1}{2}pK_b - \tfrac{1}{2}\log C_{BHX} \tag{3.29}$$

This method of calculation can be used to find, e.g. the pH of an ammonium chloride solution. The ammonium ion is a weak acid dissociating according to the equation $NH_4^+ + H_2O \rightleftharpoons H_3O^+ + NH_3$, with dissociation constant 5.75×10^{-10}. However, if we consider ammonium chloride as a salt of the weak base ammonia, then we can use the value of the basic dissociation constant of ammonia, K_b, which is $10^{-14}/(5.75 \times 10^{-10})$ $= 1.74 \times 10^{-5}$, and use Eq. (3.28) for the calculation.

Example 3.7. Calculate the pH of a $1.00 \times 10^{-2} M$ solution of pyridine hydrochloride $C_5H_5NH^+Cl^-$ if pK_a for the pyridinium ion is 5.3.

On the basis of Eq. (3.27) we calculate the concentration of H_3O^+

$$[H_3O^+] = \sqrt{1.0 \times 10^{-2} \times 5.01 \times 10^{-6}} = 2.24 \times 10^{-4} M$$

i.e.

$$pH = 3.65$$

These calculations can be done in a different way if we start from the dissociation contant of pyridine as a base $pK_b = 14 - 5.3 = 8.7$.

By using Eq. (3.28), we obtain the same result, because

$$pH = 7 - \tfrac{1}{2} \times 8.7 + \tfrac{1}{2} \times 2.00 = 3.65$$

The results obtained from these calculations are correct because the degree of dissociation is small. However, the calculations should always be checked, especially when very dilute solutions are being considered.

The method described above can also be used to calculate the pH of salt solutions of certain metals which have hydrated cations that behave as weak acids.

Example 3.8. Calculate the pH of a $0.100M$ solution of zinc chloride.

The main reaction in solution is the acid dissociation of the hydrated zinc cation

$$Zn(H_2O)_4^{2+} + H_2O \rightleftharpoons Zn(H_2O)_3OH^+ + H_3O^+$$

For this acid–base equilibrium the acid dissociation constant is

$$K_a = \frac{[Zn(H_2O)_3OH^+][H_3O^+]}{[Zn(H_2O)_4^{2+}]} = 6.3 \times 10^{-10}$$

The concentration of the H_3O^+ ion is calculated on the assumption that in the equilibrium state the concentration of the cation acid $Zn(H_2O)_4^{2+}$ is equal to the original concentration of the salt. Thus,

$$[H_3O^+] = \sqrt{6.3 \times 10^{-10} \times 1.00 \times 10^{-1}} = 7.9 \times 10^{-6}M$$

from which

$$pH = 5.10$$

In this example, use of the simplified equation is justified because, as can easily be verified, $C_{HA}/K_a = 1.6 \times 10^8$, so the error resulting from the approximation is negligible.

The magnitude of the dissociation constant of a weak acid characterizes the equilibrium of the dissociation irrespective of the concentration of the acid in the solution. In the case of a fixed total concentration of the acid the process of dissociation can also be characterized by the *degree of dissociation*, which is the ratio of the number (or concentration) of dissociated molecules to the total number (or concentration) of molecules of acid in the solution. The degree of dissociation depends of course on the concentration. If the concentration of dissociated molecules is equal to the concentration of the anion, then the degree of dissociation is given by the equation

$$x = \frac{[A^-]}{C_{HA}} \tag{3.30}$$

whence the concentration of the anion is

$$[A^-] = C_{HA}x \tag{3.31}$$

If the molecules of the acid are the only source of anions and if the reactions in the solution do not disturb the dissociation of the acid, then the concentration of anions is equal to the concentration of hydrogen ions, and thus

$$[H_3O^+] = C_{HA}x \tag{3.32}$$

The concentration of undissociated molecules is then equal, according to Eq. (3.15), to the difference between the initial concentration of acid and the concentration of either of the ions produced in the dissociation. Substituting these quantities in the equation for the acid dissociation constant, we obtain

$$K_a = \frac{(C_{HA}x)^2}{C_{HA}-C_{HA}x} = \frac{C_{HA}x^2}{1-x} \tag{3.33}$$

This equation, sometimes called *Ostwald's dilution law*, describes the change in the degree of dissociation as an acid is diluted. When the degree of dissociation is small (< 0.05) the denominator is $1-x \cong 1$, and Eq. (3.33) simplifies to

$$K_a = C_{HA}x^2 \tag{3.34}$$

which is applicable to weak electrolytes dissociated to a small degree. It allows calculation of the degree of dissociation of an acid of known concentration and dissociation constant.

Example 3.9. Calculate the degree of dissociation of $1.0 \times 10^{-3} M$ hydrocyanic acid, HCN ($K_a = 4.4 \times 10^{-10}$).

Using the simplified expression (3.34), we obtain

$$x_A = \sqrt{\frac{K_a}{C_{HA}}} = \sqrt{\frac{4.4 \times 10^{-10}}{1.0 \times 10^{-3}}} = 6.6 \times 10^{-4}$$

or, expressed as a percentage

$$x_A(\%) = 0.066\%$$

Such a small value of the degree of dissociation justifies the use of the simplified formula, because $1 - 6.6 \times 10^{-4} \cong 1$.

When the degree of dissociation is greater than 5% (i.e. > 0.05) it is necessary to solve the quadratic Eq. (3.33), which may be written as

$$C_{HA}x_A^2 + K_a x_A - K_a = 0 \tag{3.35}$$

If we solve this for x_A we obtain

$$x_A = \frac{-K_a + \sqrt{K_a^2 + 4K_a C_{HA}}}{2C_{HA}} \tag{3.36}$$

Example 3.10. Calculate the degree of dissociation of $0.050M$ dichloro-acetic acid, $(K_a = 8.0 \times 10^{-2})$.

On the basis of Eq. (3.36) we calculate

$$x = \frac{-8.0 \times 10^{-2} + \sqrt{64.0 \times 10^{-4} + 4 \times 8.0 \times 10^{-2} \times 5.0 \times 10^{-2}}}{2 \times 5.0 \times 10^{-2}}$$

$$= 7.0 \times 10^{-1}$$

or, as a percentage

$$x(\%) = 70\%$$

If we solved this example on the basis of the simplified expression (3.34) we would obtain

$$x = \sqrt{8.0 \times 10^{-2} \times 5.0 \times 10^{-2}} = 6.3 \times 10^{-2}$$

or

$$x(\%) = 6.3\%$$

The very great difference from the correct answer gives an excellent demonstration of the necessity to use Eq. (3.36).

3.4 SOLUTIONS OF WEAK BASES

Calculation of the pH of solutions of weak bases is done in a way analogous to that used for solutions of weak acids. If we assume that the only source of hydroxide ions is the reaction

$$B + H_2O \rightleftharpoons BH^+ + OH^- \tag{3.37}$$

for which the equilibrium constant is the basic dissociation constant K_b, then

$$K_b = \frac{[BH^+][OH^-]}{[B]} \tag{3.38}$$

Since no other reaction gives BH^+ ions and OH^- ions and since these ions do not take part in any other reactions, their concentrations must be equal

$$[BH^+] = [OH^-] \tag{3.39}$$

and

$$[B] = C_B - [OH^-] \tag{3.40}$$

Substitution of (3.39) and (3.40) in the expression for the dissociation constant, (3.38), gives

$$K_b = \frac{[OH^-]^2}{C_B - [OH^-]} \tag{3.41}$$

which is a quadratic equation that can be solved for $[OH^-]$

$$[OH^-] = \frac{-K_b + \sqrt{K_b^2 + 4K_b C_B}}{2} \tag{3.42}$$

From the ion product of water we can now calculate the concentration of H_3O^+ ions

$$[H_3O^+] = \frac{2K_w}{-K_b + \sqrt{K_b^2 + 4K_b C_B}} \tag{3.43}$$

If base B is a weak base, then the concentration of the dissociated fraction is very small in comparison with the total concentration. We can thus use a simplified expression according to which

$$[OH^-] = \sqrt{K_b C_B} \tag{3.44}$$

i.e.

$$[H_3O^+] = \frac{K_w}{\sqrt{K_b C_B}} \tag{3.45}$$

or in logarithmic form

$$pH = pK_w - \tfrac{1}{2}pK_b + \tfrac{1}{2}\log C_B \tag{3.46}$$

The ultimate criterion for deciding whether it is possible to use the simplified formula is the comparison (as for acids) of the concentration and the dissociation constant. If

$$\frac{C_B}{K_b} \geqslant 400 \tag{3.47}$$

we can use the simplified expressions (3.44), (3.45) and (3.46).

The formulae for the concentration of H_3O^+ ions in a solution of a weak base and the concentration of OH^- ions in a solution of a weak acid have a similar form. The only difference is the appearance of the basic dissociation constant instead of the acid dissociation constant and, of course, of the base concentration instead of the acid concentration.

Example 3.11. Calculate the pH of a $0.20M$ solution of methylamine, $(K_b = 5.0 \times 10^{-4})$.

In this example C_B/K_b is exactly 400, so we are considering a limiting case. To make sure that the simplified assumptions are correct, we will calculate by both methods.

According to Eq. (3.45) the concentration of H_3O^+ ions is

$$[H_3O^+] = \frac{10^{-14}}{\sqrt{5.0 \times 10^{-4} \times 2.0 \times 10^{-1}}} = 1 \times 10^{-12} M$$

so

$$pH = 12.00$$

We first calculate the concentration of OH^- ions by means of the general expression (3.42)

$$[OH^-] = \frac{-5.0 \times 10^{-4} + \sqrt{25.0 \times 10^{-8} + 4 \times 5.0 \times 10^{-4} \times 2.0 \times 10^{-1}}}{2}$$

$$= \frac{-5.0 \times 10^{-4} + \sqrt{2.5 \times 10^{-7} + 4.0 \times 10^{-4}}}{2}$$

$$= \frac{-5.0 \times 10^{-4} + \sqrt{4.0 \times 10^{-4}}}{2}$$

$$= \frac{-5.0 \times 10^{-4} + 2.0 \times 10^{-2}}{2} = 9.75 \times 10^{-3} M$$

Hence

$$[H_3O^+] = \frac{10^{-14}}{9.75 \times 10^{-3}} = 1.02 \times 10^{-12} M$$

and

$$pH = 11.99$$

The difference between the results obtained is inside the limits of pH measurement error, and a larger error certainly exists in the value given for the dissociation constant of methylamine. Thus, even for this limiting case we may use the simplified formula, which gives a quicker calculation. In this example, the calculations were based on the thermodynamic value of the ion product of water. This is not quite proper, as in given conditions its value is close to 1.3×10^{-14}. This has, however, no effect on the comparison of the two modes of calculation as long as the same value is used in both.

As for weak acids, the formulae given for weak bases can be used either for a solution of a molecular base or for a charged base, i.e. for what used to be called hydrolysis. Thus, if we have a solution of a salt which has an anion with the properties of a weak base (e.g. acetate or cyanide), then all expressions are applicable. It should be added that sometimes, instead of the basic dissociation constant of the ion (e.g. acetate or cyanide), the dissociation constant of the conjugate acid (e.g. acetic acid or hydrocyanic acid) can be used. Then, in place of Eq. (3.45) we obtain

$$[H_3O^+] = \frac{K_w}{\sqrt{\dfrac{K_w C_B}{K_a}}} = \sqrt{\frac{K_w K_a}{C_B}} \qquad (3.48)$$

$$pH = \tfrac{1}{2}pK_w + \tfrac{1}{2}pK_a + \tfrac{1}{2}\log C_B \qquad (3.49)$$

where C_B denotes the concentration of the salt.

Example 3.12. Calculate the pH of a $0.10M$ solution of sodium acetate (the acid dissociation constant of acetic acid is 1.74×10^{-5} and the ion product of water is 1.74×10^{-14}).

We first calculate the basic dissociation constant of the acetate ion

$$K_b = \frac{1.74 \times 10^{-14}}{1.74 \times 10^{-5}} = 1.0 \times 10^{-9}$$

$$pK_b = 9.00$$

Then the value of pH from Eq. (3.46) is

$$pH = 13.76 - \tfrac{1}{2} \times 9.00 + \tfrac{1}{2} \times 1.00 = 9.76$$

The same result is obtained if we use Eq. (3.49) and start from the dissociation constant of acetic acid or from the value $pK_a = 4.76$. We then obtain

$$pH = \tfrac{1}{2} \times 13.76 + \tfrac{1}{2} \times 4.76 + \tfrac{1}{2} \times 1.00 = 9.76$$

3.5 GENERAL CASES OF pH CALCULATIONS FOR SOLUTIONS OF WEAK ELECTROLYTES

When we discussed the dissociation of weak acids or weak bases in solution, we considered the dissociation of these electrolytes as the only reaction. We disregarded the fact that H_3O^+ or OH^- ions can also be produced from the dissociation of water. As with strong acids and strong bases (cf. Section 3.2), very dilute solutions have pH values close to neutrality. If the pH is between 6 and 8, the dissociation of water makes the formulae given in Sections 3.3 and 3.4 fail.

Thus, if we must assume that hydrogen ions in a solution of a weak acid originate from both the dissolved acid HA and water, then

$$[H_3O^+] = [A^-] + [OH^-] \qquad (3.50)$$

The concentrations in this equation can be found from the equilibrium constants of the acid dissociation. Thus, for an acid we have

$$K_a = \frac{[H_3O^+][A^-]}{[HA]} = \frac{[H_3O^+][A^-]}{C_{HA} - [A^-]} \qquad (3.51)$$

from which

$$[A^-] = \frac{C_{HA} K_a}{K_a + [H_3O^+]} \qquad (3.52)$$

and for the dissociation of water we have

$$[OH^-] = \frac{K_w}{[H_3O^+]} \qquad (3.53)$$

When these expressions are substituted in Eq. (3.50) we obtain

$$[H_3O^+] = \frac{C_{HA}K_a}{K_a + [H_3O^+]} + \frac{K_w}{[H_3O^+]} \qquad (3.54)$$

ann rearrangement with respect to $[H_3O^+]$ gives a cubic equation

$$[H_3O^+]^3 + K_a[H_3O^+]^2 - (K_w - C_{HA}K_a)[H_3O^+] - K_aK_w = 0 \qquad (3.55)$$

It is not easy to solve this equation for the general case. However, in most cases it is possible to introduce simplifications which reduce the problem to the solution of a quadratic equation in $[H_3O^+]$. Examples of simplifying assumptions, given further on, will illustrate the methods. We first transform Eq. (3.54) with respect to C_{HA}, which will permit us to find ranges in which particular terms of the equation can be neglected. The expression obtained,

$$C_{HA} = \left(1 + \frac{[H_3O^+]}{K_a}\right)\left([H_3O^+] - \frac{K_w}{[H_3O^+]}\right) \qquad (3.56)$$

can readily be represented as a graph of pH vs. $-\log C_{HA}$. Figure 3.1 is such a graph for acetic acid ($pK_a = 4.76$). In addition, we show the graph of pH vs. $-\log C_{HA}$ for a strong acid (HCl): this corresponds to the data given in Table 3.1.

Let us begin by considering the case of a weak acid with dissociation constant larger than 10^{-6}. At relatively high concentrations, the concentrations of the dissociated ions can be neglected in comparison with the

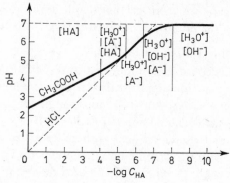

Fig. 3.1. Dependence of the pH of an acetic acid solution on its concentration.

total concentration of the acid, and the concentration of H_3O^+ arising from the dissociation of water can also be neglected. When we make these simplifications, we can say that $[H_3O^+] = K_aC_{HA}$. For lower concentrations, where $C_{HA}/K_a < 400$, we can no longer neglect the ions formed from the dissociation. Consequently, the pH calculation requires the solution of the complete quadratic equation (see Section 3.3). If we continue diluting the solution, the dissociation advances so far that the concentration of undissociated molecules becomes negligible in relation to the total concentration of the acid. Then, we can assume complete dissociation and calculate pH in the same way as for strong acids, where $[H_3O^+] = C_{HA}$ (see Section 3.1). After further dilution, we can treat the system as if it were a very dilute solution of a strong acid, so that dissociation of water begins to play an important part. Thus, $[H_3O^+]$ is again calculated from the quadratic equation

$$[H_3O^+]^2 - C_{HA}[H_3O^+] - K_w = 0 \tag{3.57}$$

(This equation applies in cases where the pH does not differ greatly from 7, i.e. where the concentration of the acid is close to $10^{-7}M$). The final case concerns solutions that are so dilute that practically the only factor to be taken into consideration is the dissociation of water, i.e. pH = 7, irrespective of the concentration of the acid.

The relationships are rather different if the value of the acid dissociation constant is always much smaller than the concentration of hydrogen ions. An example is hydrocyanic acid, which has $pK_a = 9.12$. In more concentrated solutions the degree of dissociation of HCN is very small and the concentration of the anion can be neglected in comparison with the total concentration of the acid. If we start diluting a solution of this acid, then, before the acid has dissociated sufficiently to affect the total concentration of the acid, the proportion of the H_3O^+ ions from the dissociation of water will have increased so much that it will have to be included in the calculations. Then the total concentration of H_3O^+ will be given by

$$[H_3O^+] = \sqrt{K_aC_{HA} + K_w} \tag{3.58}$$

where the terms under the square root sign correspond to the contributions of acid dissociation and of water dissociation. Further dilution will of course bring about a situation in which H_3O^+ ions will be supplied only by water.

Table 3.2 lists the equations for calculation of $[H_3O^+]$ in all the above-mentioned cases and shows the ions with concentrations that cannot be neglected; we also give the simplifying assumptions for Eq. (3.56). It should be emphasized that Eq. (3.56) can also be used for strong acids, for which, of course, $[H_3O^+]$ is always smaller than the dissociation constant K_a.

Table 3.2

Simplifying assumptions and equations used for the calculation of H_3O^+ concentration

— Decrease of acid concentration →

Particles with non-negligible concentration	HA	H_3O^+, A^-, HA	H_3O^+, A^-	H_3O^+, A^-, OH^-	H_3O^+, OH^-
Strong acids					
Simplifying assumptions	—	—	$1 \gg \dfrac{[H_3O^+]}{K_a}$; $\dfrac{K_w}{[H_3O^+]} \ll [H_3O^+]$	$1 \gg \dfrac{[H_3O^+]}{K_a}$; $\dfrac{K_w}{[H_3O^+]} \approx [H_3O^+]$	$1 \gg \dfrac{[H_3O^+]}{K_a}$; $\dfrac{K_w}{[H_3O^+]} = [H_3O^+]$
C_{HA}			$[H_3O^+]$	$[H_3O^+] - \dfrac{K_w}{[H_3O^+]}$	$\to 0$
$[H_3O^+]$			C_{HA}	$\dfrac{C_{HA} + \sqrt{C_{HA}^2 + 4K_w}}{2}$	$\sqrt{K_w}$
Weak acids, $K_a \gg 10^{-7}$					
Simplifying assumptions	$1 \ll \dfrac{[H_3O^+]}{K_a}$; $\dfrac{K_w}{[H_3O^+]} \ll [H_3O^+]$	$1 \approx \dfrac{[H_3O^+]}{K_a}$; $\dfrac{K_w}{[H_3O^+]} \ll [H_3O^+]$	$1 \gg \dfrac{[H_3O^+]}{K_a}$; $\dfrac{K_w}{[H_3O^+]} \ll [H_3O^+]$	$1 \gg \dfrac{[H_3O^+]}{K_a}$; $\dfrac{K_w}{[H_3O^+]} \approx [H_3O^+]$	$1 \gg \dfrac{[H_3O^+]}{K_a}$; $\dfrac{K_w}{[H_3O^+]} = [H_3O^+]$
C_{HA}	$\dfrac{[H_3O^+]^2}{K_a}$	$\left(1 + \dfrac{[H_3O^+]}{K_a}\right)[H_3O^+]$	$[H_3O^+]$	$[H_3O^+] - \dfrac{K_w}{[H_3O^+]}$	$\to 0$
$[H_3O^+]$	$\sqrt{K_a C_{HA}}$	$\dfrac{-K_a + \sqrt{K_a^2 + 4K_a C_{HA}}}{2}$	C_{HA}	$\dfrac{C_{HA} + \sqrt{C_{HA}^2 + 4K_w}}{2}$	$\sqrt{K_w}$
Very weak acids, $K_a \ll 10^{-7}$					
Simplifying assumptions	$1 \ll \dfrac{[H_3O^+]}{K_a}$; $\dfrac{K_w}{[H_3O^+]} \ll [H_3O^+]$	—	—	$1 \ll \dfrac{[H_3O^+]}{K_a}$; $\dfrac{K_w}{[H_3O^+]} \approx [H_3O^+]$	$1 \ll \dfrac{[H_3O^+]}{K_a}$; $\dfrac{K_w}{[H_3O^+]} = [H_3O^+]$
C_{HA}	$\dfrac{[H_3O^+]^2}{K_a}$			$\dfrac{[H_3O^+]}{K_a}\left([H_3O^+] - \dfrac{K_w}{[H_3O^+]}\right)$	$\to 0$
$[H_3O^+]$	$\sqrt{K_a C_{HA}}$			$\sqrt{K_a C_{HA} + K_w}$	$\sqrt{K_w}$

When the dissociation constant of the acid is close to 10^{-7} and the solution considered is very dilute, no simplifications are possible and only the solution of the third degree equation in $[H_3O^+]$ can give a sufficiently exact answer.

Example 3.13. Calculate the concentration of H_3O^+ ions in a $10^{-4}M$ solution of phenol (dissociation constant 10^{-10}).

Phenol is a very weak acid and the dissociation of water greatly affects the equilibria in its dilute solutions. Consequently the concentration of H_3O^+ ions is equal to the sum of the concentrations of phenolate anions and OH^- ions, i.e.

$$[H_3O^+] = [A^-] + [OH^-]$$

or after substitution of the values of the equilibrium constants,

$$[H_3O^+] = \frac{K_a[HA]}{[H_3O^+]} + \frac{K_w}{[H_3O^+]}$$

If $[HA] = C_{HA}$, we obtain on rearrangement

$$[H_3O^+] = \sqrt{K_a C_{HA} + K_w} = \sqrt{10^{-10} \times 10^{-4} + 10^{-14}}$$
$$= \sqrt{2 \times 10^{-14}} = 1.4 \times 10^{-7}M$$

so the pH $= 6.85$.

If the dissociation of water were omitted in this calculation the result obtained would be pH $= 7.0$.

To make sure that it is not necessary to use the general equation of the third degree we may calculate the concentration of undissociated phenol molecules at pH 6.85. From the expression for the dissociation constant of phenol

$$K_a = \frac{[H_3O^+](C_{HA} - [HA])}{[HA]}$$

we can calculate $[HA]$

$$[HA] = \frac{[H_3O^+]C_{HA}}{K_a + [H_3O^+]} = \frac{1.4 \times 10^{-7} \times 10^{-4}}{10^{-10} + 1.4 \times 10^{-7}} = 10^{-4}M$$

Thus it may be assumed with an error of less than 0.1% that the total (analytical) concentration of the acid is equal to the concentration of undissociated molecules.

If we calculate the concentration of a very dilute solution of a weak acid, then it is usually accepted that we may neglect the concentration of the undissociated molecules if this concentration is less than 5% of the total acid concentration. The degree of dissociation would then be greater than 0.95, i.e.

$$\frac{K_a}{C_{HA}} \geqslant 18$$

This condition is rarely satisfied. However, it allows simplification of the calculations for very dilute but not too weak acids.

Example 3.14. Calculate the pH of a $1.0 \times 10^{-5} M$ solution of benzoic acid ($K_a = 6.3 \times 10^{-5}$).

For this very dilute solution, $K_a/C_{HA} = 6.3$, so we cannot assume either that the total concentration is equal to the concentration of the undissociated part or that the dissociation of the acid is complete. However, to a first approximation we can use a formula in which dissociation of water is neglected. Thus

$$[H_3O^+] = \frac{-K_a + \sqrt{K_a^2 + 4K_a C_{HA}}}{2}$$

and after substituting the numerical values we have

$$[H_3O^+] = \frac{-6.3 \times 10^{-5} + \sqrt{4.0 \times 10^{-9} + 4 \times 6.3 \times 10^{-5} \times 10^{-5}}}{2}$$

$$= 8.9 \times 10^{-6} M$$

i.e.
$$pH = 5.05$$

Now it is possible to verify that $K_w/[H_3O^+] \ll [H_3O^+]$, because $10^{-14}/8.9 \times 10^{-6} = 1.1 \times 10^{-9} \ll 8.9 \times 10^{-6}$. Since $[H_3O^+] = 8.9 \times 10^{-6}$ and $K_a = 6.3 \times 10^{-5}$ are of similar order, the simplifying assumptions are valid, and the formula used for calculating the pH was indeed applicable.

Example 3.15. Calculate the pH of a $5 \times 10^{-5} M$ solution of chloroacetic acid, ($K_a = 1.3 \times 10^{-3}$).

As the ratio of the dissociation constant to the total concentration of the acid is $K_a/C_{HA} = 1.3 \times 10^{-3}/5 \times 10^{-5} = 26$, we may regard this acid as completely dissociated. We calculate the concentration of H_3O^+ ions as for a strong acid, i.e.

$$[H_3O^+] = C_{HA} = 5 \times 10^{-5} M \quad \text{and} \quad pH = 4.3$$

If in this example we used the full formula for weak acids to calculate $[H_3O^+]$ we would obtain

$$[H_3O^+] = \frac{-1.3 \times 10^{-3} + \sqrt{1.7 \times 10^{-6} + 4 \times 1.3 \times 10^{-3} \times 5 \times 10^{-5}}}{2}$$

$$= \frac{-1.3 \times 10^{-3} + \sqrt{1.96 \times 10^{-6}}}{2} = 5 \times 10^{-5} M$$

and so pH = 4.3. This result, which is equal to the value obtained previously, has been obtained in a much more complicated way.

The pH values of solutions of weak bases are calculated similarly. The formulae are the same as for weak acids but the symbols are changed appropriately.

Acids [H_3O^+]   Bases [OH^-]

pH   pOH

$$K_a = \frac{[H_3O^+][A^-]}{[HA]} \qquad K_b = \frac{[BH^+][OH^-]}{[B]}$$

C_{HA} C_B

Thus, for bases, the general form of Eq. (3.55) is

$$[OH^-]^3 + K_b[OH^-]^2 - (K_w - C_B K_b)[OH^-] - K_b K_w = 0 \qquad (3.59)$$

and after transformation

$$C_B = \left(1 + \frac{[OH^-]}{K_b}\right)\left([OH^-] - \frac{K_w}{[OH^-]}\right) \qquad (3.60)$$

The concentrations of hydroxide ions calculated for bases are finally used to calculate concentrations of hydrogen ions.

Since the formulae involved have analogous forms, the simplifications used for acids will be valid also for bases. Thus, if $C_A/K_b \geqslant 400$, we assume that the dissociation of the base is small compared with the concentration of undissociated molecules, and if $K_b/C_B > 18$ we follow the same reasoning as in the case of dilute solutions of strong bases. Also, if the pH is below 8, we must take into account the concentration of OH^- ions arising from the dissociation of water.

Example 3.16. Calculate the pH of a $2 \times 10^{-5} M$ solution of pyridine if the acid dissociation constant of the pyridinium ion is 5×10^{-6}.

We first calculate the basic dissociation constant of pyridine

$$K_b = \frac{K_w}{K_a} = \frac{10^{-14}}{5 \times 10^{-6}} = 2 \times 10^{-9}$$

The ratio of the total concentration to the value of the dissociation constant is

$$\frac{C_B}{K_b} = \frac{2 \times 10^{-5}}{2 \times 10^{-9}} = 10^4$$

so the simplified formula may be used if it is assumed that, in relation to the total concentration of base, the concentration of the dissociated fraction is very small ($\ll 5\%$). However, since the base is very dilute, we should consider whether the dissociation of water can be neglected. From a formula analogous to the one used for acids we obtain

$$[OH^-] = \sqrt{K_b C_B + K_w} = \sqrt{2 \times 10^{-9} \times 2 \times 10^{-5} + 10^{-14}}$$
$$= \sqrt{5 \times 10^{-14}} = 2.2 \times 10^{-7} M$$

i.e.

$$pOH = 6.65 \quad \text{or} \quad pH = 14 - 6.65 = 7.35$$

It can readily be shown that neglect of dissociation of water would lead to a result which differs by 0.05, i.e. pH = 7.30.

The dependence of the pH of solutions of acids and bases on their total concentration is shown in a diagram given by Flood (Fig. 3.2). The

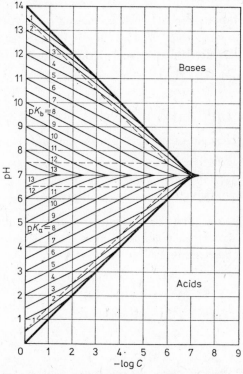

Fig. 3.2. Flood diagram—the dependence of the pH of acid and base solutions on concentration for various values of the dissociation constant.

diagram given in Fig. 3.1, for acetic acid, is a particular case of this dependence. In Flood's diagram the thicker (limiting) lines correspond to strong electrolytes (acid and base). The curves inside the area limited by these lines correspond to acids with values of pK_a from 1 to 13 (above the line for pH = 7.0) or to bases with values of pK_b from 1 to 13 (under the line for pH = 7.0). The two triangles limited by dashed lines correspond to the ranges of the concentrations of weak acids. Inside these ranges the simplest equation, $[H_3O^+] = \sqrt{KC}$ may be used. Thus, they are regions of limited dissociation. Between the dashed lines, close to pH = 7, there is a range in which it is necessary to take into account dissociation of water as a source of H_3O^+ and OH^- ions. In the vicinity of the region of strong acids or bases the calculations must allow for the presence of both ions and molecules of the acid or base.

3.6 BUFFER SOLUTIONS

If we mix two solutions, one of which contains a weak acid and the other one the conjugate base, then the state of equilibrium between the conjugate pair in the mixture is determined by the reaction

$$HA + H_2O \rightleftharpoons H_3O^+ + A^- \tag{3.61}$$

or the reaction

$$HA + OH^- \rightleftharpoons H_2O + A^- \tag{3.62}$$

The equilibrium constant of the first of these reactions is the acid dissociation constant of the acid HA

$$K_a = \frac{[H_3O^+][A^-]}{[HA]} \tag{3.63}$$

This expression implies that the concentration of H_3O^+ ions in a solution depends on the ratio of the concentration of the acid to the concentration of the conjugate base because

$$[H_3O^+] = K_a \frac{[HA]}{[A^-]} \tag{3.64}$$

or in logarithmic form

$$pH = pK_a + \log \frac{[A^-]}{[HA]} \tag{3.65}$$

The pH of a solution can thus be calculated from the concentrations of the acid and the conjugate base (e.g. NH_4^+ and NH_3, CH_3COOH and

CH_3COO^-, HCO_3^- and CO_3^{2-}) at equilibrium, and the acid dissociation constant.

The concentrations of acid and base at equilibrium are usually calculated by assuming (as a first approximation) that they are equal to the initial concentrations (i.e. the concentrations taken for preparing the solution) of acid and conjugate base, i.e.

$$[HA] = C_{HA} \quad \text{and} \quad [A^-] = C_A \tag{3.66}$$

On substitution of this in equation (3.6) we obtain

$$pH = pK_a + \log \frac{C_A}{C_{HA}} \tag{3.67}$$

A solution containing a fairly large concentration of an acid and its conjugate base is called a *buffer solution*. Buffer solutions are very widely used in many branches of chemistry and biochemistry, medicine and other fields of science and in technology, because they allow the pH of a solution to be maintained, with quite good accuracy, at a constant value, even on addition of small quantities of strong acids or bases or on dilution.

This behaviour of buffer solutions can be explained by use of Eqs. (3.61) and (3.62). If a solution containing undissociated acid HA has OH^- ions added to it, then they will at once react according to reaction (3.62). As a result the proportion $[A^-]/[HA]$ will increase somewhat, but the pH will not increase much. Similarly, if a buffer solution has H_3O^+ ions added to it, then reaction (3.61) will proceed from right to left. The H_3O^+ ions will not remain in solution, but will combine with the base A^-, to form weakly dissociated acid HA. The change in pH will be due to the small decrease in the value of the ratio $[A^-]/[HA]$ as a result of the consumption of A^- ions and the production of the acid HA.

Dilution of a buffer solution does not usually change the pH value. It changes the values of the concentrations but not their ratio.

Example 3.17. Show that the addition of 10 mmoles of strong acid (HCl) or 10 mmoles of strong base (NaOH) and also dilution of the solution to double its volume have an insignificant influence (or no influence at all) on the pH of 1 litre of a buffer solution consisting of $0.2M$ formic acid and $0.5M$ sodium formate. Compare the pH changes of this buffer solution with the pH changes for similar treatment of a solution of strong acid with the same initial pH.

For calculation of the pH of the buffer solution we use Eq. (3.67). The dissociation constant of formic acid is 2.5×10^{-4} ($pK_a = 3.60$). Hence

$$\text{pH} = \text{p}K_a + \log\frac{C_A}{C_{HA}} = 3.60 + \log\frac{0.5}{0.2} = 3.60 + 0.40 = 4.00$$

If we wanted to prepare a solution of a strong acid with the same pH, then the concentration of H_3O^+ ions would have to be equal to the total concentration of that acid. Thus, $\log C_{HA} = -4.00$ and $C_{HA} = 1 \times 10^{-4}M$.

Let us now calculate the changes of pH when the strong acid is added, i.e. 10 mmoles of H_3O^+ ions.

In the buffer, as a result of the reaction of H_3O^+ ions with the base (the formate ion) undissociated formic acid is formed, of which the final concentration, if we assume the reaction to be complete, is

$$[\text{HCOOH}] = 0.20 + \frac{10}{1000} = 0.21M$$

The concentration of the formate ion will decrease by the amount by which the concentration of the acid has increased, so

$$[\text{HCOO}^-] = 0.50 - 0.01 = 0.49M$$

The pH of the solution after the acid is added will be

$$\text{pH} = 3.60 + \log\frac{0.49}{0.21} = 3.60 + 0.37 = 3.97$$

Thus, the change in pH is 0.03.

However, if we add the same quantity of the acid to a litre of $1 \times 10^{-4}M$ hydrochloric acid (pH 4.00), then the solution obtained will be

$$1 \times 10^{-4} + 0.01 \cong 0.01M \text{ HCl}$$

so the concentration of H_3O^+ ions will be $10^{-2}M$ and pH = 2. Thus, the change will be 2.00 pH units.

We reason similarly to find the effect of the addition of the same quantity of a strong base. The concentration of formic acid will decrease by $0.01M$, i.e.

$$[\text{HCOOH}] = 0.20 - 0.01 = 0.19M$$

The total concentration of formate ions will then be

$$[\text{HCOO}^-] = 0.50 + 0.01 = 0.51M$$

Then, from Eq. (3.67)

$$\text{pH} = 3.60 + \log\frac{0.51}{0.19} = 3.60 + 0.42 = 4.02$$

This change is, as before, comparatively small, namely 0.02 pH unit.

However, if we added the same quantity of a strong base to a litre of

$1 \times 10^{-4} M$ hydrochloric acid, then the pH value would be determined by the concentration of the non-neutralized base, i.e.

$$[OH^-] = 0.01 - 1 \times 10^{-4} \cong 0.01 M \quad \text{and} \quad pH = 12$$

These large changes in pH values in an unbuffered solution show the great importance of buffer solutions for stabilization of pH values.

Similar effects are observed with dilution. In the case of a buffer, dilution does not affect the pH value, because the identical change in the acid and base concentrations has no effect on the value of the concentration ratio. However, dilution of a $1 \times 10^{-4} M$ solution of HCl to double its volume will result in

$$[H_3O^+] = \frac{1 \times 10^{-4}}{2} = 5 \times 10^{-5} M \quad \text{so} \quad pH = 4.30$$

Thus the change in pH will be 0.30.

Buffer solutions act most effectively when the concentrations of the acid and the base are equal. Such a buffer is the least sensitive to an addition of a strong acid or a strong base—its *buffer capacity* is maximal. In these optimum conditions the pH of the solution is equal to the value of pK_a. It follows from Eq. (3.67) that if the ratio of the base concentration to the acid concentration changes from 1:1 to 10:1 or 1:10, then in both cases the change of pH will be 1. This range of changes of the concentration ratio is usually assumed to be the maximum for buffer solutions. When the change of the concentration ratio is higher, the solution is said to have lost its capacity for buffer action. Thus, the region of effective buffer action is

$$pH = pK_a \pm 1 \tag{3.68}$$

To be able to prepare buffer solutions with arbitrary values of pH we must have at our disposal a fairly large assortment of species with pK_a values spaced at regular (but not too wide) intervals. For example, a buffer made from chloroacetic acid and sodium chloroacetate has its optimum capacity at pH = 2.7, because the pK_a of chloroacetic acid has this value. In a less acid region a formate buffer is effective, prepared from sodium formate and formic acid ($pK_a = 3.60$). In a still higher range of pH we can use an acetate buffer, because for acetic acid, $pK_a = 4.76$. In alkaline solutions the following buffer solutions are often used: an ammonium buffer, containing a mixture of ammonium ions ($pK_a = 9.37$) and ammonia, and a borate buffer composed of boric acid ($pK_a = 9.1$) and its sodium salt.

Solutions of salts of polyprotic acids are particularly useful and often used as buffers. For example, orthophosphoric acid and its mono-sub-stituted salt KH_2PO_4 are in the equilibrium

$$H_3PO_4 + H_2O \rightleftharpoons H_2PO_4^- + H_3O^+ \tag{3.69}$$

which is determined by the value of the first dissociation constant of ortho-phosphoric acid, $pK_{a1} = 2.0$. This should be the value of the pH of a buffer made of these components in equimolar quantities. When we mix sol-utions of the mono- and di-substituted salts, e.g. KH_2PO_4 and Na_2HPO_4, then in accordance with the equilibrium

$$H_2PO_4^- + H_2O \rightleftharpoons HPO_4^{2-} + H_3O^+ \tag{3.70}$$

the second dissociation constant of orthophosphoric acid will determine the pH of the solution, because

$$pH = pK_{a2} + \log \frac{[HPO_4^{2-}]}{[H_2PO_4^-]} \tag{3.71}$$

where $pK_{a2} = 6.9$. In a solution of a di- and a tri-substituted phosphate, e.g. Na_2HPO_4 and Na_3PO_4, the equilibrium

$$HPO_4^{2-} + H_2O \rightleftharpoons PO_4^{3-} + H_3O^+ \tag{3.72}$$

determines the pH, which in this case is the most basic, because

$$pH = pK_{a3} + \log \frac{[PO_4^{3-}]}{[HPO_4^{2-}]} \tag{3.73}$$

where $pK_{a3} = 11.7$. Thus, starting from orthophosphoric acid and its salts it is possible to prepare three different buffer solutions, which show buffer properties in three different ranges of pH.

Analogously, we can use the buffer properties of systems which orig-inate from other diprotic and polyprotic acids.

Equation (3.65), which determines the pH value of a buffer solution, is an exact equation. However, the introduction of total concentrations instead of equilibrium concentrations may be a source of error. These follow from the fact that acid dissociation, however weak, does occur, and causes a decrease in the concentration of undissociated molecules and an increase in the concentration of ions. Thus, the simplified formula cannot be used for every dilute solution, nor when the concentrations of H_3O^+ and OH^- ions are comparable with other concentrations. In these cases the equilibrium concentration of an acid differs from the total concentration by the concentration of the molecules which have dissociated. The concen-tration of dissociated molecules is equal to the equilibrium concentration of the anions, which in turn corresponds to the difference between the concentration of the hydrogen ions and the concentration of the hydroxide ions. Thus,

$$C_{HA} = [HA] + [A^-] \quad \text{or} \quad [HA] = C_{HA} - [A^-] \tag{3.74}$$

and since .

$$[H_3O^+] = [A^-] + [OH^-] \tag{3.75}$$

we have

$$[A^-] = [H_3O^+] - [OH^-] \tag{3.76}$$

and finally

$$[HA] = C_{HA} - ([H_3O^+] - [OH^-]) \tag{3.77}$$

Reasoning analogously, in the case of a base we obtain

$$[A^-] = C_A + ([H_3O^+] - [OH^-]) \tag{3.78}$$

After substitution of the results obtained, Eq. (3.63) takes the form

$$K_a = \frac{[H_3O^+]\{C_A + ([H_3O^+] - [OH^-])\}}{C_{HA} - ([H_3O^+] - [OH^-])} \tag{3.79}$$

This equation is rather complicated and not easy to solve. However, in a great majority of cases it is sufficient to use the simplified Eq. (3.65). It can also be used to judge when it is necessary to introduce more complicated expressions.

Example 3.18. Determine the pH value of a buffer solution made from equimolar quantities of sulphamic acid (NH_2SO_3H) and its sodium salt, if the concentrations are $0.10M$ and the pK_a of sulphamic acid is 0.65.

Because of the small value of pK_a, it may be assumed that, in the buffer obtained, the concentration of H_3O^+ ions is large in comparison with the total concentrations of the acid and the base (sulphamate anion), and thus cannot be neglected in calculations. However, it is admissible to neglect the concentration of OH^- ions. Thus, we use formula (3.79), omitting the $[OH^-]$ terms, and obtain

$$K_a = \frac{[H_3O^+](C_A + [H_3O^+])}{C_{HA} - [H_3O^+]} \tag{3.80}$$

and after transformation

$$[H_3O^+]^2 + [H_3O^+](C_A + K_a) - K_a C_{HA} = 0 \tag{3.81}$$

We substitute in this equation $C_A = 0.10$, $C_{HA} = 0.10$ and $K_a = 0.22$, obtaining

$$[H_3O^+]^2 + [H_3O^+](0.10 + 0.22) - 0.22 \times 0.10 = 0$$

so

$$[H_3O^+] = \frac{-0.32 + \sqrt{0.102 + 0.088}}{2} = 0.058 M$$

and

$$pH = 1.24$$

If we had solved this problem by use of the simplified formula, we would have obtained the wrong result: $[H_3O^+] = 0.22$ and $pH = 0.66$.

Example 3.19. Find the concentration of sodium sulphamate needed if the pH of a solution with $0.10M$ total concentration of sulphamic acid is to be 1.15 ($pK_a = 0.65$).

By Eq. (3.80), the same as that used in the previous example, we calculate C_A

$$C_A = \frac{-[H_3O^+]^2 + K_a(C_{HA} - [H_3O^+])}{[H_3O^+]}$$

Then

$$C_A = \frac{-5 \times 10^{-3} + 0.22(0.1 - 0.07)}{0.07} = \frac{-0.005 + 0.0066}{0.07}$$

$$= 0.023 M$$

If we had used the simplified form of the formula, we would have obtained the wrong result: $C_A = 0.31M$.

Solutions with a large concentration of strong acids or strong bases can also be regarded as buffer solutions, since addition of small quantities of an acid or a base does not bring about a large change of pH. In the case of strong acids we should consider the system as consisting of H_3O^+ as the acid and H_2O as the conjugate base and, in the case of strong bases, of H_2O as the acid and OH^- as the conjugate base. This permits uniform treatment of all systems which are known to act as buffers.

The concept of buffers can be extended to other types of system. Thus, a redox "buffer", or "poised" solution is a solution containing components which make it possible to maintain a relatively constant redox potential. *Buffers for metal ions* can be formed by systems where the displacement of a complexation equilibrium of a metal makes it possible to keep the concentration (activity) of free metal ions at a constant, generally low, level, e.g. $10^{-10}M$. Analogous buffers can be prepared for some anions or ligands.

3.7 BUFFER CAPACITY

We have shown that the most effective buffer properties are found when the concentration of the acid and the concentration of the base are equal and the pH is equal to the pK_a of the acid. In the process of adding a strong acid to a solution of this kind, the base of the buffer system is consumed and at the same time the conjugate acid is produced. Addition of a strong base produces the reverse reaction—the acid is consumed and the concentration of the conjugate base is increased. The pH of a buffer solution should not change by more than a unit.

A quantitative measure of the resistance of a buffer solution to the addition of strong acid or strong base is the *buffer capacity*. It is an expression of the rate at which the addition of strong acid or base changes the pH. Thus, buffer capacity is defined by the equation

$$\beta = \frac{dC}{dpH} \tag{3.82}$$

where dC denotes the number of moles of the strong base added (which caused a numerically equal increase in the concentration of the buffer base, the increase taking place at the expense of the conjugate acid). Addition of the same number of moles of a strong acid will cause an effect equal in absolute value, but in the opposite direction.

The total concentration of the base in the solution is

$$C_A = [A^-] + [OH^-] - [H_3O^+] \tag{3.83}$$

We assume that the total concentration of buffer components is

$$C = [HA] + [A^-] \tag{3.84}$$

After substituting the values of [HA] expressed by means of the dissociation constant of the acid, we obtain

$$C = \frac{[H_3O^+][A^-]}{K_a} + [A^-] \tag{3.85}$$

and then

$$[A^-] = \frac{CK_a}{K_a + [H_3O^+]} \tag{3.86}$$

We substitute this value in Eq. (3.83) to obtain

$$C_A = \frac{K_w}{[H_3O^+]} - [H_3O^+] + \frac{CK_a}{K_a + [H_3O^+]} \tag{3.87}$$

and differentiate with respect to $[H_3O^+]$

$$\frac{dC}{d[H_3O^+]} = -\frac{K_w}{[H_3O^+]^2} - 1 - \frac{CK_a}{(K_a + [H_3O^+])^2} \tag{3.88}$$

However, buffer capacity is a derivative of concentration with respect to pH and not with respect to concentration of H_3O^+ ions. Thus

$$\beta = \frac{dC}{dpH} = \frac{dC}{d[H_3O^+]} \frac{d[H_3O^+]}{dpH} = -2.303[H_3O^+]\frac{dC}{d[H_3O^+]}$$

(3.89)

On substitution of Eq. (3.88) in Eq. (3.89), we obtain

$$\beta = 2.303\left(\frac{K_w}{[H_3O^+]} + [H_3O^+] + \frac{CK_a[H_3O^+]}{(K_a + [H_3O^+])^2}\right)$$

(3.90)

This equation allows us to calculate the buffer capacity of any system containing a weak acid HA and the conjugate base, A^-. If the solution also contains other systems with buffer properties, then we must replace the last term inside the brackets by the sum of such terms calculated separately for each of the systems. Thus, we obtain the general expression

$$\beta = 2.303\left(\frac{K_w}{[H_3O^+]} + [H_3O^+] + \sum \frac{C_i K_{ai}[H_3O^+]}{(K_{ai} + [H_3O^+])^2}\right)$$

(3.91)

which is also applicable to buffer systems originating from polyprotic acids.

In Eqs. (3.90) and (3.91) there are two terms independent of the nature of the buffer used and corresponding to the concentrations of OH^- ions and of H_3O^+ ions. They determine the increase in buffer capacity in a strongly alkaline region and in a strongly acid region, and confirm the buffering character of concentrated solutions of strong acids and bases. On a diagram showing the dependence of β on pH (Fig. 3.3) these terms make the curve of buffer capacity increase rapidly at both ends of the pH scale. The buffer

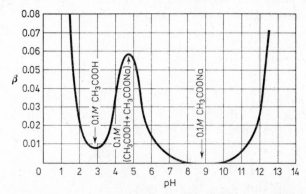

Fig. 3.3. Buffer capacity of a 0.1M acetic acid system as a function of pH. The minima on the diagram correspond to pure acetic acid and to pure sodium acetate and the maximum is an equimolar buffer solution.

action of any weak acid–weak base system, on the other hand, is shown by means of a bell-shaped curve with a maximum at $K_a = [H_3O^+]$. The height of the maximum depends on the total concentration of the buffer solution. If there are several buffer systems in a solution, then there are several maxi-

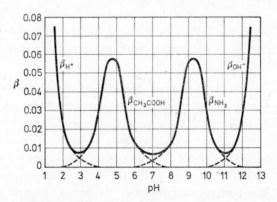

Fig. 3.4. The buffer capacity of a mixture of two systems, $0.10M$ acetic acid and $0.10M$ ammonia, as a function of pH. The pH was changed by adding strong acid or strong base. On the diagram are marked the branches of the curves corresponding to the individual values of β (dashed line) and their sum (continuous line).

ma (Fig. 3.4) on the curve of β vs. pH. It follows from Eq. (3.91) that the total buffer capacity is equal to the sum of the buffer capacities of the individual systems

$$\beta = \beta_{OH^-} + \beta_{H_3O^+} + \sum \beta_{HA/A^-} \tag{3.92}$$

Thus the diagram of β can be obtained by summing the diagrams for the individual systems.

Example 3.20. By using the concept of buffer capacity, calculate the change in the pH of a buffer solution consisting of $0.2M$ formic acid and $0.05M$ sodium formate, if we add to 1 litre of the solution 10 mmoles of strong acid. Assume K_a for formic acid is 2.5×10^{-4} and the ion product of water 2.5×10^{-14} ($pK_w = 13.60$). The pH of the solution, according to the calculations of Example 3.17, is 4.00.

According to Eq. (3.90), the buffer capacity is

$$\beta = 2.303 \left[2.5 \times 10^{-10} + 1.0 \times 10^{-4} + \frac{0.7 \times 2.5 \times 10^{-4} \times 1.0 \times 10^{-4}}{(2.5 \times 10^{-4} + 1.0 \times 10^{-4})^2} \right]$$

$$= 2.303 \times 1.43 \times 10^{-1} = 0.33$$

From the definition of buffer capacity (Eq. (3.82)) we can calculate the change of pH; by assuming that the differentials are approximately equal to small finite increments, we obtain

$$\Delta pH = \frac{\Delta C}{\beta} = \frac{10^{-2}}{0.33} = 0.03$$

This result is within the limits of error expected to result from the approximations used and is in accordance with the result obtained in Example 3.17.

The expressions derived indicate that as its dilution increases, the buffer capacity of a system decreases and that the stability of the pH of a buffer solution is only maintained at fairly high concentrations. If C_A (or C_{HA}) in the general formula for K_a has a value which is comparable with the concentrations of H_3O^+ and OH^-, then the constancy of the ratio of acid to base is not maintained and the pH of the buffer solution (which therefore ceases to be a true buffer) tends towards 7, i.e. to the value of pH in neutral solutions.

The equation for buffer capacity (Eq. (3.90)) can be transformed by replacing the dissociation constant K_a by the concentrations of ions and molecules. Then, after rearrangement, we obtain the equation

$$\beta = 2.303 \left(\frac{K_w}{[H_3O^+]} + [H_3O^+] + \frac{[HA][A^-]}{[HA] + [A^-]} \right) \tag{3.93}$$

which sometimes makes it easier to calculate the buffer capacity.

Example 3.21. Calculate the buffer capacity of a solution with pH $= 5.30$ which contains 1.0×10^{-2} mole of pyridine and 5.0×10^{-3} mole of HCl per litre.

In this solution there is a buffer consisting of equimolar quantities of pyridine and pyridine hydrochloride, the concentration of each component being $5.0 \times 10^{-3} M$. Consequently, the pH of the solution is equal to the value of the pK_a of the pyridinium ion. We calculate the buffer capacity from Eq. (3.93).

$$\beta = 2.303 \left[2.0 \times 10^{-9} + 5.0 \times 10^{-6} + \frac{(5.0 \times 10^{-3})^2}{1.0 \times 10^{-2}} \right]$$

$$= 2.303 \times 2.5 \times 10^{-3} = 5.8 \times 10^{-3} = 0.0058$$

Example 3.22. Calculate the required total concentration of a buffer solution with pH $= 5.0$, consisting of acetic acid and sodium acetate, in order that an addition of 5 mmoles of hydrochloric acid to 200 ml of the solution should cause a pH change of less than 0.05 unit. The dissociation constant of acetic acid is approximately $K_a = 1.8 \times 10^{-5}$.

It is first necessary to calculate the required buffer capacity of the solution. We start from the formula

$$\beta = \frac{\Delta C}{\Delta pH}$$

The change in the concentration is $\Delta C = (5/200)M = 0.025M$, so the change $\Delta pH = 0.05$ would occur when $\beta = 0.025/0.05 = 0.5$.

For a buffer solution with pH = 5 we can neglect the influence of $[H_3O^+]$ and $[OH^-]$ on the buffer capacity and write

$$\beta \cong 2.3 \frac{CK_a[H_3O^+]}{(K_a + [H_3O^+])^2}$$

so

$$C = \frac{\beta(K_a + [H_3O^+])^2}{2.3K_a[H_3O^+]} = \frac{0.5(1.8 \times 10^{-5} + 1.0 \times 10^{-5})}{2.3 \times 1.8 \times 10^{-5} \times 1.0 \times 10^{-5}} = 0.95M$$

Thus, if the total concentration of the buffer (i.e. the sum of the concentrations of acetic acid and sodium acetate) is not smaller than $0.95M$, then the pH does not change by more than 0.05 unit. Note, that in this example we have assumed the dissociation constant of acetic acid which is probably not exactly valid at such high ionic strength.

For practical calculations it is necessary to remember that Eqs. (3.90) and (3.91) were derived on the assumption that the changes of concentration (dC) were small and also that the changes in pH (dpH) were small. If this assumption is not valid, there can be a large calculation error. If in Example 3.22 we had wanted to calculate the buffer capacity of a solution for which the pH does not change more than by 0.5 unit when 50 mmoles of hydrochloric acid are added, application of the same reasoning would lead to incorrect results. Indeed, the addition of this quantity of acid would change the ratio C_{HA}/C_A from 1.79 ($\log C_{HA}/C_A = 0.25$) to 9.56 ($\log C_{HA}/C_A = 0.98$), which corresponds to a change in pH of 0.73, i.e. bigger than the one assumed.

Also, for such large changes of pH the buffer capacity on addition of an acid will differ from that on addition of a base, except when $C_{HA} = C_A$. To calculate in the general case the changes of the pH of a buffer solution on addition of an amount ΔC_{HA} of a strong acid, it is necessary to use general equations for the pH of a buffer solution. Thus, the change in pH will be

$$\Delta pH = \log \frac{C_{HA}}{C_A} - \log \frac{C_{HA} + \Delta C_{HA}}{C_A - \Delta C_{HA}}$$

or in the case of an addition of a strong base ΔC_A

$$pH = \log \frac{C_{HA}}{C_A} - \log \frac{C_{HA} - \Delta C_A}{C_A + \Delta C_A}$$

These equations can of course only be used if during the addition of an acid, $C_A > \Delta C_{HA}$, and if during the addition of a base, $C_{HA} > \Delta C_A$. If these conditions are not satisfied, we are really just considering a solution of a strong and a weak acid or a solution of a strong and a weak base, and we may consider the buffer capacity to be exceeded.

Example 3.23. A buffer solution with pH 9.0 has been obtained by acidification of $1M$ ammonia solution with $5M$ hydrochloric acid. Calculate whether the addition of 10 ml of $5M$ hydrochloric acid to 1 litre of this solution will cause a change of pH bigger than 0.20 unit. K_a for NH_4^+ is arbitrarily taken as 3×10^{-10}.

We first calculate the concentration of NH_3 (a base) and NH_4^+ (an acid) in the buffer solution. It is easy to calculate the ratio of these concentrations from the known pK_a and known pH of the solution

$$\log \frac{C_{HA}}{C_A} = pK_a - pH = 9.52 - 9.00 = 0.52$$

Hence $C_{HA}/C_A = 3.31$.

The total concentration of $[NH_4^+]$ and $[NH_3]$ has decreased in relation to the initial concentration of the base as a result of dilution during neutralization. To calculate the degree of dilution of the solution, we assume that the number of moles of ammonia before the addition of hydrochloric acid was $1 \times V_0$ moles, and the number of moles of the acid added, equal to the number of moles of the NH_4^+ produced, is $5 \times V$, where V_0 and V denote the respective volumes of the solutions of NH_3 and HCl. Thus, in the final buffer solution

$$\frac{[NH_4^+]}{[NH_3]} = \frac{5V}{1V_0 - 5V} = 3.31$$

whence

$$\frac{V}{V_0} = 0.154$$

i.e. the dilution of the initial solution is

$$\frac{V_0}{V_0 + V} = 0.867$$

which is numerically equal to the sum of $[NH_4^+] + [NH_3]$. Thus, the concentrations of the individual species in the solution are

$$[NH_4^+] = 0.666M \quad \text{and} \quad [NH_3] = 0.201M$$

After an addition of 10 ml of $5M$ hydrochloric acid these concentrations will change, the concentration of NH_4^+ will increase by $0.050M$ and the concentration of NH_3 will decrease by the same amount. Thus, the change in pH of the solution is

$$\Delta pH = \log 3.31 - \log \frac{0.666 + 0.050}{0.201 - 0.050} = -0.16$$

i.e. it is smaller than the value assumed in the example.

3.8 PREPARATION OF BUFFER SOLUTIONS

The correct preparation of buffer solutions is very important. As already mentioned, the greatest buffer capacity is shown by a buffer with a pH value numerically equal to the pK_a of the acid in it. For this reason, the first stage in the preparation of a buffer solution is the choice of suitable components with pK_a values as close to the required pH as possible; certainly the difference should not be more than 1 unit of pH. If the difference is any greater, i.e. if $|pH - pK_a| > 1$, the buffer effect is relatively small. For choosing the components of buffer solutions it is useful to have a table of dissociation constants of acids listed in order of increasing values of pK_a. Among the possible choices, we should then consider the availability of the components in adequate purity. We should also make sure that the buffer components do not react with the other components of the solution. Such reactions can often occur, e.g. the components of pyridine buffer can react with many metal ions to form stable pyridine complexes, and orthophosphoric acid, which produces otherwise satisfactory buffer solutions, can precipitate many metals as sparingly soluble orthophosphates.

The next stage is to calculate the required ratio of the base and acid concentrations. This calculation requires use of the basic equation for determining the pH of a buffer solution, Eq. (3.67). When the ratio of the concentrations is known, the concentrations of the individual components can be calculated from the required value of the buffer capacity of the system. When the total concentration of the buffer and the ratio of the concentrations of the two components are known, we can plan the practical method of preparing the solution.

A buffer solution can be prepared by several methods. It is easiest to take exactly measured quantities of the components (a weak acid and its conjugate base), mix them and dilute to volume. For example, a buffer may be prepared by mixing calculated volumes of a solution of NH_3 and a solution of NH_4Cl, or by dissolving in water the calculated weights of KH_2PO_4 and Na_2HPO_4.

This method is only practicable if the two components are available in a pure state. However, if we want to prepare a pyridine buffer and have pyridine but no pyridine hydrochloride, the buffer can be made by preparing a suitable solution of pyridine and neutralizing it with a solution of hydrochloric acid. The volume of acid to be added can be calculated, or the neutralization can be continued until the solution reaches the desired pH value (measured by means of a pH-meter).

Similarly, the starting substance can be a free acid, which we neutralize with a solution of a strong base such as sodium hydroxide, either by adding an exactly calculated volume, or adding it until a definite pH value is obtained.

These three methods usually produce essentially the same buffer solution. The only difference may be the presence of a neutral salt in the solution, if we start from a cationic acid or an anionic base. However, this salt need be taken into account only in more exact calculations, when activity coefficients must be considered.

Example 3.24. Decide how to prepare a buffer solution with pH = 10.0 and buffer capacity $\beta = 0.03$.

To prepare the buffer several acids might be used, e.g.

hydrogen cyanide	$pK_a = 9.2$
ammonium ion	$pK_a = 9.37$
aminoacetic acid	$pK_a = 9.7$
phenol	$pK_a = 9.8$
hydrogen carbonate ion	$pK_a = 10.1$
ethylammonium ion	$pK_a = 10.8$

With reference to the pK_a, suitable substance would be phenol; however, a phenolate buffer is unstable, so it is more convenient to use hydrogen carbonate, which is readily available as a sufficiently pure sodium salt.

To determine the required ratio of the concentrations of the acid and its conjugate base we use the equation

$$\log \frac{[A^-]}{[HA]} = pH - pK_a$$

from which

$$\log \frac{[CO_3^{2-}]}{[HCO_3^-]} = 10.0 - 10.1 = -0.1$$

and

$$\frac{[HCO_3^-]}{[CO_3^{2-}]} = 10^{0.1} = 1.26$$

To calculate the required total concentration of buffer constituents for buffer capacity 0.3, we can disregard $[H_3O^+]$ and $[OH^-]$ compared with the concentration of the buffer constituents, and write

$$\beta = 2.3 \frac{CK_a[H_3O^+]}{(K_a + [H_3O^+])^2}$$

and hence

$$C = \frac{\beta(K_a + [H_3O^+])^2}{2.3K_a[H_3O^+]}$$

and

$$C = \frac{0.3(8 \times 10^{-11} + 10^{-10})^2}{2.3 \times 8 \times 10^{-11} \times 10^{-10}} \simeq 0.53M$$

Since the concentration ratio of HCO_3^- to CO_3^{2-} is to be equal to 1.26, we set $[CO_3^{2-}] = x$ and $[HCO_3^-] = 1.26x$, so

$$x + 1.26x = 0.53$$

and finally

$$x = \frac{0.53}{2.26} = 0.235M = [CO_3^{2-}]$$

and

$$1.26x = 0.295M = [HCO_3^-]$$

Thus, by dissolving 0.235 mole of Na_2CO_3 and 0.295 mole of $NaHCO_3$ in 1 litre of water, we obtain a buffer solution with the desired properties.

To avoid the laborious calculations of buffer composition, it is possible to use tables published in many textbooks, which give the composition of typical buffer solutions. Table 3.3 gives examples of several such solutions. The buffer solution with the pH given in the last column is obtained by mixing the two components in the correct proportions. Obviously, such solutions cannot be used in regions far from the pH value equal to pK_a if a large buffer capacity is required. The calculations of the pH values given

Table 3.3

Compositions of some buffer solutions used in analytical practice

Acetate buffer			Phosphate buffer			Borate buffer		
0.2M CH$_3$COOH x ml	0.2M CH$_3$COONa y ml	pH	1/15M KH$_2$PO$_4$ x ml	1/15M Na$_2$HPO$_4$ y ml	pH	0.05M Na$_2$B$_4$O$_7$ x ml	0.1M HCl y ml	pH
9.5	0.5	3.42	9.5	0.5	5.59	5.5	4.5	7.93
9.0	1.0	3.72	9.0	1.0	5.91	6.0	4.0	8.27
8.0	2.0	4.05	8.0	2.0	6.24	7.0	3.0	8.67
7.0	3.0	4.27	7.0	3.0	6.47	8.0	2.0	8.89
6.0	4.0	4.45	6.0	4.0	6.64	9.0	1.0	9.07
5.0	5.0	4.63	5.0	5.0	6.81	10.0		9.22
4.0	6.0	4.80	4.0	6.0	6.98		0.1M	
3.0	7.0	4.99	3.0	7.0	7.17		NaOH	
2.0	8.0	5.23	2.0	8.0	7.38		y ml	
1.0	9.0	5.57	1.0	9.0	7.73			
0.5	9.5	5.89	0.5	9.5	8.04	9.0	1.0	9.32
						8.0	2.0	9.48
						7.0	3.0	9.65
						6.0	4.0	9.94
						5.5	4.5	10.28

Table 3.4

Compositions of the standard buffer solutions used for calibration purposes

Component	Concentration, g/l	Concentration, M	pH at 20°C	pH at 25°C
Potassium tetroxalate KH$_3$(C$_2$O$_4$)$_2 \cdot$ 2H$_2$O	12.70	0.05	1.68	1.68
Potassium hydrogen tartrate KHC$_4$H$_4$O$_6$	saturated at 25°C	~ 0.034		3.56
Potassium hydrogen phthalate KHC$_8$H$_4$O$_4$	10.21	0.05	4.00	4.01
Disodium hydrogen phosphate Na$_2$HPO$_4$	3.55	0.025 ⎫	6.88	6.86
Potassium dihydrogen phosphate KH$_2$PO$_4$	3.40	0.025 ⎬		
Borax Na$_2$B$_4$O$_7 \cdot$ 10H$_2$O	3.81	0.01	9.22	9.18
Calcium hydroxide Ca(OH)$_2$	saturated at 25°C	~ 0.0203	12.63	12.45

Note: the alkaline buffers should be prepared in CO$_2$-free water.

in this table take into account ionic strength, and so the values correspond to activities of hydrogen ions.

Precise measurements of pH by means of pH meters make use of certain standard buffer solutions that define a conventional pH scale. These have been recommended on the basis of careful research, and are prepared from substances of very high purity; they have a definite ionic strength. To achieve the highest precision, the standard buffer used for calibration of a pH meter should have a pH as close as possible to the pH of the unknown solution. This is particularly important when pH measurements are to be made in an alkaline solution with a glass electrode. Compositions of standard buffer solutions and their pH values at 20 and 25°C are given in Table 3.4.

3.9 SOLUTIONS OF POLYPROTIC ACIDS AND BASES

Quantitative consideration of equilibria in solutions of polyprotic acids and bases is much more complicated than it is for solutions of monoprotic acids and bases. Polyprotic acids and bases dissociate in a stepwise manner, hence their dissociation is described by as many different dissociation constants as there are protons which can dissociate. In the solution of a diprotic acid there is an equilibrium between the acid molecule H_2A, the ions HA^- and A^{2-}, and the H_3O^+ cation generated by each of the two dissociation steps. According to the Brønsted–Lowry theory, the H_2O molecule, the HA^- ion, and the H_3O^+ ion behave in this case as acids, while HA^-, A^{2-} and the H_2O molecule (in its role as a proton acceptor) function as bases. If an acid can undergo a three-step dissociation in aqueous solution (e.g. H_3PO_4) then the number of components in the solution increases and the system becomes even more complicated. It turns out, however, that polyprotic acid solutions may be treated, to a close approximation, as solutions of diprotic acids. This is also true for polyprotic bases. It should also be remembered that if we want to treat the orthophosphoric acid solution H_3PO_4 as a triprotic acid dissociating into $H_2PO_4^-$, HPO_4^{2-}, and PO_4^{3-}, then sodium orthophosphate Na_3PO_4, which in solution forms the PO_4^{3-} ion, should be treated as a triprotic base which during a three-step dissociation forms successively HPO_4^{2-}, $H_2PO_4^-$ and H_3PO_4.

Polyprotic acids and bases are weak electrolytes, except for a few cases where the first dissociation step corresponds to complete dissociation. In most cases, however, the dissociation constants of the successive dissociation steps differ greatly, sometimes by several orders of magnitude.

For this reason, only the first two dissociation steps need be taken into account in the solutions of polyprotic acids and bases. Further dissociation is so small that the concentrations of the species which are being generated may be neglected. Thus, to calculate the pH of a polyprotic acid solution only a diprotic acid needs to be considered, and of a polyprotic base, a diprotic base.

If in the general case we denote an acid by the symbol H_2A, then the equilibria in a solution are expressed by the equations

$$H_2A + H_2O \rightleftharpoons H_3O^+ + HA^- \tag{3.94}$$

$$HA^- + H_2O \rightleftharpoons H_3O^+ + A^{2-} \tag{3.95}$$

for which the equilibrium constants are the corresponding dissociation constants

$$K_{a1} = \frac{[H_3O^+][HA^-]}{[H_2A]} \tag{3.96}$$

$$K_{a2} = \frac{[H_3O^+][A^{2-}]}{[HA^-]} \tag{3.97}$$

When the solutions are very dilute, it is necessary, as for monoprotic acids, to consider also the dissociation of water. In that case, there are three equilibria, determined by three equilibrium constants. The general equation is thus of the fourth order with respect to $[H_3O^+]$. Fortunately, it is usually sufficient to consider only equilibria (3.94) and (3.95), which lead to a third order equation.

The total concentration of acids in the solution is

$$C_a = [H_2A] + [HA^-] + [A^{2-}] \tag{3.98}$$

and the concentration of H_3O^+ ions is

$$[H_3O^+] = [HA^-] + 2[A^{2-}] \tag{3.99}$$

The coefficient 2 appears because in the reaction of forming the A^{2-} ion from the H_2A molecule, two H_3O^+ ions are liberated. In both equations, $[H_2A]$ and $[A^{2-}]$ can be expressed as functions of $[H_3O^+]$ and $[HA^-]$, and we obtain

$$C_a = \frac{[H_3O^+][HA^-]}{K_{a1}} + [HA^-] + \frac{K_{a2}[HA^-]}{[H_3O^+]} \tag{3.100}$$

and

$$[H_3O^+] = [HA^-] + 2\frac{K_{a2}[HA^-]}{[H_3O^+]} \tag{3.101}$$

Dividing (3.100) by (3.101) leads to

$$\frac{C_a}{[H_3O^+]} = \frac{[H_3O^+]^2 + K_{a1}[H_3O^+] + K_{a1}K_{a2}}{K_{a1}([H_3O^+] + 2K_{a2})} \tag{3.102}$$

This equation can be rearranged into a third order equation with respect to $[H_3O^+]$

$$[H_3O^+]^3 + K_{a1}[H_3O^+]^2 + (K_{a2} - C_a)K_{a1}[H_3O^+]$$
$$- 2C_aK_{a1}K_{a2} = 0 \tag{3.103}$$

The solution of this is rather complicated: it can be solved by a trial and error procedure, in which approximate probable values of $[H_3O^+]$ are substituted into the equation. It can be solved iteratively, by use of a computer program. Several simplified methods can also be used.

For many acids the difference between the dissociation constants K_{a1} and K_{a2} is so large that the second dissociation step may be neglected. This approximation holds true not only in cases of such acids as sulphurous acid ($K_{a1} = 1.74 \times 10^{-2}$, $K_{a2} = 6.3 \times 10^{-8}$) or hydrogen sulphide ($K_{a1} = 6.3 \times 10^{-8}$, $K_{a2} = 6.3 \times 10^{-14}$) where the dissociation constants of the first step and the second step differ by several orders of magnitude, but also when the dissociation constants are much closer, e.g. in the case of succinic acid ($K_{a1} = 6 \times 10^{-5}$, $K_{a2} = 2.5 \times 10^{-6}$). Thus, in Eq. (3.103) we can omit the terms containing K_{a2} to obtain the simpler equation

$$[H_3O^+]^2 + K_{a1}[H_3O^+] - K_{a1}C_a = 0 \tag{3.104}$$

which is identical to the equation for a weak monoprotic acid. This equation can be solved either exactly, or by use of the same approximations as we used for weak monoprotic acids.

Example 3.25. Calculate the H_3O^+ concentration and the pH of $0.1 M$ oxalic acid $H_2C_2O_4$ if the pK_a values for the successive dissociation steps are 1.1 and 4.0 respectively.

Since the difference between pK_{a1} and pK_{a2} is about 3 units, we can neglect the second dissociation step and apply Eq. (3.104), and thus

$$[H_3O^+]^2 + 8.0 \times 10^{-2}[H_3O^+] - 8.0 \times 10^{-2} \times 10^{-1} = 0$$

so

$$[H_3O^+] = \frac{-8.0 \times 10^{-2} + \sqrt{6.4 \times 10^{-3} + 32.0 \times 10^{-3}}}{2} = 0.058$$

$$= 5.8 \times 10^{-2} M$$

and

$$pH = 1.24$$

The pH value obtained with the use of the more exact Eq. (3.103) would differ from this by less than 0.1 unit, so it is reasonable to treat this system as if it were a monoprotic acid.

Example 3.26. Distilled water is usually in equilibrium with air containing carbon dioxide. The consequent total concentration of CO_2 in the solution is $10^{-4}M$. Calculate the pH of the solution, if the carbonic acid dissociation constants are $K_{a1} = 4 \times 10^{-7}$ and $K_{a2} = 5 \times 10^{-11}$.

When CO_2 is dissolved in water, the solution becomes acidic as a result of acidic dissociation of the carbonic acid formed. The pH of the solution will thus be less than 7. The second dissociation constant need not be included in the calculations, since it is smaller than the first one by about 4 orders of magnitude. The system can be regarded as a solution of very dilute monoprotic acid, so we must consider both the dissociation of the acid and the dissociation of water. So

$$[H_3O^+] = [HCO_3^-] + [OH^-]$$

and

$$[H_3O^+] = \frac{K_{a1}[H_2CO_3]}{[H_3O^+]} + \frac{K_w}{[H_3O^+]}$$

Since the carbonic acid is only slightly dissociated, we have $[H_2CO_3] \cong C_a$ and thus

$$[H_3O^+]^2 = 4 \times 10^{-7} \times 1 \times 10^{-4} + 10^{-14} = 4 \times 10^{-11} + 10^{-14}$$

This equation indicates that, in spite of the small dissociation constant of carbonic acid, the dissociation of water may be neglected, hence

$$[H_3O^+] = 6.3 \times 10^{-6}M \quad \text{and} \quad pH = 5.2$$

We have solved this problem with the assumption that the dissociation constant of carbonic acid 4×10^{-7}, which corresponds to the total equilibrium

$$(CO_2 + H_2CO_3) + H_2O = H_3O^+ + HCO_3^-$$

(because almost all the CO_2 exists as a hydrate in the solution, in equilibrium with the carbonic acid). If we were given the concentration of H_2CO_3 molecules instead of the total concentration of $CO_2 + H_2CO_3$, we should use the dissociation constant $K_a = 1.6 \times 10^{-4}$.

A slightly better approximation can be obtained as follows: Since the HA^- ions which are formed during the first dissociation step undergo only negligible further dissociation, it can be stated that $[H_3O^+] = [HA^-]$. From this assumption, it follows that the A^{2-} anion concentration is equal to the second dissociation constant

$$K_{a2} = \frac{[H_3O^+][A^{2-}]}{[HA^-]} = [A^{2-}] \qquad (3.105)$$

Since the total concentration of H_3O^+ ions is equal to the sum of the concentrations of the H_3O^+ ions formed in the first step, $[H_3O^+]_1$, and of the H_3O^+ ions formed in the second step as a result of the dissociation of HA^-, we have

$$[H_3O^+] = [H_3O^+]_1 + [A^{2-}] = [H_3O^+] + K_{a2} \qquad (3.106]$$

More exact results can be obtained if we replace, in the equation of the second dissociation constant (3.105), the value of $[H_3O^+]$ by the sum of the concentrations of H_3O^+ ions formed in both steps, $[H_3O^+]_1 + + [H_3O^+]_2$, and the HA^- concentration by the concentration calculated from the first constant ($[H_3O^+]_1$) decreased by the fraction which has undergone dissociation, i.e. ($[H_3O^+]_1 - [H_3O^+]_2$). This substitution gives

$$K_{a2} = \left(\frac{[H_3O^+]_1 + [H_3O^+]_2}{[H_3O^+]_1 - [H_3O^+]_2} \right) [H_3O^+]_2 \qquad (3.107)$$

This method of calculation proves particularly helpful when the first dissociation step corresponds to a strong acid and the second step to a weak acid. Sulphuric acid is an example of such an acid.

Example 3.27. Calculate the H_3O^+ concentration in $0.005M$ sulphuric acid, if $K_{a2} = 1.2 \times 10^{-2}$.

Application of the approximate Eq. (3.106) leads to the result:

$$[H_3O^+] = [H_3O^+]_1 + K_{a2}$$
$$[H_3O^+] = 0.005 + 0.012 = 0.017M$$

This result is obviously erroneous, because even complete dissociation of both the protons of sulphuric acid would lead to an H_3O^+ concentration of $0.01M$ at most. Equation (3.107), leads to a nearly correct answer

$$K_{a2} = 0.012 = \left(\frac{0.005 + [H_3O^+]_2)}{0.005 - [H_3O^+]_2)} \right) [H_3O^+]_2$$

or

$$[H_3O^+]_2^2 + 0.017[H_3O^+]_2 - 6 \times 10^{-5} = 0$$

so

$$[H_3O^+]_2 = \frac{-1.7 \times 10^{-2} + \sqrt{2.89 \times 10^{-4} + 2.4 \times 10^{-4}}}{2}$$

$$= 3.0 \times 10^{-3} M$$

Consequently the total H_3O^+ concentration is

$$[H_3O^+] = 0.005 + 0.003 = 0.008M = 8.0 \times 10^{-3} M$$

and

$$pH = 2.10$$

A result obtained in this way is probably nearly correct; it agrees with an intuitive estimate of the H_3O^+ concentration and with experimental data.

Example 3.28. Calculate the pH and the H_3O^+ concentration in $1 \times 10^{-3} M$ ethylenediaminetetra-acetic acid solution (H_4Y). Dissociation constants: $K_{a1} = 8.5 \times 10^{-3}$, $K_{a2} = 1.8 \times 10^{-3}$, $K_{a3} = 7.0 \times 10^{-7}$, $K_{a4} = 4.6 \times 10^{-11}$.

We shall treat ethylenediaminetetra-acetic acid as a dibasic acid. This asssumption is justified because the third dissociation constant is much smaller than the second and thus also than the first. However, the small difference between the first and the second dissociation constants forces us to exercise caution in applying approximations. In this case we cannot assume that only the first dissociation step is significant.

Let us try applying Eq. (3.106). The concentration of H_3O^+ from the first dissociation step is calculated as for a monoprotic acid

$$[H_3O^+]_1^2 + 8.5 \times 10^{-3}[H_3O^+]_1 - 8.5 \times 10^{-3} \times 1 \times 10^{-3} = 0$$

On solving the quadratic equation, we obtain

$$[H_3O^+]_1 = 9.0 \times 10^{-4} M$$

According to Eq. (3.106) the total H_3O^+ concentration is obtained by adding to this the numerical value of the second acid dissociation constant

$$[H_3O^+] = 9.0 \times 10^{-4} + 1.8 \times 10^{-3} = 2.7 \times 10^{-3} M$$

which is nonsensical, because even if both the first two protons were completely dissociated, the H_3O^+ concentration could not be 2.7 times the molar concentration of the acid.

We shall therefore use Eq. (3.107), which should allow a correct calculation of $[H_3O^+]_2$. Then

$$1.8 \times 10^{-3} = \left(\frac{9.0 \times 10^{-4} + [H_3O^+]_2}{9.0 \times 10^{-4} - [H_3O^+]_2} \right) [H_3O^+]_2$$

so

$$[H_3O^+]_2^2 + 2.7 \times 10^{-3}[H_3O^+]_2 - 1.62 \times 10^{-6} = 0$$
$$[H_3O^+]_2^2 = 1.4 \times 10^{-4} M$$

On adding the concentrations of H_3O^+ from the two dissociation steps, we obtain the total concentration

$$[H_3O^+] = 9.0 \times 10^{-4} + 1.4 \times 10^{-4} = 1.04 \times 10^{-3} M$$

and

$$pH = 2.98$$

This result appears to be a reasonable one.

However, because the dissociation constants have very similar values $(K_{a1}:K_{a2} = 4.5)$, this result must also be viewed critically. The most reasonable way to check it is to find out whether the value of $[H_3O^+]$ obtained satisfies the general Eq. (3.103)

$$[H_3O^+]^3 + K_{a1}[H_3O^+]^2 + (K_{a2} - C_a)K_{a1}[H_3O^+] = 2C_a K_{a1} K_{a2}$$

On substitution of the numerical value $[H_3O^+] = 1.04 \times 10^{-3} M$, we obtain

$$(1.04 \times 10^{-3})^3 + 8.5 \times 10^{-3}(1.04 \times 10^{-3})^2$$
$$+ (1.8 \times 10^{-3} - 1.0 \times 10^{-3}) \times 8.5 \times 10^{-3} \times 1.04 \times 10^{-3}$$
$$= 2 \times 1.0 \times 10^{-3} \times 8.5 \times 10^{-3} \times 1.8 \times 10^{-3}$$

This leads to values of 1.74×10^{-8} for the left-hand side and 3.06×10^{-8} for the right-hand side, which indicates that the true value of $[H_3O^+]$ must be somewhat bigger than $1.04 \times 10^{-3} M$. If we try a slightly bigger value, e.g. $[H_3O^+] = 1.20 \times 10^{-3} M$ (pH = 2.92), the left-hand side is equal to 2.21×10^{-8}, which is closer to the true solution. The next test, $[H_3O^+] = 1.40 \times 10^{-3} M$ (pH = 2.85) gives an almost perfect answer because the value of the left-hand side polynomial is 2.90×10^{-8}. An even better result is given by $[H_3O^+] = 1.44 \times 10^{-3} M$ (pH = 2.84), the value of the left-hand side polynomial then being 3.04×10^{-8}. In this case the difference in the pH values is smaller than 0.01, which is within the accuracy of measurements.

In work with polyprotic acid solutions, it is often important to know the relative proportions of all the chemical species that make up the total concentration of the acid. Calculation of the concentrations is particularly important when a given pH is to be obtained. A typical problem concerns calculation of the concentration of the sulphide ion, S^{2-}, that is reacting to precipitate metal ions in a solution with a given pH. Another is the calculation of changes in the concentration of each species during a titration.

Let us consider a triprotic acid H_3A with the following dissociation constants:

$$K_{a1} = \frac{[H_3O^+][H_2A^-]}{[H_3A]}$$

$$K_{a2} = \frac{[H_3O^+][HA^{2-}]}{[H_2A^-]}$$

$$K_{a3} = \frac{[H_3O^+][A^{3-}]}{[HA^{2-}]}$$

For each product of the acid–base reactions we can write an equation determining its mole fraction

$$x_3 = \frac{[H_3A]}{C_{H_3A}}, \qquad x_2 = \frac{[H_2A^-]}{C_{H_3A}}$$

$$x_1 = \frac{[HA^{2-}]}{C_{H_3A}}, \qquad x_0 = \frac{[A^{3-}]}{C_{H_3A}} \tag{3.108}$$

since

$$C_{H_3A} = [H_3A] + [H_2A^-] + [HA^{2-}] + [A^{3-}]$$

the reciprocals of the mole fractions can be expressed as

$$\frac{1}{x_3} = \frac{[H_3A] + [H_2A^-] + [HA^{2-}] + [A^{3-}]}{[H_3A]}$$

$$\frac{1}{x_2} = \frac{[H_3A] + [H_2A^-] + [HA^{2-}] + [A^{3-}]}{[H_2A^-]}$$

$$\frac{1}{x_1} = \frac{[H_3A] + [H_2A^-] + [HA^{2-}] + [A^{3-}]}{[HA^{2-}]} \tag{3.109}$$

$$\frac{1}{x_0} = \frac{[H_3A] + [H_2A^-] + [HA^{2-}] + [A^{3-}]}{[A^{3-}]}$$

If we now express each equilibrium concentration as a function of $[H_3O^+]$ through the corresponding values of the dissociation constants, then, after rearrangement, we obtain

$$\frac{1}{x_3} = 1 + \frac{K_{a1}}{[H_3O^+]} + \frac{K_{a1}K_{a2}}{[H_3O^+]^2} + \frac{K_{a1}K_{a2}K_{a3}}{[H_3O^+]^3}$$

$$\frac{1}{x_2} = \frac{[H_3O^+]}{K_{a1}} + 1 + \frac{K_{a2}}{[H_3O^+]} + \frac{K_{a2}K_{a3}}{[H_3O^+]^2}$$

$$\frac{1}{x_1} = \frac{[H_3O^+]^2}{K_{a1}K_{a2}} + \frac{[H_3O^+]}{K_{a2}} + 1 + \frac{K_{a3}}{[H_3O^+]} \tag{3.110}$$

$$\frac{1}{x_0} = \frac{[H_3O^+]^3}{K_{a1}K_{a2}K_{a3}} + \frac{[H_3O^+]^2}{K_{a2}K_{a3}} + \frac{[H_3O^+]}{K_{a3}} + 1$$

These equations permit calculation of the contribution of each species to the total concentration of polyprotic acid at any given pH. Obviously in different pH regions certain terms may be neglected, because the values are very small in relation to the others.

To calculate the concentration of each species by use of these equations, it is only necessary (Eq. (3.108)) to divide the total concentration of polyprotic acid by the appropriate value of $1/x$.

Example 3.29. Calculate the concentrations of H_2S, HS^- and S^{2-} in a hydrogen sulphide solution with total concentration $0.02M$, in the presence of $0.1M$ hydrochloric acid. The dissociation constants for H_2S are $K_{a1} = 1.0 \times 10^{-7}$, $K_{a2} = 1.25 \times 10^{-13}$.

To calculate the concentrations we use the reciprocals of the mole fractions

$$\frac{1}{x_2} = 1 + \frac{K_{a1}}{[H_3O^+]} + \frac{K_{a1}K_{a2}}{[H_3O^+]^2}$$

$$\frac{1}{x_1} = \frac{[H_3O^+]}{K_{a1}} + 1 + \frac{K_{a2}}{[H_3O^+]}$$

$$\frac{1}{x_0} = \frac{[H_3O^+]^2}{K_{a1}K_{a2}} + \frac{[H_3O^+]}{K_{a2}} + 1$$

Substitution of numerical values leads to the equations

$$\frac{1}{x_2} = 1 + \frac{1.0 \times 10^{-7}}{1 \times 10^{-1}} + \frac{1.0 \times 10^{-7} \times 1.25 \times 10^{-13}}{1 \times 10^{-2}} = 1$$

$$\frac{1}{x_1} = \frac{1 \times 10^{-1}}{1.0 \times 10^{-7}} + 1 + \frac{1.25 \times 10^{-13}}{1 \times 10^{-1}} = 1.0 \times 10^6$$

$$\frac{1}{x_0} = \frac{1 \times 10^{-2}}{1.0 \times 10^{-7} \times 1.25 \times 10^{-13}} + \frac{1 \times 10^{-1}}{1.25 \times 10^{-13}} + 1$$

$$= 1.25 \times 10^{18}$$

The concentrations can then be calculated

$$[H_2S] = 2 \times 10^{-2}/1 = 2.0 \times 10^{-2}M$$

$$[HS^-] = 2 \times 10^{-2}/1.0 \times 10^6 = 2.0 \times 10^{-8}M$$

$$[S^{2-}] = 2 \times 10^{-2}/1.25 \times 10^{18} = 1.6 \times 10^{-20}M$$

Because of the very small values of the dissociation constants and because the total concentration of hydrogen sulphide may be taken to be equal to the concentration of H_2S molecules, the results obtained from the calculation above are similar to the result of an approximate calculation of $[S^{2-}]$ from the product of the dissociation constants, and of $[HS^-]$ from the first dissociation constant alone

$$K_{a1}K_{a2} = \frac{[H_3O^+]^2[S^{2-}]}{[H_2S]}, \qquad [S^{2-}] = K_{a1}K_{a2}\frac{C_{H_2S}}{[H_3O^+]^2}$$

$$K_{a1} = \frac{[H_3O^+][HS^-]}{[H_2S]}, \quad [HS^-] = K_{a1}\frac{C_{H_2S}}{[H_3O^+]}$$

In many cases, however, such approximations cannot be used.

Example 3.30. Calculate the concentrations of the neutral molecules and both ions of tartaric acid $H_2C_4H_4O_6$ (H_2A) when the total concentration is $1 \times 10^{-2}M$, the pH has been brought to 4.0. The values of pK_1 and pK_2 are 3.0 and 4.3.

The mole-fraction reciprocals are calculated from the general equations

$$\frac{1}{x_{H_2A}} = 1 + \frac{10^{-3.0}}{10^{-4.0}} + \frac{10^{-3.0} \times 10^{-4.3}}{10^{-8.0}}$$
$$= 1 + 10^{1.0} + 10^{0.7} = 16$$

$$\frac{1}{x_{HA^-}} = \frac{10^{-4.0}}{10^{-3.0}} + 1 + \frac{10^{-4.3}}{10^{-4.0}} = 0.1 + 1 + 0.5 = 1.6$$

$$\frac{1}{x_{A^{2-}}} = \frac{10^{-8.0}}{10^{-3.0} \times 10^{-4.3}} + \frac{10^{-4.0}}{10^{-4.3}} + 1 = 0.2 + 2 + 1 = 3.2$$

The actual concentrations are thus

$$[H_2A] = \frac{1 \times 10^{-2}}{16} = 6.25 \times 10^{-4}M$$

$$[HA^-] = \frac{1 \times 10^{-2}}{1.6} = 6.25 \times 10^{-3}M$$

$$[A^{2-}] = \frac{1 \times 10^{-2}}{3.2} = 3.125 \times 10^{-3}M$$

The total is $1 \times 10^{-2}M$, which confirms that the mole-fraction reciprocals were calculated correctly.

The methods given for calculating $[H_3O^+]$ and the concentrations of individual species in polyprotic acid solutions may also be used for polyprotic bases. Just as for monoprotic acids and bases, acid dissociation constants should be replaced by basic dissociation constants and H_3O^+ concentrations by OH^- concentrations in the equations.

Example 3.31. Calculate the pH of a $5 \times 10^{-3}M$ aqueous solution of ethylenediamine, if the acid dissociation constants of the appropriate conjugate acids $(NH_3CH_2CH_2NH_3)^{2+}$ and $(NH_2CH_2CH_2NH_3)^+$ under the conditions of the experiment are $10^{-7.5}$ and $10^{-10.6}$.

First we calculate the dissociation constants of the base: for the first dissociation step we get $K_{b1} = 10^{-14}/10^{-10.6} = 10^{-3.4}$, and for the second $K_{b2} = 10^{14}/10^{-7.5} = 10^{-6.5}$. Since the constants differ by about three orders of magnitude, we calculate the OH^- concentration, to a first approximation, from the first basic dissociation constant only

$$[OH^-]_1^2 + K_{b1}[OH^-]_1 - K_{b1}C_b = 0$$

$$[OH^-]_1 = \frac{-4.0 \times 10^{-4} + \sqrt{16.0 \times 10^{-8} + 4 \times 4.0 \times 10^{-4} \times 5 \times 10^{-2}}}{2}$$

$$= 2.7 \times 10^{-4} M$$

To calculate the total concentration of OH^- ions generated from the two dissociation steps we use the approximate equation to obtain

$$[OH^-] = [OH^-]_1 + K_{b2} = 2.7 \times 10^{-4} + 3.2 \times 10^{-7}$$

$$= 2.7 \times 10^{-4} M$$

Thus, only the OH^- ions from the first dissociation step need be taken into account. We calculate $[H_3O^+]$ and the pH of this solution

$$[H_3O^+] = \frac{10^{-14}}{2.7 \times 10^{-4}} = 3.7 \times 10^{-9} M$$

and

$$pH = 8.43.$$

As has already been stated, the same procedure may be used for molecular and ionic acids or bases. Consequently, we may use similar methods to calculate the pH of salts of polyprotic acids and bases such as sodium carbonate or sodium orthophosphate.

Example 3.32. Calculate the pH in $5 \times 10^{-3} M$ sodium arsenate (Na_3AsO_4) solution. Under the conditions of the experiment, the successive constants for basic dissociation of the AsO_4^{3-} ion are: $K_{b1} = 2.5 \times 10^{-4}$, $K_{b2} = 1.0 \times \times 10^{-7}$, $K_{b3} = 1.6 \times 10^{-12}$.

We first calculate the contribution of the OH^- ions produced from the first dissociation step, just as for a monoprotic base,

$$[OH^-]^2 + 2.5 \times 10^{-4}[OH^-] - 2.5 \times 10^{-4} \times 5.0 \times 10^{-3} = 0$$

$$[OH^-] = \frac{-2.5 \times 10^{-4} + \sqrt{6.25 \times 10^{-8} + 5.0 \times 10^{-6}}}{2}$$

$$= 1.0 \times 10^{-3} M$$

When we compare the resulting value with the second dissociation constant, we can see that further dissociation steps will not contribute significantly to production of OH^- ions, since

$$[OH^-] = 1.0 \times 10^{-3} + 1.0 \times 10^{-7} \cong 1.0 \times 10^{-3} M = 10^{-3.0} M$$

Finally, the pH value is

$$pH = 14.0 - 3.0 = 11.0$$

assuming that the value of the ion product of water is 1×10^{-14}.

Polyprotic acid ions appear in many of the reactions of analytical chemistry. For example, the S^{2-} ion participates in the dissolution reactions of sparingly soluble metal sulphides, and in complexation reactions use is often made of the properties of the tetraprotic ion of ethylenediaminetetraacetic acid (EDTA), which forms stable complexes with many metals. In such reactions the analyst is interested in the concentration at various pH values of the actual species that reacts with the metal ions. This species is usually the anion. For calculating these concentrations, side-reaction coefficients are of great practical value (see Section 1.5). These greatly facilitate the calculation of equilibrium constants for complex reactions, known as conditional constants. The side-reaction coefficients for the final product of dissociation of a polyprotic acid is defined as the reciprocal of the mole fraction. Since the only factor determining the value of the mole fractions is the concentration of H_3O^+, (i.e. the pH of the solution) it is convenient to list calculated values in the form of tables or graphs.

For example, for a tetraprotic anion, such as the anion of EDTA, the side-reaction coefficient $\alpha_{EDTA(H)}$ is expressed by the equation

$$\alpha_{EDTA(H)} = 1 + \frac{[H_3O^+]}{K_{a4}} + \frac{[H_3O^+]^2}{K_{a3}K_{a4}} + \frac{[H_3O^+]^3}{K_{a2}K_{a3}K_{a4}}$$
$$+ \frac{[H_3O^+]^4}{K_{a1}K_{a2}K_{a3}K_{a4}} \tag{3.111}$$

where K_{a1}, K_{a2}, K_{a3} and K_{a4} denote the successive acid dissociation constants of the acid. Naturally, for an acid with a smaller number of protons the equation is simpler.

Example 3.33. Calculate the side-reaction coefficients for the tartrate anion for a range of pH values if the successive dissociation constants of tartaric acid are $10^{-3.0}$ and $10^{-4.3}$,

In this case, the equation for calculation is

$$\alpha_{tart(H)} = 1 + \frac{[H_3O^+]}{K_{a2}} + \frac{[H_3O^+]^2}{K_{a1}K_{a2}}$$

With it, we calculate the side-reaction coefficients for successive pH values.

When pH $= 0$

$$\alpha_{tart(H)} = 1 + \frac{10^0}{10^{-4.3}} + \frac{10^0}{10^{-7.3}} = 1 + 10^{4.3} + 10^{7.3} = 10^{7.3}$$

(In this calculation, as in any other, in a sum of widely differing terms we can neglect terms which are less than 5% of the largest term.)

When pH $= 1$

$$\alpha_{tart(H)} = 1 + \frac{10^{-1}}{10^{-4.3}} + \frac{10^{-2}}{10^{-7.3}} = 1 + 10^{3.3} + 10^{5.3} = 10^{5.3}$$

When pH $= 2$

$$\alpha_{tart(H)} = 1 + \frac{10^{-2}}{10^{-4.3}} + \frac{10^{-4}}{10^{-7.3}} = 1 + 10^{2.3} + 10^{3.3}$$

$$= 2 \times 10^2 + 2 \times 10^3 = 2.2 \times 10^3 = 10^{3.34}$$

When pH $= 3$

$$\alpha_{tart(H)} = 1 + \frac{10^{-3}}{10^{-4.3}} + \frac{10^{-6}}{10^{-7.3}} = 1 + 10^{1.3} + 10^{1.3}$$

$$= 1 + 2 \times 10^1 + 2 \times 10^1 = 41 = 10^{1.61}$$

When pH $= 4$

$$\alpha_{tart(H)} = 1 + \frac{10^{-4}}{10^{-4.3}} + \frac{10^{-8}}{10^{-7.3}} = 1 + 10^{0.3} + 10^{-0.7}$$

$$= 1 + 2 + 0.2 = 3.2 = 10^{0.50}$$

When pH $= 5$

$$\alpha_{tart(H)} = 1 + \frac{10^{-5}}{10^{-4.3}} + \frac{10^{-10}}{10^{-7.3}} = 1 + 10^{-0.7} + 10^{-2.7}$$

$$= 1 + 0.2 = 1.2 = 10^{0.08}$$

If the pH is 6 or more, then all fractional terms of the sum are much less than 1 and the value of the side-reaction coefficient is $1 = 10^0$.

Since the values of side-reaction coefficients vary greatly with the pH of the solution, they are usually given in logarithmic form. Table 3.5 gives such data for some anions. Since in many cases, graphical representation is sufficiently exact, we often come across graphs of $\log \alpha$ vs. pH. Several such graphs are given in Fig. 3.5.

To calculate the actual concentration of an anion by the use of side-reaction coefficients, the value of the total concentration C has to be divided by the value of the coefficient α

Table 3.5

Values of log α for some conjugate bases of weak acids, at various pH values

Acid	pH														
	0	1	2	3	4	5	6	7	8	9	10	11	12	13	14
								log α							
Acetic acid	4.8	3.8	2.8	1.8	0.8	0.3	0.1								
Arsenic acid	20.6	17.7	14.8	12.5	10.4	8.4	6.5	4.7	3.6	2.5	1.5	0.6	0.1		
Carbonic acid	16.3	14.3	12.3	10.3	8.3	6.3	4.5	3.1	2.0	1.0	0.3				
Ethylenediaminetetra-acetic acid	21.4	17.4	13.7	10.8	8.6	6.6	4.8	3.4	2.3	1.4	0.5	0.1			
Hydrogen sulphide	19.5	17.5	15.5	13.5	11.5	9.5	7.6	5.9	4.6	3.6	2.6	1.6	0.7	0.1	
Orthophosphoric acid	20.7	17.7	15.0	12.7	10.6	8.6	6.7	5.0	3.7	2.7	1.7	0.8	0.2		
Oxalic acid	5.1	3.4	2.1	1.1	0.3	0.1									
Tartaric acid	7.3	5.3	3.3	1.6	0.5	0.1									
Ammonium ion	9.4	8.4	7.4	6.4	5.4	4.4	3.4	2.4	1.4	0.5	0.1				

$$[X^-] = \frac{C}{\alpha} \tag{3.112}$$

Example 3.34. Calculate the concentration of the anion of oxalic acid $C_2O_4^{2-}$ in a solution with pH = 2 when the total oxalate concentration (i.e. $[H_2C_2O_4] + [HC_2O_4^-] + [C_2O_4^{2-}] = C$) is 0.10$M$.

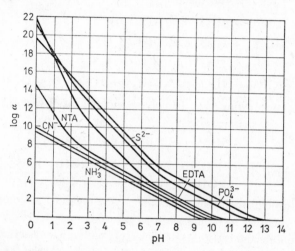

Fig. 3.5. Dependence of side-reactions coefficients, as logα, on pH for some bases (EDTA—the anion of ethylenediaminetetra-acetic acid, NTA—the anion of nitrilotriacetic acid).

We first look up the value of the logarithm of the side-reaction coefficient, α. Then

$$\log[C_2O_4^{2-}] = \log C - \log \alpha = -1 - 2.1 = -3.1$$

This result means that

$$[C_2O_4^{2-}] = 10^{-3.1} = 8.0 \times 10^{-4} M$$

3.10 ACID–BASE EQUILIBRIA INVOLVING METAL IONS

One of the facts effectively explained by the Brønsted–Lowry theory is the acid reaction of many metal ion salts. As mentioned earlier, this is due to the dissociation of protons from the water molecules solvating the metal ions. In addition to this qualitative explanation, the Brønsted–Lowry theory makes it possible to calculate equilibrium states in solutions. This is of great practical value since such systems are often encountered in analytical chemistry. However, they are rather complicated, and it is usually insufficient to treat them as mono- or even polyprotic acid systems. In the literature different approaches to these systems may be found. Acid–base equilibrium constants may be treated formally either as acid dissociation constants (or protonation constants) or as stability constants of hydroxide complexes. Thus, for the reactions

$$M^{2+} + H_2O \rightleftharpoons MOH^+ + H^+$$
$$MOH^+ + H_2O \rightleftharpoons M(OH)_2 + H^+$$
$$M(OH)_2 + H_2O \rightleftharpoons M(OH)_3^- + H^+ \qquad (3.113)$$
$$M(OH)_3^- + H_2O \rightleftharpoons M(OH)_4^{2-} + H^+$$

the equilibrium constants are the corresponding acid dissociation constants

$$K_{a1} = \frac{[MOH^+][H^+]}{[M^{2+}]}, \quad K_{a2} = \frac{[M(OH)_2][H^+]}{[MOH^+]}$$

$$\qquad (3.114)$$

$$K_{a3} = \frac{[M(OH)_3^-][H^+]}{[M(OH)_2]}, \quad K_{a4} = \frac{[M(OH)_4^{2-}][H^+]}{[M(OH_3)^-]}$$

These acidic dissociation reactions may be regarded as composite complexation reactions in which protons are liberated. It is customary to mark the equilibrium constants of such reactions with an asterisk. Therefore instead of writing K_{a1}, K_{a2}, K_{a3} and K_{a4} we may write $*K_1, *K_2, *K_3$ and $*K_4$. Also for the overall reaction

$$M^{2+} + nH_2O \rightleftharpoons M(OH)_n^{2-n} + nH^+$$

the equilibrium constant is denoted by the symbol $*\beta_n = *K_1 \cdot *K_2 \ldots *K_n$.

The same reactions may also be represented as simple complex formation, characterized by the stability constants of the complexes formed in the reactions

$$M^{2+} + OH^- \rightleftharpoons MOH^+$$
$$MOH^+ + OH^- \rightleftharpoons M(OH)_2$$
$$M(OH)_2 + OH^- \rightleftharpoons M(OH)_3^- \qquad (3.115)$$
$$M(OH)_3^- + OH^- \rightleftharpoons M(OH)_4^{2-}$$

in the form

$$K_1 = \frac{[MOH^+]}{[M^{2+}][OH^-]}, \quad K_2 = \frac{[M(OH)_2]}{[MOH^+][OH^-]}$$

$$\qquad (3.116)$$

$$K_3 = \frac{[M(OH)_3^-]}{[M(OH)_2][OH^-]}, \quad K_4 = \frac{[M(OH)_4^{2-}]}{[M(OH)_3^-][OH^-]}$$

The interdependence of the two kinds of constants may be presented in the following manner:

$$*K_1 = K_1 K_w, \quad *K_2 = K_2 K_w$$
$$*K_3 = K_3 K_w, \quad *K_4 = K_4 K_w \tag{3.117}$$

As can be seen from these equations, in each of the two ways of representing the reactions the water molecules which solvate the ions taking part in the reactions are omitted from the reaction equations.

In many cases, the differences between the successive equilibrium constants are not large: for example for Ag^+ the values of the successive acid dissociation constants are actually equal: $*K_1 = 1.3 \times 10^{-12}$, $*K_2 = 1.3 \times \times 10^{-12}$; this means that there is a possibility of more than one reaction taking place simultaneously. Therefore, to calculate the concentration of free, uncomplexed ions of the metal that result from a given total concentration, it is convenient to use the side-reaction coeffcient $\alpha_{M(OH)}$. This coefficient shows the relationship between the concentration of free (hydrated) metal ions plus all the forms resulting from proton dissociation ($= [M']$), and the actual concentration of free metal ions

$$\alpha_{M(OH)} = \frac{[M']}{[M]} = \frac{[M] + [MOH] + [M(OH)_2] + \ldots}{[M]} \tag{3.118}$$

(charges on the ions have been omitted). Calculation of $\alpha_{M(OH)}$ coefficients is simple if all the equilibrium constants (or at least all the significant ones) are known

$$\alpha_{M(OH)} = 1 + K_1[OH^-] + K_1 K_2[OH^-]^2 + \ldots \tag{3.119}$$

or

$$\alpha_{M(OH)} = 1 + *K_1[H_3O^+]^{-1} + *K_1 *K_2[H_3O^+]^{-2} + \ldots \tag{3.120}$$

This treatment of equilibria in metal ion solutions does not allow for the formation of sparingly soluble hydroxides. These arise when the positive charge of a metal ion is completely neutralized by the negative charges of hydroxide ions. Equilibrium with the precipitate is described by the solubility product. Most often the equation for the solubility product is written in terms of uncomplexed metal ions only, i.e.

$$K_{so} = [M^{n+}][OH^-]^n \tag{3.121}$$

which is the equilibrium constant of the reaction

$$M(OH)_n \rightleftharpoons M^{n+} + nOH^- \tag{3.122}$$

Another formulation of the solubility product is possible, in which it is the equilibrium constant of a protolytic reaction

$$M(OH)_n + nH_3O^+ \rightleftharpoons M^{n+} + 2nH_2O \tag{3.123}$$

expressed by the equation

$$*K_{so} = \frac{[M^{n+}]}{[H_3O^+]^n} = \frac{K_{so}}{K_w^n} \tag{3.124}$$

By use of solubility products, the concentration of metal ions in equilibrium with the precipitate for any given pH (or pOH) value can be calculated.

Thus, in the general case

$$[M^{n+}] = K_{so}[OH^-]^{-n} = K_{so}K_w^{-n}[H_3O^+]^n = *K_{so}[H_3O^+]^n \tag{3.125}$$

or in logarithmic form

$$pM = pK_{so} - npK_w + npH = p*K_{so} + npH \tag{3.126}$$

The way in which the pH precipitation of a metal hydroxide depends on the total concentration of metal is governed by two sets of relationships. Thus, it is the value of $[M^{n+}]$ obtained from the solubility product that must be substituted into the expression for $\alpha_{M(OH)}$

$$[M'] = \alpha_{M(OH)}[M^{n+}] = \alpha_{M(OH)}*K_{so}[H_3O^+]^n \tag{3.127}$$

or

$$[M'] = (1 + K_{a1}[H_3O^+]^{-1} + K_{a1}K_{a2}[H_3O^+]^{-2} + ...)*K_{so}[H_3O^+]^n \tag{3.128}$$

From this expression we can easily calculate $[M']$, but in the general case the determination of the pH at which precipitation of metal hydroxide begins requires solution of an equation of higher order with respect to $[H_3O^+]$. Therefore, we usually use a graphical approximation based on the $\log[M']$ vs. pH graph (Fig. 2.3).

The accuracy of readings taken from the graph usually seems to be at least as good as the accuracy of the equilibrium constants used to prepare the graph.

Example 3.34. The logarithms of the stability constants of the hydroxide complexes of gallium(III) are $\log\beta_1 = 11.4$, $\log\beta_2 = 22.1$, $\log\beta_3 = 31.7$ and $\log\beta_4 = 39.4$ Calculate the corresponding acid dissociation constants of $Ga(H_2O)_4^{3+}$. Express the changes in the gallium ion concentrations as a function of pH, using the side-reaction coefficient.

We calculate the successive stability constants from the ratios of the corresponding overall constants (or the differences of their logarithms), and then we calculate the successive acid dissociation constants by multiplying the K_n values by the ion product of water

$$\log K_1 = \log\beta_1 = 11.4, \quad \log *K_1 = 11.4 - 14.0 = -2.6$$

$$\log K_2 = \log\beta_2 - \log\beta_1 = 10.7, \qquad \log^*K_2 = 10.7 - 14.0 = -3.3$$

$$\log K_3 = \log\beta_3 - \log\beta_2 = 9.6, \qquad \log^*K_3 = 9.6 - 14.0 = -4.4$$

$$\log K_4 = \log\beta_4 - \log\beta_3 = 7.7, \qquad \log^*K_4 = 7.7 - 14.0 = -6.3$$

From the resulting $^*K_n$ values the pH regions in which a particular ionic species is dominant can readily be calculated. In the region pH $<$ 2.6 the Ga^{3+} ion is dominant; in the region $2.6 <$ pH < 3.3 the $GaOH^{2+}$ ion is dominant; for the region $3.3 <$ pH < 4.4 it is the ion $Ga(OH)_2^+$; at $4.4 <$ pH < 6.3 the undissociated $Ga(OH)_3$ molecule is present in highest concentration, and when pH > 6.3, $Ga(OH)_4^-$ dominates.

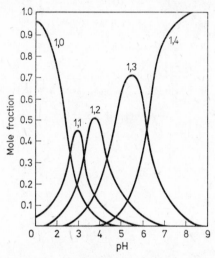

Fig. 3.6. Mole fractions of the individual complexes as a function of the pH for gallium(III)–OH^- system. The total Ga(III) concentration is $10^{-5}M$ (from C. F. Baes Jr. and R. E. Mesmer, *The Hydrolysis of Cations*, Wiley, New York, 1976; by permission of the copyright holders).

These findings are true for a solution which, at pH $\gtrsim$ 3, is in equilibrium with the $Ga(OH)_3$ [or $GaO(OH)$] precipitate. For this reason, on the y-axis of the graph showing the predominance areas of the various ionic forms (Fig. 3.6) we plot the mole fractions of only the species found in the solution. The relation presented here holds only for dilute solutions of Ga(III) salts, i.e. $10^{-5}M$. In more concentrated solutions of most metal-ion salts, polynuclear complexes are formed.

To calculate the side-reaction coefficient $\alpha_{Ga(OH)}$ we start from the general expression:

$$\alpha_{Ga(OH)} = 1 + 10^{-2.6}[H_3O^+]^{-1} + 10^{-5.9}[H_3O^+]^{-2}$$

$$+ \; 10^{-10.3}[H_3O^+]^{-3} + 10^{-16.6}[H_3O^+]^{-4}$$

Let us calculate the $\alpha_{Ga(OH)}$ values for several pH values:

$$pH = 2.5 \quad \text{or} \quad [H_3O^+] = 10^{-2.5}$$

$$\alpha_{Ga(OH)} = 1 + 10^{-0.1} + 10^{-0.9} + 10^{-2.8} + 10^{-6.6} = 1.9$$
$$= 10^{0.28}$$

$$pH = 3.5 \quad \text{or} \quad [H_3O^+] = 10^{-3.5}$$

$$\alpha_{Ga(OH)} = 1 + 10^{0.9} + 10^{1.1} + 10^{0.2} + 10^{-2.6} = 23.2 = 10^{1.37}$$

$$pH = 4.5 \quad \text{or} \quad [H_3O^+] = 10^{-4.5}$$

$$\alpha_{Ga(OH)} = 1 + 10^{1.9} + 10^{3.1} + 10^{3.2} + 10^{2.4} = 3180 = 10^{3.50}$$

$$pH = 5.5 \quad \text{or} \quad [H_3O^+] = 10^{-5.5}$$

$$\alpha_{Ga(OH)} = 1 + 10^{2.9} + 10^{5.1} + 10^{6.2} + 10^{6.4} = 4.2 \times 10^6$$
$$= 10^{6.63}$$

Figure 3.7 is a graph of the $\alpha_{Ga(OH)}$ values as a function of pH.

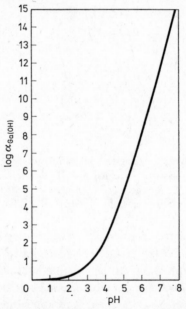

Fig. 3.7. The dependence of $\log \alpha_{Ga(OH)}$ on pH, showing the growing contribution of gallium(III) hydroxide complexes (neglecting polynuclear complexes).

Example 3.35. Calculate the successive acid dissociation constants of $Ni(H_2O)_4^{2+}$ if the stability constants of the hydroxide complexes are $\log \beta_1 = 3.8$, $\log \beta_2 = 8.8$, $\log \beta_3 = 12.0$ and $\log \beta_4 = 12.0$. Find the pH region of predominance for each complex and draw the graph of log $\alpha_{Ni(OH)}$ as a function of pH.

Before the successive (stepwise) acid dissociation constants can be calculated, the successive (stepwise) complex stability constants must be found

$$\log K_1 = \log\beta_1 = 3.8, \qquad \log{}^*K_1 = 3.8-14 = -10.2$$
$$\log K_2 = \log\beta_2 - \log\beta_1 = 5.0, \qquad \log{}^*K_2 = 5.0-14 = -9.0$$
$$\log K_3 = \log\beta_3 - \log\beta_2 = 3.2, \qquad \log{}^*K_3 = 3.2-14 = -10.8$$
$$\log K_4 = \log\beta_4 - \log\beta_3 = 0.0, \qquad \log{}^*K_4 = 0.0-14 = -14.0$$

If the values of $^*K_n$ increase regularly, the determination of the predominance region for each complex is obvious (Example 3.34). However, in this case the situation is more complicated, and although at the pH corresponding to the value of a given constant there should be equal concentrations of the two species concerned, it does not necessarily follow that these are the predominant species. For example, the value of $\log{}^*K_2$ indicates that at pH 9.0 there will be equal concentrations of $Ni(OH)_2$ and $NiOH^+$, but the dominant species at this pH is in fact $Ni(H_2O)_4^{2+}$, as can readily be seen by calculating the ratios $[Ni(H_2O)_4^{2+}]/[NiOH^+] = [H_3O^+]/{}^*K_1$ and $[Ni(H_2O)_4^{2+}]/[Ni(OH)_2] = [H_3O^+]^2/{}^*K_1{}^*K_2$ at this pH. The values are the same, $10^{1.2}$, showing equal concentrations for the two minor species. On the other hand the concentrations of $Ni(H_2O)_4^{2+}$ and $Ni(OH)_2$

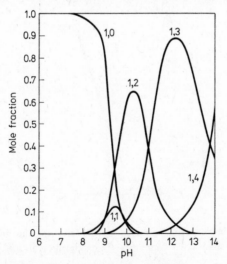

Fig. 3.8. Mole fractions of the individual complexes as a function of pH for the nickel(II)–OH$^-$ system. The total Ni(II) concentration is $10^{-5}M$ (from C. F. Baes Jr. and R. E. Mesmer, *The Hydrolysis of Cations*, Wiley, New York, 1976; by permission of the copyright holders).

are equal when $[H_3O^+]^2 = {}^*\beta_2 = {}^*K_1{}^*K_2$, i.e. $[H_3O^+] = \sqrt{10^{-19.2}}$ $= 10^{-9.6}$ or pH = 9.6.

Graphical methods of representing equilibria will be discussed later, in Section 3.14, but the Ni^{2+}–OH^- system, i.e. the protolytic equilibria of the $Ni(H_2O)_4^{2+}$ ion, is illustrated in Fig. 3.8.

Calculation of the side-reaction coefficient (of a protolytic reaction causing the lowering of the effective concentration (activity) of the hydrated metal ion) is done as before, by using the equation

$$\alpha_{Ni(OH)} = 1 + {}^*\beta_1[H_3O^+]^{-1} + {}^*\beta_2[H_3O^+]^{-2}$$
$$+ {}^*\beta_3[H_3O^+]^{-3} + {}^*\beta_4[H_3O^+]^{-4}$$
$$= 1 + 10^{-10.2}[H_3O^+]^{-1} + 10^{-19.2}[H_3O^+]^{-2}$$
$$+ 10^{-30.0}[H_3O^+]^{-3} + 10^{-44.0}[H_3O^+]^{-4}$$

As an example, the values of $\alpha_{Ni(OH)}$ are calculated for pH 10

$$\alpha_{Ni(OH)} = 1 + 10^{-0.2} + 10^{0.8} + 10^0 + 10^{-4.0} = 7.9 = 10^{0.9}$$

and for pH 12

$$\alpha_{Ni(OH)} = 1 + 10^{1.8} + 10^{4.8} + 10^{6.0} + 10^{4.0} = 10^{6.0}$$

The graph of $\alpha_{Ni(OH)}$ as a function of pH is shown in Fig. 3.9.

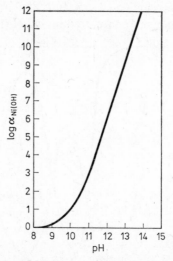

Fig. 3.9. The dependence of $\log\alpha_{Ni(OH)}$ on pH, showing the growing contribution of nickel(II) hydroxide complexes (neglecting polynuclear complexes).

Example 3.36. Calculate the concentration of hydrated Cu^{2+} ions in equilibrium with a $Cu(OH)_2$ precipitate at pH 6 when the solubility product of $Cu(OH)_2$ is $K_{s0} = 10^{-19.1}$.

At pH 6 the concentration of OH^- is $10^{-8}M$ and therefore

$$[Cu^{2+}] = \frac{10^{-19.1}}{(10^{-8})^2} = 10^{-3.1}M = 7.94 \times 10^{-4}M$$

Thus, if the Cu^{2+} concentration exceeds this value, precipitation of cupric hydroxide will begin.

It should be noted, that the value given here for the solubility product refers to the fresh precipitate considered in this example. After a time this precipitate changes into CuO, for which the solubility product is $K_{so} = 10^{-20.1}$.

Example 3.37. Calculate the maximum pH of $10^{-2}M$ zinc ion solution from which zinc hydroxide (solubility product $K_{so} = 1.26 \times 10^{-16}$) does not precipitate. To a first approximation, we neglect the formation of soluble hydroxide complexes of zinc.

By use of Eq. (3.126), we calculate

$$pH \leqslant \frac{1}{n}(pM - pK_{so}) + pK_w$$

and on substitution of numerical values, we obtain

$$pH \leqslant \frac{1}{2}(2 - 15.9) + 14 = 7.05$$

Next, the value of the side-reaction coefficient $\alpha_{Zn(OH)}$ must be checked for this pH, in order to determine whether, in addition to free zinc ions, the solution contains any significant amount of hydroxo-complexes. Therefore, considering first the first step leading to $ZnOH^+$, for which $K_1 = 10^{4.8}$, we calculate

$$\alpha_{Zn(OH)} = 1 + 10^{4.8}\frac{10^{-14}}{10^{-7.3}} = 1 + 10^{-1.9} = 1.01$$

This value indicates that the contribution of the first complex at this pH value is about 1%, and thus can be neglected. In practice, at a total zinc concetration of $10^{-2}M$ zinc hydroxide is found to start to precipitate when the pH is 7.1.

This treatment of acid–base equilibria of metals is not yet completely exact, since the formation of polynuclear complexes has not yet been considered. Polynuclear complexes are complexes in which the complex ion (molecule) contains more than one metal ion. They occur particularly with multivalent metal ions and diverse formulae are found, e.g. $Sc_2(OH)_2^{4+}$, $Sc_3(OH)_4^{5+}$, $Al_2(OH)_2^{4+}$, $Bi_6(OH)_{12}^{6+}$.

Nevertheless, in many instances the data on the formation of such complexes should be looked at sceptically, since sometimes the only proof of their existence is the result of complicated calculations on experimental data which may be subject to an experimental error of great significance. The formation of polynuclear complexes is a process which runs parallel to the formation of mononuclear hydroxide complexes; concentrations of polynuclear species tend to increase as the concentration of metal ions increases. For ferric ions the process may occur as follows:

$$Fe^{3+} \longrightarrow FeOH^{2+} \longrightarrow Fe(OH)_2^+ \longrightarrow Fe_2(OH)_2^{4+}$$

The formation of polynuclear complexes greatly complicates the calculations of ionic equilibria. In the case of mononuclear systems the side-reaction coefficient $\alpha_{M(OH)}$ is a function of the hydrogen ion concentration only, but formation of polynuclear complexes gives rise to a new term which depends on the concentration of metal ion uncomplexed by OH^- ions. For the case of Fe(III) ions we can thus write

$$\alpha_{Fe(OH)} = \frac{[Fe']}{[Fe^{3+}]}$$

$$= \frac{[Fe^{3+}] + [FeOH^{2+}] + [Fe(OH)_2^+] + 2[Fe_2(OH)_2^{4+}]}{[Fe^{3+}]}$$

$$(3.129)$$

If we express the concentrations in terms of the corresponding equilibrium constants (the stability constants of the hydroxide complexes), we obtain

$$\alpha_{Fe(OH)} = 1 + {}^*\beta_1[H_3O^+]^{-1} + {}^*\beta_2[H_3O^+]^{-2}$$
$$+ 2{}^*\beta_{22}[Fe^{3+}][H_3O^+]^{-2} \qquad (3.130)$$

Thus for different Fe^{3+} concentrations the value of $\alpha_{Fe(OH)}$ are also different. The relationships are clearly seen in Fig. 3.10. When the iron(III) concentration is less than $10^{-4}M$, the dependence of $\log \alpha_{Fe(OH)}$ on pH is unchanged. In this case the last term of Eq. (3.130) is always smaller than the preceding term, which determines the value of $\alpha_{Fe(OH)}$, since if

$${}^*\beta_2 = 2{}^*\beta_{22}[Fe^{3+}] \qquad (3.131)$$

then

$$[Fe^{3+}] = 2 \times 10^{-4}M$$

The curve corresponding to this concentration is called the *mononuclear wall*.

If we take into account in the system all protolytic reactions involving metal ions, we obtain the following scheme:

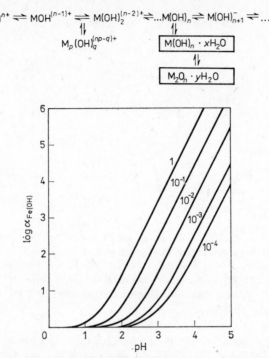

$$M^{n+} \rightleftharpoons MOH^{(n-1)+} \rightleftharpoons M(OH)_2^{(n-2)+} \rightleftharpoons ...M(OH)_n \rightleftharpoons M(OH)_{n+1}^- \rightleftharpoons ...$$

$$M_p(OH)_q^{(np-q)+} \qquad \boxed{M(OH)_n \cdot xH_2O}$$

$$\boxed{M_2O_n \cdot yH_2O}$$

Fig. 3.10. The dependence of $\log \alpha_{Fe(OH)}$ on pH for various Fe^{3+} concentrations. The numbers represent molar concentrations.

In this scheme, the formulae enclosed in boxes represent precipitates. In laboratory practice, besides the reactions shown, other accompanying processes often take place. For example, owing to the presence of atmospheric carbon dioxide, which saturates all solutions if they are not protected, metal carbonates may precipitate, for example cadmium carbonate.

The complexity of protolytic reactions taking place in solutions with the participation of metal ions makes the calculation of the equilibria very laborious even when other substances which complex metal ions are absent. Nowadays this difficulty can be overcome by use of a computer program; this can not only obtain numerical values but also draw curves characterizing the system. An example of such a graph is given in Fig. 3.11 for the system Sc^{3+}–OH^-. In the legend graph, the values of $\log^* \beta$ are given for the mononuclear complexes $ScOH^{2+}$, $Sc(OH)_2^+$, and $Sc(OH)_3$ and the polynuclear complexes $Sc_2(OH)_2^{4+}$, $Sc_3(OH)_4^{5+}$, $Sc_3(OH)_5^{4+}$. In addition, $\log^* K_{s0}$ for the sparingly soluble $Sc(OH)_3$ is given.

The vertical lines on the graph define the pH values at which the corresponding complex predominates in the solution. Thus the line marked

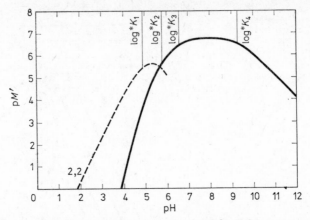

Fig. 3.11. The dependence of pM′ on pH for the Sc^{3+}–OH^- system, calculated and plotted by a computer. The corresponding equilibrium data are: $\log^*\beta_1$ = −4.9, $\log^*\beta_2$ = −10.7, $\log^*\beta_3$ = −17.3, $\log^*\beta_4$ = −26.6, $\log^*\beta_{22}$ = −6.0, $\log^*\beta_{53}$ = −17.2, $\log^*K_{so}$ = 10.5 (from J. Kragten, *Atlas of Metal–Ligand Equilibria in Aqueous Solutions*, Horwood, Chichester, 1978; by permission of the copyright holders).

$\log^*K_1$ corresponds to the borderline between the predominance region of the Sc^{3+} ion and the predominance region for the $ScOH^{2+}$ ion. The line marked 2,2 is the boundary of the region in which the ion $Sc_2(OH)_2^{4+}$ is present at > 1% of the total, approached from low pH values. Contribution of other polynuclear complexes is negligible. The unmarked curve corresponds to the region where $Sc(OH)_3$ is precipitated and is in equilibrium with the various $Sc_p(OH)_q^{(3p-q)+}$ species in the solution.

3.11 MIXTURES OF ACIDS AND MIXTURES OF BASES

In analytical practice we must often analyse solutions containing a mixture of two or more acids (or bases) which differ with respect to both strength and concentration. To evaluate such composite mixtures, several approximations may be used as in the case of polyprotic acids or bases. However, it should be remembered that the system is more complicated because the effect of each component on the H_3O^+ concentration is determined not only by the value of its dissociation constant but also by its concentration ratio to the other components. As in other calculations, we can obviously neglect the participation of a weaker or less concentrated acid, if the H_3O^+ concentration due to its dissociating contributes less than 5% of the total.

In a mixture of a weak acid and a strong acid, it can be assumed that the H_3O^+ concentration is equal to the sum of the total concentration of the strong acid (complete dissociation) and the concentration of the anion (the conjugate base of the weak acid), which corresponds to the contribution of the weaker acid to the total H_3O^+ concentration. If we denote by X the concentration of H_3O^+ ions resulting from dissociation of the weak acid (present with concentration C_2), then the total H_3O^+ concentration will be equal to $C_1 + X$; the dissociation constant of the weak acid is given by

$$K_a = \frac{(C_1 + X)X}{(C_2 - X)} \tag{3.132}$$

(C_1 is the concentration of completely dissociated strong acid).

Example 3.38. Calculate the pH of a solution that has $0.01M$ hydrochloric acid concentration and $0.05M$ formic acid concentration. The acid dissociation constant of formic acid is $K_a = 10^{-3.7}$.

For $0.01M$ strong acid, $[H_3O^+] = 10^{-2}M$. The total concentration of H_3O^+ is

$$[H_3O^+] = 1 \times 10^{-2} + X$$

From the formic acid dissociation constant

$$K_a = \frac{(1 \times 10^{-2} + X)X}{(5 \times 10^{-2} - X)} = 2.0 \times 10^{-4}$$

We solve the quadratic equation

$$X^2 + X(1 \times 10^{-2} + 2.0 \times 10^{-4}) - 5 \times 10^{-2} \times 2.0 \times 10^{-4} = 0$$

and obtain

$$X = 9.0 \times 10^{-4}M$$

The total H_3O^+ concentration is then equal to

$$[H_3O^+] = 1 \times 10^{-2} + 9.0 \times 10^{-4} = 1.09 \times 10^{-2}M$$

and

$$pH = 1.96$$

If the solution contains a mixture of two weak acids, the situation is more complicated, since with a relatively small difference between the dissociation constants there is a mutual influence between the two components of the solution. The total H_3O^+ concentration in a solution containing several acids is equal to the sum of the concentrations of all the conjugate bases, e.g. for a mixture of two acids

$$[H_3O^+] = [A_1^-] + [A_2^-] \qquad (3.133)$$

The individual concentrations of the conjugate bases may be determined from the dissociation constants K_{a1} and K_{a2} and from the total concentrations of each acid

$$C_1 = [A_1^-] + [HA_1] \qquad (3.134)$$

$$C_2 = [A_2^-] + [HA_2] \qquad (3.135)$$

The concentrations of the conjugate bases will thus be

$$[A_1^-] = \frac{K_{a1}C_1}{[H_3O^+] + K_{a1}}, \qquad [A_2^-] = \frac{K_{a2}C_2}{[H_3O^+] + K_{a2}} \qquad (3.136)$$

On substitution of these values in Eq. (3.133), we obtain

$$[H_3O^+] = \frac{K_{a1}C_1}{[H_3O^+] + K_{a1}} + \frac{K_{a2}C_2}{[H_3O^+] + K_{a2}} \qquad (3.137)$$

In the case of very dilute acid solutions, the effect of dissociation of water must also be considered. An additional term, $K_w/[H_3O^+]$ is required in Eq. (3.137). However, we will disregard such cases for the present. Equation (3.137) is of the third degree in $[H_3O^+]$, so its solution may be difficult. Usually, without much error, we may assume that $[H_3O^+] \gg K_{a1}$ and $[H_3O^+] \gg K_{a2}$, in which case we obtain a quadratic equation which is easy to solve

$$[H_3O^+] = \frac{K_{a1}C_1}{[H_3O^+]} + \frac{K_{a2}C_2}{[H_3O^+]} \qquad (3.138)$$

$$[H_3O^+]^2 = K_{a1}C_1 + K_{a2}C_2 \qquad (3.139)$$

This equation can be generalized for a system containing any number of acids, provided that their concentrations and dissociation constants are known.

The result obtained from Eq. (3.139) is approximate. If it turns out that the value of $[H_3O^+]$ obtained is comparable to the value of at least the biggest dissociation constant, then it should be treated as a first approximation and substituted in the denominators of Eq. (3.137) to allow a more exact value of $[H_3O^+]$ to be calculated.

Example 3.39. Calculate the pH of a $0.5M$ ammonium chloride solution which is saturated with hydrogen sulphide to a concentration of $0.05M$. The dissociation constants of hydrogen sulphide are $K_{a1} = 1.25 \times 10^{-7}$ and $K_{a2} = 1.6 \times 10^{-13}$, and the dissociation constant of the ammonium ion is $K_a = 4.0 \times 10^{-10}$.

Because of the large difference in its dissociation constants hydrogen sulphide may be treated as a monoprotic acid with dissociation constant

K_{a1} and consequently we may consider this system to be a mixture of two acids, H_2S and NH_4^+. From Eq. (3.139) we obtain to a first approximation

$$[H_3O^+]^2 = 1.25 \times 10^{-7} \times 5 \times 10^{-2} + 4.0 \times 10^{-10} \times 5 \times 10^{-1}$$

$$= 6.25 \times 10^{-9} + 2.0 \times 10^{-10} = 6.45 \times 10^{-9}$$

so

$$[H_3O^+] = 8.0 \times 10^{-5}M \quad \text{and} \quad pH = 4.10$$

The value $[H_3O^+] = 8.0 \times 10^{-5}M$ is over 600 times the value of the hydrogen sulphide dissociation constant, and this indicates that the approximate assumptions made were correct. Should the effect of ammonium chloride (4%) in this calculation be totally omitted, the H_3O^+ concentration found would be 7.9×10^{-5} (pH = 4.10).

The method given above can also be used for mixtures of bases if the dissociation constants of acids are replaced by the dissociation constants of bases, $[H_3O^+]$ is replaced by $[OH^-]$ and analytical concentrations of acids are replaced by the corresponding concentrations of bases.

Example 3.40. Calculate the pH of a solution obtained by mixing equal volumes of $0.1M$ NH_3 solution and $0.1M$ Na_2CO_3 solution. The basic dissociation constants for the carbonate ion are 2.5×10^{-4} and 6.3×10^{-7}, and the basic dissociation constant of ammonia is 6.0×10^{-5}, the ion product of water is 2.5×10^{-14}.

The small value of the second dissociation constant of the CO_3^{2-} ion allows us to treat it as a monoprotic base. Therefore

$$[OH^-]^2 = 2.5 \times 10^{-4} \times 10^{-1} + 6.0 \times 10^{-5} \times 10^{-1}$$

$$= 2.5 \times 10^{-5} + 6.0 \times 10^{-6} = 3.1 \times 10^{-5}$$

$$[OH^-] = 5.6 \times 10^{-3}M$$

so

$$pOH = 2.25 \quad \text{and} \quad pH = 11.35$$

As can be seen from the calculation, the value of the first dissociation constant of the carbonate ion is about 3% of the value of the calculated H_3O^+ ion concentration. To find out how much the final result would be changed if the value obtained were substituted in the denominator of the equation corresponding to Eq. (3.136), we can perform the following calculation:

$$[OH^-] = \frac{K_{b1}C_1}{[OH^-] + K_{b1}} + \frac{K_{b2}C_2}{[OH^-] + K_{b2}}$$

$$[OH^-] = \frac{2.5 \times 10^{-4} \times 10^{-1}}{5.6 \times 10^{-3} + 2.5 \times 10^{-4}} + \frac{6.0 \times 10^{-5} \times 10^{-1}}{5.6 \times 10^{-3} + 6.0 \times 10^{-5}}$$

$$= 5.3 \times 10^{-3} M$$

and thus

$$pOH = 2.27 \quad \text{and} \quad pH = 11.33$$

This result, different from the preceding one by 0.02 pH unit only, is within the limits of allowable error. There was therefore no need for the additional, more exact calculations.

3.12 MIXTURES OF ACIDS AND BASES

If solutions of a sufficiently strong acid and a strong base are mixed, then a neutralization reaction will take place. The resulting products will determine the pH of the solution. The value will depend on the strengths of the acid and the base and on their relative and absolute concentrations. For example, mixing a strong acid and a strong base may yield an acid solution (excess of acid), a neutral solution (equivalent amounts) or a basic solution (excess of base). Calculation of the pH in each of these cases is simple and can be done on the basis of the expressions already discussed (see Sections 3.1 and 3.2).

Mixing equivalent amounts of a strong acid and a weak base or a weak acid and a strong base yields a system which may be treated as a solution of a weak acid (conjugate to the weak base) or as a solution of a weak base (conjugate to the weak acid). Calculation of the pH for such a system has also been discussed (see Section 3.3, 3.4 and 3.5). When these reagents are mixed in non-equivalent quantities, we may have either mixtures of acids and mixtures of bases (excess of the strong electrolyte over the weak electrolyte—see Section 3.10) or buffer solutions (excess of the weak electrolyte over the strong electrolyte).

We have not yet discussed the case of a mixture of a weak acid and a weak base. If a problem of this kind has to be solved, simplifications should be sought, but if none is possible, the problem should be tackled in the following manner.

When a weak acid HA with dissociation constant $K_a = [A^-][H_3O^+]/[HA]$ and total concentration C_{HA} is dissolved in water together with a weak base B with dissociation constant $K_b = [BH^+][OH^-]/[B]$ and total concentration C_B, then proton transfer reactions between the acids and bases may take place, viz. between acid HA and base B, between acid HA and base H_2O, and between acid H_2O and base B. As a result of these reactions, the total concentration of the species resulting from loss of protons, i.e. OH^- and A^-, must equal the total concentration of the species produced by gain of protons, i.e. H_3O^+ and BH^+. Therefore

$$[H_3O^+] + [BH^+] = [OH^-] + [A^-] \tag{3.140}$$

From the dissociation constants and total concentrations of HA and B

$$C_{HA} = [HA] + [A^-], \quad C_B = [BH^+] + [B] \tag{3.141}$$

we obtain

$$[A^-] = \frac{K_a C_{HA}}{K_a + [H_3O^+]} \tag{3.142}$$

and

$$[BH^+] = \frac{K_b C_B}{K_b + [OH^-]} \tag{3.143}$$

Substitution of Eqs. (3.142) and (3.143) in Eq. (3.140) and rearrangement will lead to a quadratic equation in $[H_3O^+]$ and $[OH^-]$

$$[H_3O^+]^2 K_b + [H_3O^+]\{K_b(C_B + K_a) + K_w\} - \{K_a K_b(C_{HA}$$
$$- C_B) - K_w(K_a - K_b)\} - [OH^-]\{K_a(C_{HA}$$
$$+ K_b) + K_w\} - [OH^-]^2 K_a = 0 \tag{3.144}$$

By replacing $[OH^-]$ by $K_w/[H_3O^+]$ we obtain a fourth degree equation in $[H_3O^+]$, which is difficult to solve in the general form.

However, under specified conditions it is possible to introduce several simplifications, which reduce the equation to a quadratic form. When the solution is definitely acidic, $[H_3O^+] \gg [OH^-]$ so we can omit in Eq. (3.144) the last two terms, containing $[OH^-]$. We thus obtain the equation

$$[H_3O^+]^2 K_b + [H_3O^+]\{K_b(C_B + K_a) + K_w\}$$
$$- [K_a K_b(C_{HA} - C_B) - K_w(K_a - K_b)] = 0 \tag{3.145}$$

which is readily solvable.

In a definitely basic solution, such that $[H_3O^+] \ll [OH^-]$, it is possible to disregard the first two terms of Eq. (3.144). We then obtain an equation similar to Eq. (3.145) but containing terms characterizing an acid instead of a base and vice versa

$$[OH^-]^2 K_a + [OH^-]\{K_a(C_{HA} + K_b) + K_w\}$$
$$- \{K_a K_b(C_B - C_{HA}) - K_w(K_b - K_a)\} = 0 \tag{3.146}$$

In the pH region close to neutrality, the H_3O^+ concentration is small compared with the total acid concentration ($C_{HA} \gg [H_3O^+]$) and the concentration of the base is much greater than the OH^- concentration ($C_B \gg [OH^-]$). In such cases only the three middle terms of Eq. (3.144) need be taken into account. After rearrangement and substitution for $[OH^-]$ we obtain

$$[H_3O^+]^2 \{K_b(C_B + K_a) + K_w\} - [H_3O^+] \{K_aK_b(C_{HA} - C_B)$$
$$- K_w(K_a - K_b)\} - K_w\{K_a(C_{HA} + K_b) + K_w\} = 0 \qquad (3.147)$$

These three equations allow us to solve practically all problems concerning mixtures of weak acids and weak bases.

Example 3.41. Calculate the hydrogen ion concentration of a solution containing a mixture of $0.002M$ hydrazine hydrochloride ($K_a = 10^{-8.0}$) and $0.001M$ ammonia ($K_b = 10^{-4.76}$).

Both compounds in this system are weak electrolytes. The hydrazinium ion $N_2H_5^+$ is an acid, ammonia is a base; consequently the solution should have a pH close to neutrality, and Eq. (3.147) can be used. On substitution of the relevant values of the concentrations and dissociation constants, we obtain

$$[H_3O^+]^2 \{10^{-4.76}(10^{-3} + 10^{-8.0}) + 10^{-14}\}$$
$$- [H_3O^+] \{10^{-8.0} \times 10^{-4.76}(10^{-2.7} - 10^{-3.0})$$
$$- 10^{-14}(10^{-8.0} - 10^{-4.76})\}$$
$$- 10^{-14} \{10^{-8.0}(10^{-2.7} + 10^{-4.76}) + 10^{-14}\} = 0$$

Disregarding small terms and completing the calculation, we obtain

$$[H_3O^+]^2 \times 10^{-7.76} - [H_3O^+] \times 10^{-15.76} - 10^{-24.7} = 0$$
$$[H_3O^+]^2 - [H_3O^+] \times 10^{-8.0} - 10^{-16.94} = 0$$

so

$$[H_3O^+] = 1.11 \times 10^{-8}M \quad \text{and} \quad pH = 7.96$$

A result with a small error may be obtained in a much simpler manner if we treat the system as a buffer solution composed of equimolar amounts of the hydrazinium ion and hydrazine. For such a system the pH value should be equal to pK_a, i.e. 8.0.

In Example 3.41 ammonia, as a base, is a much stronger electrolyte than the hydrazinium ion. When the strengths of the two components are comparable, the displacement of the acid dissociation equilibrium caused by the addition of a base is smaller, and treatment of such a system as a buffer will result in a much greater error. Then, such drastic simplifications are not advisable.

Example 3.42. Calculate the pH in a solution of a weak acid at $10^{-2}M$ concentration $(K_a = 10^{-5})$ and a base at $5 \times 10^{-3}M$ concentration $(K_b = 10^{-9})$.

The values of the acid and base dissociation constants allow us to guess that $[H_3O^+]$ will be much smaller than the total acid concentration and $[OH^-]$ will be much smaller than the total concentration of the base. We should therefore use Eq. (3.147). On substitution of numerical values, we obtain

$$[H_3O^+]^2 \{10^{-9}(5 \times 10^{-3} + 10^{-5}) + 10^{-14}\} - [H_3O^+] \times$$
$$\times \{10^{-5} \times 10^{-9}(10^{-2} - 5 \times 10^{-3}) - 10^{-14}(10^{-5} - 10^{-9})\}$$
$$- 10^{-14} \{10^{-5}(10^{-2} + 10^{-9}) + 10^{-14}\} = 0$$

On disregarding small terms and completing the calculation we obtain

$$[H_3O^+]^2 - [H_3O^+] \times 10^{-5} - 2 \times 10^{-10} = 0$$

Then, on solving the quadratic equation, we get

$$[H_3O^+] = 2 \times 10^{-5}M \quad \text{and} \quad pH = 4.7$$

We see that in this example the weak base with $pK_b = 9$ cannot cause a sufficient displacement of the acid dissociation equilibrium to allow the system to be treated as a buffer prepared from the acid and its salt. If we used the method of calculating pH for buffers we would obtain pH = 5.0, and the consequent error of 0.3 pH unit is too large even for approximate analytical calculations.

In analytical problems we often encounter equimolar mixtures of a weak acid and a weak base. Such cases correspond to solutions of salts of weak acids and weak bases, e.g. solutions of ammonium acetate or ammonium cyanide. In such systems, $C_{HA} = C_B = C$. Further simplification is achieved by the use of Eq. (3.147), i.e. it is assumed that the solution is neither very acidic nor very basic. If the solution of the salt is not too dilute, then $C \gg [H_3O^+]$ and $C \gg [OH^-]$. Morever, the assumption that K_w is significantly smaller than $K_{a(HB)}$ and $K_{b(B)}$ is equivalent to stating that $C \gg K_{b(A)}$ and $C \gg K_{a(HB)}$ (where $K_{b(A)}$ is the dissociation constant of base A conjugate to acid HA, and $K_{a(HB)}$ is the dissociation constant of acid HB conjugate to base B). For such cases, on neglecting the small terms in Eq. (3.147), we obtain

$$[H_3O^+]^2 K_{b(B)}(C + K_{a(HA)}) - K_w K_{a(HA)}(C + K_{b(B)}) = 0$$

and therefore

$$[H_3O^+] = \sqrt{\frac{K_w K_{a(HA)}(C + K_{b(B)})}{K_{b(B)}(C + K_{a(HA)})}} \qquad (3.148)$$

When, in addition, the concentration is much higher than the values of $K_{b(B)}$ and $K_{a(HA)}$, we obtain

$$[H_3O^+] = \sqrt{\frac{K_w K_{a(HA)}}{K_{b(B)}}} = \sqrt{K_{a(HB)} K_{a(HA)}} \qquad (3.149)$$

This formula may be derived more directly if we assume that the neutralization reaction

$$HA + B \rightleftharpoons HB^+ + A^- \qquad (3.150)$$

runs almost completely ($> 95\%$) to the right and the contribution of water ions in Eq. (3.140) is not significant. In this case $[HB^+] = [A^-] = C$ and $[B] = [HA]$. From the equations for the dissociation constants we obtain

$$[HA] = \frac{[H_3O^+][A^-]}{K_{a(HA)}} = [B] = \frac{[HB^+][OH^-]}{K_{b(B)}} \qquad (3.151)$$

from which

$$\frac{[H_3O^+][A^-]}{K_{a(HA)}} = \frac{[HB^+][OH^-]}{K_{b(B)}}$$

so

$$\frac{[H_3O^+][A^-]}{[OH^-][HB^+]} = \frac{K_{a(HA)}}{K_{b(B)}}$$

and

$$\frac{[H_3O^+]^2}{K_w} = \frac{K_{a(HA)}}{K_{b(B)}}$$

and finally

$$[H_3O^+] = \sqrt{\frac{K_w K_{a(HA)}}{K_{b(B)}}} \qquad (3.152)$$

i.e. an expression identical with Eq. (3.149).

Example 3.43. Calculate the pH of a solution of ammonium nitrite at concentration $C = 5 \times 10^{-2} M$, given that $K_{a(HNO_2)} = 10^{-3.2}$, $K_{b(NH_3)} = 10^{-4.4}$ and $K_w = 10^{-13.8}$.

A preliminary evaluation of the problem indicates that $C \gg K_{a(HNO_2)}$ and $C \gg K_{b(NH_3)}$). Since the concentration is thus much higher than the values of the dissociation constants, it may be assumed that the pH is not too far from 7, which allows us to use Eq. (3.149). We obtain

$$[H_3O^+] = \sqrt{\frac{10^{-13.8} \times 10^{-3.2}}{10^{-4.4}}} = 10^{-6.3} = 5.0 \times 10^{-7}M$$

therefore pH = 6.30.

However, if we were considering a solution with a concentration of $5 \times 10^{-4}M$, we could not neglect the values of the dissociation constants in the brackets in the more exact Eq. (3.148). We would then have to calculate the pH in the following way:

$$[H_3O^+] = \sqrt{\frac{10^{-13.8} \times 10^{-3.2} \times (5 \times 10^{-4} + 4.0 \times 10^{-5})}{10^{-4.4} \times (5 \times 10^{-4} + 6.3 \times 10^{-4})}}$$

$$= \sqrt{\frac{10^{-17.0} \times 5.4 \times 10^{-4}}{10^{-4.4} \times 1.13 \times 10^{-3}}} = \sqrt{12 \times 10^{-14}}$$

$$= 3.5 \times 10^{-7}M$$

hence pH = 6.46.

The difference between the results obtained by the two methods of calculation, ΔpH = 0.16, indicates a rather significant error, which should not be neglected. The error could be somewhat smaller when the more correct thermodynamic constants were used in calculations. Thus, although the concentration C does not appear in Eq. (3.149), which was used to calculate the first version of the problem $(C = 5 \times 10^{-2}M)$, knowledge of its value is necessary for a decision about whether the approximate simple equation may be used.

Example 3.44. Calculate the pH of $0.05M$ ammonium dichloroacetate solution; the dissociation constant of ammonia is 3.7×10^{-5}, the dissociation constant of dichloroacetic acid is 8.0×10^{-2} and the ion product of water is 1.2×10^{-14}.

In this case we shall use Eq. (3.148) since the total concentration is comparable to the value of the dissociation constant of the acid

$$[H_3O^+] = \sqrt{\frac{10^{-13.92} \times 10^{-1.1}(5 \times 10^{-2} + 3.7 \times 10^{-5})}{3.7 \times 10^{-5}(5 \times 10^{-2} + 8 \times 10^{-2})}}$$

$$= \sqrt{\frac{10^{-15.02} \times 10^{-1.3}}{10^{-4.43} \times 1.3 \times 10^{-1.0}}}$$

hence

$$[H_3O^+] = \sqrt{10^{-11.0}} = 10^{-5.50}M = 3.1 \times 10^{-6}M$$

and

$$pH = 5.50$$

In all cases such as those considered above, e.g. ammonium acetate, we may equally well treat the solution as a mixture of acetic acid as the weak acid and ammonia as the weak base or as a mixture of ammonium ions as the weak acid and acetate as the weak base. The result of the calculations will be the same.

3.13 SOLUTIONS OF AMPHIPROTIC SUBSTANCES

For quantitative consideration of ampholyte solutions, we may use some of the methods already discussed, at least partially. We obtain equations which are similar to the expressions derived for mixtures of weak acids and weak bases. Let us consider a system of two weak acids of different strengths (with different dissociation constants) in equimolar concentration. If we add to this system half the amount of a strong base needed for total neutralization, we obtain a solution in which the stronger acid is totally converted into its conjugate base while the weaker acid remains in only slightly dissociated form.

If, instead of a mixture of two acids, we start with a half-neutralized solution of the diprotic acid H_2A; then HA^- ions will appear in the system. They may serve as a base, by accepting a proton, or as an acid, by losing the remaining proton. A system of this kind is described by two dissociation constants

$$K_{a1} = \frac{[H_3O^+][HA^-]}{[H_2A]}, \quad K_{a2} = \frac{[H_3O^+][A^{2-}]}{[HA^-]} \tag{3.153}$$

and the total concentration of acid

$$C = [H_2A] + [HA^-] + [A^{2-}] \tag{3.154}$$

In a solution of HA^- (i.e. an ampholyte solution), the pH results from proton loss by HA^- (A^{2-} formation) and by H_2O (OH^- formation), and from proton gain by HA^- (H_2A formation) and H_2O (H_3O^+ formation). Since the number of protons lost is equal to the number of protons gained, we have

$$[H_3O^+] + [H_2A] = [A^{2-}] + [OH^-] \tag{3.155}$$

By substituting in Eq. (3.155) the values of $[H_2A]$ and $[A^{2-}]$ evaluated from the dissociation constants K_{a1} and K_{a2} and $[OH^-]$ from the ion product of water, we obtain

$$[H_3O^+] + \frac{[H_3O^+][HA^-]}{K_{a1}} = \frac{K_{a2}[HA^-]}{[H_3O^+]} + \frac{K_w}{[H_3O^+]} \qquad (3.156)$$

and on rearrangement

$$[H_3O^+] = \sqrt{\frac{K_{a2}[HA^-] + K_w}{1 + [HA^-]/K_{a1}}} = \sqrt{\frac{(K_{a2}[HA^-] + K_w)K_{a1}}{K_{a1} + [HA^-]}}$$
$$(3.157)$$

However, this comparatively simple equation requires knowledge of $[HA^-]$. If the values of the two dissociation constants are not too close, it may be assumed to a good approximation that $[H_2A]$ and $[A^-]$ in Eq. (3.154) are small in comparison with $[HA^-]$, so by setting $C = [HA^-]$ we obtain

$$[H_3O^+] = \sqrt{\frac{K_{a1}(K_{a2}C + K_w)}{K_{a1} + C}} \qquad (3.158)$$

Further simplification of Eq. (3.158) is possible in cases where the ampholyte concentration is much higher than K_{a1}, which is equivalent to the assumption that in Eq. (3.155) $[H_3O^+]$ is much smaller than $[H_2A]$. Then

$$[H_3O^+] = \sqrt{\frac{K_{a1}(K_{a2}C + K_w)}{C}} \qquad (3.159)$$

When $[OH^-]$ is omitted from Eq. (3.155) which is equivalent to saying that $K_{a2} \gg K_w$, we obtain

$$[H_3O^+] = \sqrt{\frac{K_{a1}K_{a2}C}{K_{a1} + C}} \qquad (3.160)$$

Simultaneous introduction of both simplifying assumptions leads to the simplest form of Eq. (3.158), namely

$$[H_3O^+] = \sqrt{K_{a1}K_{a2}} \qquad (3.161)$$

These expressions make it possible to calculate the pH of ampholytic solutions whether the ampholyte appears as an anion, e.g. HPO_4^{2-}, as a neutral molecule, e.g. $NH_2CH_2COOH^*$, or in a form of a cation, e.g. $NH_2CH_2CH_2NH_3^+$.

---

* A molecule of aminoacetic acid (glycine) usually appears in solution in the form of a dipolar ion with the structure $^+NH_3CH_2COO^-$; this, however, does not influence our calculations.

Calculation of the pH of solutions of amino acids, which appear in solutions in amphiprotic form, is important because of their role in bio-chemical studies. The mathematical treatment is the same as that given above; it is worth emphasizing, however, that the calculated pH value corresponds to the isoelectric point pH_{iso}. It is usually not far from 7.0 (Table 3.6) and corresponds to the maximum concentration of ampholyte ion.

In more concentrated solutions the pH value is independent of the amino acid concentration. However, when the concentration is not much higher than K_{a1}, the pH value is dependent on the concentration to a small extent.

Table 3.6

The pH values of the isoelectric point of $0.1M$ solutions of some amino acids

	pK_{a1}	pK_{a2}	pH_{iso}
α-aminoacetic acid (glycine)	2.35	9.77	6.07
α-aminopropionic acid (α-alanine)	2.35	9.87	6.11
α-aminoisovaleric acid (valine)	2.29	9.72	6.01
α-aminoisocapronic acid (leucine)	2.36	9.60	5.98

Example 3.45. The cation of protonated aminoacetic acid, formula $^{+}NH_3 \cdot CH_2 \cdot COOH$, is a diprotic acid. Calculate the pH of a solution of aminoacetic acid with a concentration of $1.0 \times 10^{-2} M$ if the values of the dissociation constants of the diprotic acid are $K_{a1} = 4.5 \times 10^{-3}$ and $K_{a2} = 1.7 \times 10^{-10}$.

Here, we cannot use the simplest Eq. (3.161) because K_{a1} is only slightly smaller than C. However, since $K_{a2}C = 1.7 \times 10^{-12}$ is 170 times larger than K_w, we can use Eq. (3.160). We then obtain

$$[H_3O^+] = \sqrt{\frac{7.65 \times 10^{-15}}{1.45 \times 10^{-2}}} = 7.26 \times 10^{-7} M$$

and pH = 6.14.

By using Eq. (3.161) we would obtain pH = 6.06, so the difference would not be great.

Example 3.46. Calculate the pH of a $0.01M$ solution of sodium hydrogen carbonate, $NaHCO_3$, if pK_{a1} and pK_{a2} for carbonic acid are 6.3 and 10.1 respectively.

Preliminary evaluation of the problem suggests the simplest Eq. (3.161) can be used for the calculations since $K_{a1} = 10^{-6.3}$ is much smaller than $C = 10^{-1.0}$ and $K_{a2} C = 10^{-11.1}$ is much greater than K_w. Therefore

$$pH = 1/2(pK_{a1} + pK_{a2}) = 1/2 \times 16.4 = 8.2$$

Example 3.47. Calculate the pH of $0.010M$ sodium hydrogen sulphide solution, NaHS, if the acid dissociation constants of hydrogen sulphide are $K_{a1} = 10^{-7.1}$, $K_{a2} = 10^{-13.1}$, and ion-product of water $K_w = 10^{-13.9}$.

The conditions of the problem indicate that $K_{a1} \ll C$ but $K_{a2} C$ and K_w are of the same order of magnitude. In this situation Eq. (3.159), obtained by omitting the H_3O^+ concentration in Eq. (3.155), should be used. Hence

$$[H_3O^+] = \sqrt{\frac{10^{-7.1}(10^{-13.1} \times 10^{-2.0} + 10^{-13.9})}{10^{-2}}}$$

$$= \sqrt{10^{-5.1}(7.9 \times 10^{-16} + 1.26 \times 10^{-14})}$$

$$= \sqrt{10^{-5.1} \times 10^{-13.9}} = 10^{-9.5}M$$

and pH = 9.5.

If we were to use, not too critically, the simplest equation, we would obtain pH = 10.1 and therefore we would make an error quite significant even for approximate calculations.

3.14 GRAPHICAL METHODS FOR ILLUSTRATING ACID–BASE SYSTEMS

The methods which we used in the preceding sections for calculation of the pH and the composition of solutions required the solution of higher order equations. Sometimes, to obtain the solution it was necessary to make simplifying assumptions, and the validity of these always had to be checked. Such treatment leads to exact results, but is often time-consuming and, besides, requires some intuition about the likely state of equilibrium in a solution with a composition that may be very complicated. In many cases the procedure is made much easier by the use of graphs to represent the equilibria in the system. These often make it possible to obtain satisfactorily exact results very quickly.

There are two main methods of graphical representation of equilibria in acid–base systems, *distribution diagrams* and *logarithmic diagrams*. Since the position of an acid–base equilibrium is always a function of pH, the pH value is an independent variable in both types of graphs.

Distribution diagrams show the dependence of the concentrations of particular species (expressed as mole fractions or percentages of total concentration) on the pH of the solution. If we know the value of the dissociation constant of the acid

$$K_a = \frac{[H_3O^+][A^-]}{[HA]} \tag{3.162}$$

and its total (analytical) concentration

$$C_{HA} = [HA] + [A^-] \tag{3.163}$$

we can easily calculate the mole fraction of the acid HA and the conjugate base A^-, as a function of pH.

The mole fraction of the conjugate base A^-

$$x_{A^-} = \frac{[A^-]}{C_{HA}} = \frac{K_a}{K_a + [H_3O^+]} \tag{3.164}$$

is called, as already mentioned, the *degree of dissociation of the acid*.

The mole fraction of the acid in a solution containing HA and A^- is determined by the expression

$$x_{HA} = \frac{[HA]}{C_{HA}} = \frac{[H_3O^+]}{K_a + [H_3O^+]} = \bar{n} \tag{3.165}$$

and is known as the *degree of formation of the acid* from its ions. The degree of formation of acids may also be described as the average number of protons bound per molecule or ion of base (Brønsted and Lowry). In the case of monoprotic acids the maximum value of the degree of formation is 1. The degree of formation of these acids is denoted by x_{HA}, although we could equally well use the symbol $\bar{n}$, which is mainly used with regard to polyprotic acids.

The diagram of the degree of dissociation as a function of pH is sometimes called the *dissociation curve*, and the diagram of the degree of formation, the *formation curve*. Obviously, the sum of the degree of dissociation and the degree of formation must always be equal to unity for any pH value.

Figure 3.12 shows the dependence of x_{HA} on pH, i.e. the HA acid formation function; this is an example of a distribution diagram. For each pH value the segments parallel to the y-axis determine the relative contributions of HA and A^- to the total concentration of the acid. The segment between the x-axis and the formation curve is proportional to x_{HA}, and the segment between the formation curve and an ordinate value of 1.0 is equal to x_{A^-}. If we invert the line of the formation curve we

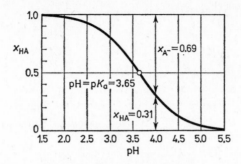

Fig. 3.12. Distribution diagram for formic acid (pK_a = 3.75). The vertical segments are proportional to the mole fractions x_{HA} and x_{A^-}.

obtain the dissociation curve. Figure 3.12 refers to formic acid, for which $K_a = 2.24 \times 10^{-4}$, i.e. p$K_a$ = 3.65. Hence the midpoint of the graph (x_{HA} = 0.50) falls at pH = 3.65. If, for example, we wanted to find the mole fractions of formic acid at pH = 4 we would read from a graph x_{HA} = 0.31 and x_A = 0.69. Such precision is quite satisfactory for most calculations.

Distribution diagrams for polyprotic acids (or bases) are somewhat more complicated. In order to show in such a system the concentration of a definite species in relation to the total acid concentration, $N+1$ mole fractions are needed where N is the number of protons that can dissociate from the acid. Thus, in the case of arsenic acid H_3AsO_4, we have

$$x_3 = \frac{[H_3AsO_4]}{C}, \quad x_1 = \frac{[HAsO_4^{2-}]}{C}$$

$$x_2 = \frac{[H_2AsO_4^-]}{C}, \quad x_0 = \frac{[AsO_4^{3-}]}{C} \tag{3.166}$$

where the subscript on x shows the number of protons in the relevant species. The total concentration can be expressed, as in Section 3.9, as the sum of the concentrations of all the species, H_3AsO_4, $H_2AsO_4^-$, $HAsO_4^{2-}$, and AsO_4^{3-}. If we express the concentrations of the individual species in terms of the acid dissociation constants K_{a1}, K_{a2} and K_{a3}, we can obtain an expression for the mole fractions, in terms of the dissociation constants and the pH of the solution. Because of its simpler mathematical form, it is more convenient to write the equation for the dependence of the reciprocal of x on $[H_3O^+]$. Thus we have

$$\frac{1}{x_3} = 1 + \frac{K_{a1}}{[H_3O^+]} + \frac{K_{a1}K_{a2}}{[H_3O^+]^2} + \frac{K_{a1}K_{a2}K_{a3}}{[H_3O^+]^3}$$

$$\frac{1}{x_2} = \frac{[H_3O^+]}{K_{a1}} + 1 + \frac{K_{a2}}{[H_3O^+]} + \frac{K_{a2}K_{a3}}{[H_3O^+]^2} \tag{3.167}$$

$$\frac{1}{x_1} = \frac{[H_3O^+]^2}{K_{a1}K_{a2}} + \frac{[H_3O^+]}{K_{a2}} + 1 + \frac{K_{a3}}{[H_3O^+]}$$

$$\frac{1}{x_0} = \frac{[H_3O^+]^3}{K_{a1}K_{a2}K_{a3}} + \frac{[H_3O^+]^2}{K_{a2}K_{a3}} + \frac{[H_3O^+]}{K_{a3}} + 1$$

For each x_n a separate diagram can be made, demonstrating its dependence on pH. Figure 3.13 shows four such diagrams. The values of the coordinates of point $x = 0.50$ correspond to equal concentrations of

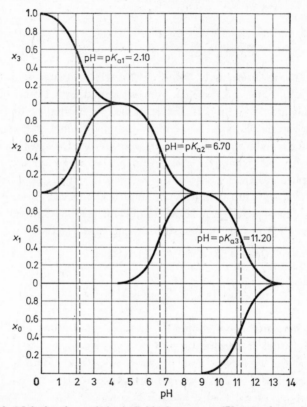

Fig. 3.13. Mole fractions of the individual protonated forms of arsenic acid as functions of pH; x_0—mole fraction of AsO_4^{3-}, x_1—$HAsO_4^{2-}$, x_2—$H_2AsO_4^{-}$, x_3—H_3AsO_4.

the acid and its conjugate base; hence the pH at that point is equal to the pK of the acid. In the case of arsenic acid, for which the values of the successive dissociation constants differ widely, there are at most two different major species in each region and their concentrations greatly exceed those of the others. This greatly simplifies the calculations.

The separate diagrams can be combined to make a single figure, which is a complete distribution diagram (Fig. 3.14). On a diagram of this kind, the distance between a given curve and the x-axis is determined by the

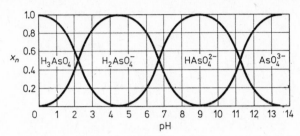

Fig. 3.14. Distribution diagram for arsenic acid ($pK_{a1} = 2.10$, $pK_{a2} = 6.70$, $pK_3 = 11.20$).

contribution of the particular species. In such diagrams, the sum of all the mole fractions is equal to one

$$x_0 + x_1 + x_2 + x_3 = 1 \tag{3.168}$$

If we introduce the concept of the degree of formation, defined, as for monoprotic acids, as the mean number of protons bound to the base AsO_4^{3-}, we obtain

$$\bar{n} = \frac{[HAsO_4^{2-}] + 2[H_2AsO_4^-] + 3[H_3AsO_4]}{C} \tag{3.169}$$

and on substituting the values of the mole fractions (Eq. (3.166))

$$\bar{n} = x_1 + 2x_2 + 3x_3 \tag{3.170}$$

The graph of $\bar{n}$ as a function of pH is called the *formation curve*.

Another method of presentation is the dissociation curve. For monoprotic acids the relationship between the formation curve and the dissociation curve is a simple one, because

$$x_A = 1 - \bar{n} \tag{3.171}$$

In the case of N-protic acids the total degree of dissociation is equal to $N - \bar{n}$, where N is the maximum number of protons attached to the base. Consequently the degree of dissociation of arsenic acid will be $3 - \bar{n}$. Thus from Eqs. (3.168) and (3.170) we obtain

$$3 - \bar{n} = 3(x_0 + x_1 + x_2 + x_3) - x_1 - 2x_2 - 3x_3$$
$$= 3x_0 + 2x_1 + x_2 \tag{3.172}$$

The formation and dissociation curves are shown in Figs. 3.15 and 3.16. On these diagrams the segments with minimum slope (almost parallel to the pH axis) correspond to the ions predominating in a given pH region. Thus, when the pH is about 4.40, it is the $H_2AsO_4^-$ ion which predominates, and when the pH is about 8.95, it is the $HAsO_4^{2-}$ ion.

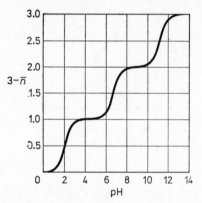

Fig. 3.15. Formation curve for arsenic acid.

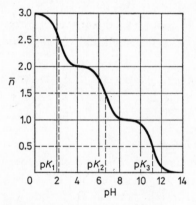

Fig. 3.16. Dissociation curve for arsenic acid.

These values of pH thus correspond to the pH of solutions of the salts NaH_2AsO_4 and Na_2HAsO_4.

The composition of acid and base solutions can also be represented by means of logarithmic diagrams. Distribution diagrams (and dissociation and formation curves) permit only the calculation of relative concentrations of individual species; the curves are independent of the total concentrations. Logarithmic diagrams, on the other hand, are made for definite concentrations. The independent variable is again pH, but the dependent variable is the logarithm of the concentration of the given chemical species. Logarithmic diagrams have the advantage that they enable us to calculate the concentrations of ions and molecules which do not predominate in the given solution.

To make a logarithmic diagram (Fig. 3.17) we start by plotting the dependence of the concentrations of the H_3O^+ and OH^- ions on pH. This dependence will be the same for all logarithmic diagrams, irrespective

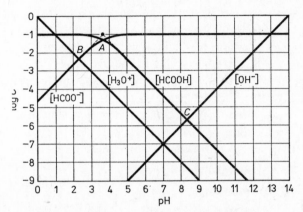

Fig. 3.17. Logarithmic diagram for the formic acid–formate ion system with total concentration $C = 0.10M$.

of the substance dissolved, provided that the solution in question is an aqueous solution for which the value of K_w is constant and equal to 10^{-14}.

The dependence of $\log[H_3O^+]$ on pH results from the definition of pH

$$\log[H_3O^+] = -pH \qquad (3.173)$$

so the slope of the straight line obtained is -1. For $\log[OH^-]$ we obtain

$$\log[OH^-] = pH - pK_w \qquad (3.174)$$

which indicates that the slope of the straight line is $+1$.

We will illustrate the next step by using as model $0.1M$ formic acid ($pK_a = 3.75$). We must draw lines for the dependence of $\log[HA]$ and $\log[A^-]$ on pH. Equations (3.164) and (3.165) can be presented in the form

$$[HA] = \frac{C_{HA}[H_3O^+]}{K_a + [H_3O^+]} \qquad (3.175)$$

$$[A^-] = \frac{C_{HA}K_a}{K_a + [H_3O^+]} \qquad (3.176)$$

To express the concentrations in logarithmic form, we must consider two cases for each of these expressions, namely when $K_a \ll [H_3O^+]$ and K_a

$\gg [H_3O^+]$. In the second case we can neglect $[H_3O^+]$ in the denominator, in comparison with K_a. Then

$$\log[A^-] = \log C_{HA} \tag{3.177}$$

$$\log[HA] = \log C_{HA} - pH + pK_a \tag{3.178}$$

It follows from this assumption, that these expressions hold when the pH of the solution is greater than pK_a, i.e. for all points to the right of point $pH = pK_a$. In this whole region the concentration of base A^- is equal, to a good approximation, to the total concentration, and thus we obtain a line segment with slope 0 (parallel to the pH axis). The acid concentration, on the other hand, is a function both of the total concentration and of pH, and the line segment has a slope of -1.

For the case when $K_a \ll [H_3O^+]$, i.e. in acid conditions, we omit the values of K_a, which are small in relation to $[H_3O^+]$, in the denominators of Eqs. (3.175) and (3.176). We then obtain

$$\log[A^-] = \log C_{HA} - pK_a + pH \tag{3.179}$$

$$\log[HA] = \log C_{HA} \tag{3.180}$$

Equation (3.179) corresponds to a line with slope $+1$ and Eq. (3.180) to a line parallel to the pH axis.

These four line segments pass through the point with coordinates $pH = pK_a$ and $\log C = \log C_{HA}$. This point is the 'characteristic point' for a given system. In our case it is determined by the conditions, i.e. $pK_a = 3.65$ and $\log C = -1.0$. We should remember, however, that for the derivation of Eqs. (3.171)–(3.180) we made simplifying assumptions, namely in the denominator we omitted either K_a or $[H_3O^+]$. Such assumptions are justifiable when the smaller term is at most 5% of the larger one. Thus it does not apply to the region of $pH = pK_a \pm 1.30$. Within this region, the graph is curved. Thus smooth curves link the segments with slopes 0 and -1 (for [HA]) and the segments with slopes $+1$ and 0 (for [A$^-$]). The curves intersect at point A, placed 0.30 logarithmic units below the characteristic point of the system. The concentrations of [HA] and [A$^-$] at that point are both equal to $C_{HA}/2$. Thus, the point corresponds to a buffer solution with concentration $[HA] = [A^-] = C_{HA}/2$.

There are two other interesting points marked on the diagram. The lines for the concentrations of H_3O^+ and A^- intersect at point B, where $[H_3O^+] = [A^-]$. This occurs in the case of a solution of a weak monoprotic acid. For $0.1M$ formic acid the pH value at that point is 2.32.

Point C is the intersection point of the lines for the concentrations of OH^- and HA. These concentrations are equal when the solution contains a weak base A^- which undergoes basic dissociation in accordance with the equation

$$A^- + H_2O \rightleftharpoons HA + OH^-$$

This point corresponds to a solution of a salt of formic acid and a strong base, for example a solution of sodium formate. The pH of $0.1M$ sodium formate can be obtained from the diagram: it is 8.37.

By means of logarithmic diagrams we can find for each value of pH (and thus also for the points of intersection) the concentrations of all the species which exist in a given solution.

The use of logarithmic diagrams, as also other graphical presentations, is approximate and the users cannot expect to obtain very precise informations on this way. For example, it is practically not possible to take into account the ionic strength effect. Therefore in diagrams the pH scale always has 14 logarithmic units, independently of the solution concentration and, in consequence, of ionic strength. On the other hand, the constants used are mostly concentration constants for $I = 0.1$.

Example 3.48. Estimate from a logarithmic diagram the concentrations of H_3O^+, HA, OH^- and A^- in a $0.1M$ solution of sodium formate.

The values of $\log C$ read from the diagram are:

$$\log[H_3O^+] = -8.32, \quad [H_3O^+] = 4.8 \times 10^{-9}M$$
$$\log[OH^-] \; = -5.68, \quad [OH^-] \; = 2.1 \times 10^{-6}M$$
$$\log[HA] \quad = -5.68, \quad [HA] \quad = 2.1 \times 10^{-6}M$$
$$\log[A^-] \quad = -1.00, \quad [A^-] \quad = 1.0 \times 10^{-1}M$$

To check the correctness of these values we may use the equations given previously.

The use of logarithmic diagrams produces correct answers when the distance between two parallel segments of the graphs at the points for which we do the calculations is not less than $\Delta \log C = 1.30$. In other cases the result is less exact, though it can indicate which ions or molecules have negligible concentrations and which should be included in the calculations. The greatest error will occur when two lines corresponding to the concentrations of different species coincide. The error is then 0.3 of a logarithmic concentration unit, so it is still small enough to allow an approximate estimate of the composition of a solution to be made.

Example 3.49. Make a logarithmic diagram for a chloroacetic acid–chloroacetate system with a total concentration of $0.010M$. The dissociation constant of chloroacetic acid is 2×10^{-3}. Calculate with the help of the diagram the pH of a $0.01M$ solution of chloroacetic acid, a $0.01M$ solution of sodium chloroacetate and a $0.01M$ chloroacetate buffer with an acid : base concentration ratio of $1:1$.

We start by plotting straight lines corresponding to $\log [H_3O^+]$ and $\log [OH^-]$, and determine the characteristic point of a system with the coordinates $pH = pK_a = -\log(2 \times 10^{-3}) = 2.70$ and $\log C = -2.0$ (Fig. 3.18). To mark the straight-line segments of $\log C$ we draw straight lines with slopes -1, 0 and $+1$ through the characteristic point of the system.

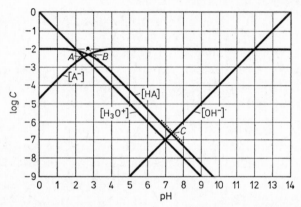

Fig. 3.18. Logarithmic diagram for the chloroacetic acid–chloroacetate ion system with total concentration $C = 0.10M$

At a distance of 0.30 of a logarithmic unit below the characteristic point of the system we mark the point of intersection of the curved segments of $\log[HA]$ and $\log[A^-]$. We then join the straight-line segments of the composition curves by arcs passing through that point. The plotting accuracy may, of course, influence the accuracy of the results obtained with the help of the diagram. The pH of $0.01M$ chloroacetic acid is then read from the diagram. It is the value of pH at point A, for which $[H_3O^+] = [A^-]$, so $pH = 2.45$. In principle, the following equation should be written for this point: $[H_3O^+] = [A^-] + [OH^-]$. However, the distance between the straight-line for $\log[OH^-]$ and the straight-line for $\log[A^-]$ is so large for $pH = 2.45$ that we can neglect the concentration of OH^- ions.

Calculation of the pH for a chloroacetate buffer with equal molar concentrations of the acid and its conjugate base is rather more difficult. A solution where $[HA] = [A^-]$ will naturally have $pH = pK_a = 2.70$ (point B). In order to obtain such a solution, however, we cannot use

equimolar quantities of chloroacetic acid and sodium chloroacetate because chloroacetic acid, which is relatively strong, will be considerably
dissociated under these conditions. Thus the pH of an equimolar buffer
will differ considerably from the value of pK_a, namely by twice the value
of the logarithm of the H_3O^+ concentration

$$[A^-] = [HA] + 2[H_3O^+]$$

Because of this, the graphical determination of pH is far from simple, and
it is more convenient to use a numerical method, from which we obtain
pH = 2.91; this result differs by about 0.2 pH unit from the result obtained by means of a diagram.

Calculation of the pH of $0.01M$ sodium chloroacetate also presents
difficulties. In the expression $[OH^-] = [HA] + [H_3O^+]$ neither of the
terms on the right-hand side can be omitted, because the distance between
the segments of the straight-lines of log [HA] and of log $[H_3O^+]$ is 0.7 of
a logarithmic unit, i.e. $[HA]/[H_3O^+] = 5$. From these expressions it
follows that:

$$[OH^-] = [HA](1 + \tfrac{1}{5}) = \tfrac{6}{5}[HA]$$

Thus, in order to determine the pH graphically, we should draw above
the line of log[HA] another line (the dotted line in Fig. 3.18) corresponding to $(\log 6/5 + \log[HA])$ and thus placed 0.08 of a logarithmic unit
above log[HA]. It is the intersection of this new line with the line for
log[OH$^-$] that gives the correct pH value for $0.01M$ sodium chloroacetate,
viz. 7.39. This value, however, differs by only 0.04 from the value which
would have been obtained if we had neglected the concentration of H_3O^+
from the dissociation of water.

The system above, $0.01M$ chloroacetic acid, is a rather rare case, in
that the direct use of logarithmic diagrams fails to give sufficiently exact
results.

Logarithmic diagrams may also be used for polyprotic acids. For each
individual dissociation constant there is a different characteristic point of
the system. For example, for $0.01M$ H_3AsO_4 there are three characteristic
points, each with $\log C = -1.0$ and of pH equal to $pK_{a1} = 2.10$, pK_{a2}
= 6.70, and $pK_{a3} = 11.20$ respectively (Fig. 3.19). However, in the case
of polyprotic acids the curves are somewhat different from those for monoprotic acids. We omit the mathematical treatment, which is fairly simple,
and state only that the slopes of the graphs of $\log C$ vs. pH, which are
equal to -1, 0 and $+1$ for a monoprotic acid, may take different absolute
values for polyprotic acids, up to a value of N, where N is the maximum
number of protons that can be lost by the acid.

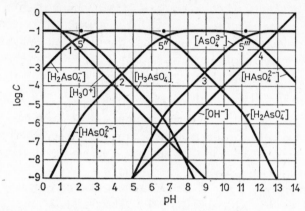

Fig. 3.19. Logarithmic diagram for the arsenic acid system with total concentration $C = 0.10M$.

Let us consider arsenic acid again. The concentration of AsO_4^{3-} depends on the first power of $[H_3O^+]$ only in the pH region between pK_{a2} and pK_{a3}. In the region of pH between pK_{a1} and pK_{a2} the concentration of AsO_4^{3-} is a function of $[H_3O^+]^2$, and thus the slope of this segment of the curve on a logarithmic diagram is $+2$. Below pH $= pK_{a1}$, $[AsO_4^{3-}]$ would be proportional to $[H_3O^+]^3$, so the slope of the appropriate segment is $+3$.

Logarithmic diagrams for polyprotic acids provide the same information as the diagrams for monoprotic acids. The amount of information available, however, is greater because of the larger number of equilibria. We shall now discuss a number of interesting points on the diagram.

Point 1. The pH of $0.1M$ H_3AsO_4 corresponds to the abscissa of this point. The reasoning is as follows:

The concentration of H_3O^+ ions in the solution of the acid is equal to the sum of the concentrations of all conjugate bases, multiplied by the appropriate coefficients which indicate the number of protons lost; thus

$$[H_3O^+] = [H_2AsO_4^-] + 2[HAsO_4^{2-}] + 3[AsO_4^{3-}] + [OH^-]$$

As all the lines with a positive slope lie much below the curve for $[H_2AsO_4^-]$, the terms corresponding to them can be disregarded. Thus

$$[H_3O^+] = [H_2AsO_4^-]$$

so the point of intersection corresponds to pH $= 1.55$.

Point 2. The pH of $0.1M$ NaH_2AsO_4 satisfies the condition

$$[H_3AsO_4] + [H_3O^+] = [OH^-] + [HAsO_4^{2-}] + 2[AsO_4^{3-}]$$

which can be simplified to

$$[H_3AsO_4] = [HAsO_4^{2-}]$$

The point of intersection for these curves read from the diagram is at pH = 4.40 (however, the proximity of the $[H_3AsO_4]$ and $[H_3O^+]$ lines suggests a somewhat higher result).

Point 3. The pH of $0.10M$ Na_2HAsO_4 satisfies the condition

$$2[H_3AsO_4] + [H_2AsO_4^-] + [H_3O^+] = [OH^-] + [AsO_4^{3-}]$$

The position of the line immediately suggest the following simpler form:

$$[H_2AsO_4^-] = [AsO_4^{3-}]$$

so the pH value at the point of intersection is 8.95.

Point 4. The pH of $0.10M$ Na_2AsO_4 is calculated from the general expression

$$3[H_3AsO_4] + 2[H_2AsO_4^-] + [HAsO_4^{2-}] + [H_3O^+] = [OH^-]$$

which can be reduced to the form

$$[HAsO_4^{2-}] = [OH^-]$$

and gives the point of intersection as pH = 12.10.

Point 5. We may indicate on the diagram the points corresponding to the equimolar buffer solutions $H_3AsO_4 + NaH_2AsO_4$ (5'), $NaH_2AsO_4 + Na_2HAsO_4$ (5'') and $Na_2HAsO_4 + Na_3AsO_4$ (5''').

For any pH we can determine the concentrations of all species present in the solution. For example, for pH = 7.50 we obtain the following concentrations:

$$\log[H_3O^+] \quad = -7.50, \quad [H_3O^+] \quad = 3.2 \times 10^{-8} M$$
$$\log[OH^-] \quad = -6.50, \quad [OH^-] \quad = 3.2 \times 10^{-7} M$$
$$\log[HAsO_4^{2-}] = -1.10, \quad [HAsO_4^{2-}] = 8.0 \times 10^{-2} M$$
$$\log[H_2AsO_4^-] = -1.80, \quad [H_2AsO_4^-] = 1.6 \times 10^{-2} M$$
$$\log[H_3AsO_4] = -7.25, \quad [H_3AsO_4] = 5.6 \times 10^{-8} M$$

The error of 4% in the total concentration of arsenic acid results from the limited accuracy of plotting.

Logarithmic diagrams can be constructed not only for one-component systems but also for mixtures. A diagram for a mixture of two acids is simply the superposition of the diagrams for each acid separately.

Example 3.50. Determine with the aid of a logarithmic diagram the pH of a solution obtained by mixing equal quantities of $0.10M$ acetic acid ($pK_a = 4.76$) and $0.025M$ formic acid ($pK_a = 3.65$).

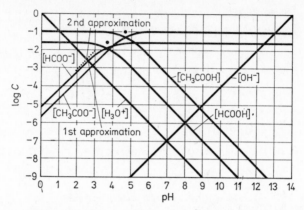

Fig. 3.20. Logarithmic diagram for the 0.10M acetic acid $+0.025M$ formic acid system.

We make a common diagram (Fig. 3.20) for the two acid systems. The two characteristic points of the system have the coordinates 4.76; -1.0, and 3.65; -1.6. To calculate the pH of the mixed solution of the two acids we have to find a point which satisfies the condition

$$[H_3O^+] = [HCOO^-] + [CH_3COO^-] + [OH^-]$$

Moving down the $[H_3O^+]$ line, we come in turn to the $[HCOO^-]$, $[CH_3COO^-]$ and $[OH^-]$ lines. This order indicates the direction in which the contributions of the individual ions to the general acidity decrease. To a first approximation we can consider only

$$[H_3O^+] = [HCOO^-]$$

so pH = 2.63. However, since the $[CH_3COO^-]$ line runs comparatively nearby, the concentrations of these ions should also be taken into account. For each value of pH the distance between the two straight-lines (in the range where they run parallel, with slope $+1$) is $\Delta \log C = 0.45$ or $[HCOO^-]/[CH_3COO^-] = 2.8$. When we include this, we obtain

$$[H_3O^+] = [HCOO^-] + [CH_3COO^-] = [HCOO^-]\left(1 + \frac{1}{2.8}\right)$$

and

$$[H_3O^+] = [HCOO^-]\frac{3.8}{2.8}$$

or

$$\log[H_3O^+] = \log[HCOO^-] + 0.13$$

Thus if we plotted on the diagram a line placed 0.13 units above the

[HCOO$^-$] line and parallel to it, the intersection of this line with the [H$_3$O$^+$] line would correspond to a pH value smaller by 0.065 units, viz. 2.56, which is the correct answer.

The use of computers in chemistry has made it much easier to do complicated calculations, such as occur in systems of ionic equilibria. Nevertheless, graphical presentation of equilibria has retained its importance as a quick and very objective method of collecting data, particularly since it is now possible to obtain the appropriate diagram directly from the computer (as mentioned in Section 3.10).

3.15 HAMMETT'S ACIDITY FUNCTION

The acidity of dilute solutions is most conveniently characterized by means of the concentration of solvated hydrogen ions. If the concentration is not too high and the difference between activity and concentration is not too large, the activity coefficients are close to unity and the difference between the concentration-based acidity scale and the activity-based scale is not large. Thus both of these scales can give sufficient accuracy for description of such properties of the system as reactivity in relation to bases, catalytic action in some reactions, and behaviour with regard to acid–base indicators.

Difficulties arise, however, when we have to consider systems in which the activity coefficients differ significantly from one, and for which we cannot determine in an independent manner the activity coefficient of the solvated proton. However, since it is possible to determine experimentally the value of the product or quotient of the activity coefficients, i.e. $f_H f_X$ for the dissociation of the molecular acid HX or f_H/f_{HB} for the cationic acid HB$^+$, we can define acidity functions, (e.g. an experimental pH scale) which contain an expression which is a combination of the activity of a hydrogen ion and one or several activity coefficients of other species. According to the system on which a given scale is based, we can regard such functions as p($a_H f_{Cl}$), p($a_H f_X$) or similar expressions giving the exact acidity.

In non-aqueous solvents and in concentrated strong acids the values of the activity coefficients differ considerably from unity and, moreover, their determination by means of potentiometric measurements is subject to a large error, caused mainly by the liquid-junction potential. Because of this, the spectrophotometric method of acidity determination is to be preferred, because it permits measurement of the concentration of both forms (acidic and basic) of the indicator from the absorption spectra.

The measurement is analogous to the spectrophotometric measurement of pH (Section 2.15) with the use of an indicator for which the value of the dissociation constant has been determined in dilute aqueous solutions. Thus from the equation

$$_aK_a = \frac{[H^+][B]}{[BH^+]} \frac{f_{H^+} f_B}{f_{BH^+}} \tag{3.181}$$

where $[H^+]$ denotes the concentration of the solvated proton, we obtain

$$[H^+] \frac{f_{H^+} f_B}{f_{BH^+}} = {_aK_a} \frac{[BH^+]}{[B]} \tag{3.182}$$

The value of the expression on the right-hand side of the equation can be calculated if the concentrations of the two forms of the indicator are determined spectrophotometrically. This value is a certain measure of the acidity of the solution in which the measurement is carried out. The exact relationship depends also on the activity coefficients of the two forms of the indicator used. However, if the indicator is an uncharged base (B), then the expression $f_{H^+} f_B / f_{BH^+}$ is very close to unity, particularly in solvents and media with a high relative permittivity. Thus the expressions

$$h_0 = {_aK_a} \frac{[BH^+]}{[B]} \tag{3.183}$$

or

$$H_0 = -\log h_0 = p_aK_a + \log \frac{[B]}{[BH^+]} \tag{3.184}$$

may be used as functions of the acidity in a medium. The function H_0 is called *Hammett's acidity function*. In dilute aqueous solutions it is equal to the pH of the solution.

Figure 3.21 shows curves which are functions of different measures of acidity for a hydrogen chloride solution with a variable concentration. These curves correspond to the ratios h_0/C_{HCl}, $[H^+]/C_{HCl}$ and a_{H^+}/C_{HCl}, where C_{HCl} denotes the analytical concentration of the acid.

Hammett's acidity function is a property of the solution that is independent of the indicator used, provided that the base B is uncharged. If we know the value of K_a for an indicator B_I, say, *p*-nitroaniline ($p_aK_a = 0.99$), we can determine the p_aK_a of another, less basic, indicator B_{II}, e.g. *o*-nitroaniline, in a solvent which is, say, a mixture of water and sulphuric acid.

Equation (3.185) is then used

$$p_aK_a^{II} = p_aK_a^{I} - \log \frac{[B_{II}][B_I H^+]}{[B_{II} H^+][B_I]} \tag{3.185}$$

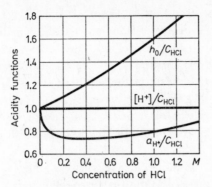

Fig. 3.21. The plot of the ratios of various acidity functions to the analytical concentration of hydrogen chloride, vs. its concentration.

on condition that, in the solvent, the appropriate ratios of the concentrations $[B_{II}]/[B_{II}H^+]$ and $[B_I]/[B_IH^+]$ can be measured. To make this possible, the p_aK_a values should not differ greatly from each other. In the case of o-nitroaniline the p_aK_a value calculated is -0.29.

By repeating this operation and working with indicators with decreasingly basic properties, we may extend the acidity scale to more and more acidic solvents, for example to mixtures with growing proportions of sulphuric acid in relation to water. A correct selection of organic in-

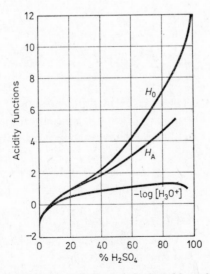

Fig. 3.22. Acidity functions H_0 and H_A and the logarithm of hydrogen ion concentration in concentrated solutions of sulphuric acid.

dicators permits the extension of Hammett's acidity scale up to an H_0 value of about 13. This value of Hammett's acidity function can be reached in solutions which no longer contain water, but contain a considerable excess of SO_3. In these conditions, as a result of dissociation of sulphuric acid, the ion $H_3SO_4^+$ appears in the solution; it has strong proton-donor properties. Figure 3.22 shows the dependence of Hammett's acidity function on the concentration of sulphuric acid and the same dependence for $\log[H_3O^+]$, which has its maximum in a solution with a concentration of about $10M$.

For measurements made in strongly acid solutions with a high content of sulphuric acid the exact knowledge of Hammett's function is indispensable; the relevant data are given in Table 3.7. However, it should be

Table 3.7

Value of Hammett's acidity function (H_0) for the system $H_2SO_4–H_2O$, at 25°C

\multicolumn{3}{c}{H_2SO_4 concentration}			\multicolumn{3}{c}{H_2SO_4 concentration}		
M	%	H_0	M	%	H_0
1.08	10	−0.31	7.1	50	−3.38
1.68	15	−0.66	9.1	60	−4.46
1.96	17.8	−0.88	10.3	65	−5.04
2.32	20	−1.01	11.5	70	−5.65
3.73	30	−1.72	12.7	75	−6.30
4.12	32.6	−1.90	14.1	80	−6.97
5.3	40	−2.41	16.6	90	−8.27
5.5	40.8	−2.48	18.6	100	−11.10

stated that though the results given by different authors for concentrations up to about $10M$ are similar, when the concentration exceeds this value discrepancies appear, and these necessitate critical evaluation of the data given in the literature.

Hammett's scale H_0 has also been determined for solutions of other acids, such as perchloric, hydrochloric, phosphoric, hydrofluoric and p-toluenesulphonic.

The significance of Hammett's acidity function is considerable because it permits comparison of the properties of very weak bases and in addition makes it possible to interpret reactions catalysed by acids in water–acid mixture.

Let us consider, for example, the hydrolysis of saccharose, for which the rate depends on the acidity of the solution. The relationship, however, is not rectilinear and varies for different strong acids, if we adopt as the measure of acidity the concentration of hydrogen ions calculated conventionally on the basis of the total acid concentration. If, instead, we estimate the acidity with the aid of Hammett's function, then the experimental points for different acids (for example $HClO_4$, H_2SO_4, HCl, HNO_3) are found to lie on a straight line, which indicates exact proportionality of the reaction rate and H_0.

Another practical application of Hammett's function is the calculation of dissociation constants of compounds which do not show basic properties in aqueous solutions. Measurements in water–sulphuric acid mixtures which have negative values of the acidity function H_0 allow determination of the acid dissociation constant of acetone, or rather of the $(CH_3COHCH_3)^+$ ion obtained as a result of the attachment of a proton to the acetone molecule. This constant is about $10^{7.2}$ ($pK_a = -7.2$), which proves that the ion is a very strong acid which cannot exist in aqueous solutions, and that acetone shows negligible basic properties in water.

It does not follow from the above that the course of Hammett's function depends on the type of compounds used as indicators for the scale determination. However, it turns out in practice that specific reactions may occur with a solvent which make it possible to obtain consistent results only for a particular type of compounds. The scale discussed above was prepared with the aid of nitroaniline indicators. Other types of compound have also been used, such as acid amides, including the amides of pyrrolecarboxylic acid or of trinitrobenzoic acid. Such an amide scale of acidity H_A has a rather different form from the H_0 scale (Fig. 3.22).

Attempts to establish acidity functions for anionic (H_-) or cationic (H_+) bases have not been so successful. The failure is due to the absence of sufficient indicators and also to the influence (difficult to eliminate in a simple manner) of the activity coefficients, which make the functions H_- and H_+ dependent on the kind of indicator used.

Hammett's acidity function makes it possible to enlarge the traditional pH scale for some systems, but it is not appropriate for all solvents. To establish Hammett's scale it is necessary to use indicators which are similar with regard to acid–base dissociation mechanism, similarly solvated and not subject to any other disturbing reactions. Therefore, it has proved impossible to use Hammett's scale for solvents of low relative permittivity (e.g. glacial acetic acid) because the acid–base reaction then leads to the formation of anion–cation ionic pairs, which precludes the preparation of the general acidity scale.

PROBLEMS

1. Calculate the pH of the following solutions:

    a. $4.25 \times 10^{-2} M$ HCl;

    b. $1.15 \times 10^{-4} M$ $HClO_4$;

    c. $9.50 \times 10^{-6} M$ HNO_3;

    d. $2.0 \times 10^{-1} M$ HNO_3.

2. Calculate the hydrogen ion concentration and pH of the following solutions. Assume that the ion product of water is 1.0×10^{-14}.

    a. $2.05 \times 10^{-3} M$ HCl;

    b. $5.0 \times 10^{-8} M$ $HClO_4$;

    c. $1.85 \times 10^{-1} M$ HNO_3;

    d. $2.00 \times 10^{-9} M$ HCl;

    e. $5.5 \times 10^{-4} M$ NaOH;

    f. $1.0 \times 10^{-10} M$ KOH;

    g. $2.5 \times 10^{-5} M$ $Ba(OH)_2$;

    h. $1.00 \times 10^{-3} M$ NaH.

3. Calculate the hydroxide ion concentration and pH in the following solutions. Assume that the ion product of water is 1.0×10^{-14}.

    a. $2.22 \times 10^{-2} M$ KOH;

    b. $5.80 \times 10^{-4} M$ HCl;

    c. $10^{-5.6} M$ HNO_3;

    d. $6.30 \times 10^{-4} M$ NaOH;

    e. $2.25 \times 10^{-8} M$ $HClO_4$;

    f. $1.50 \times 10^{-3} M$ C_2H_5ONa.

4. Calculate the hydrogen ion concentration and pH of the following solutions. Assume that the ion product of water is 1.0×10^{-14}.

    a. $1.50 \times 10^{-2} M$ acetic acid ($pK_a = 4.85$);

    b. $5.80 \times 10^{-1} M$ acetic acid ($pK_a = 4.83$);

    c. $2.25 \times 10^{-2} M$ benzoic acid ($pK_a = 4.20$);

    d. $1.10 \times 10^{-3} M$ hydrocyanic acid ($pK_a = 9.36$);

    e. $10^{-2.0} M$ phenol ($pK_a = 10.0$);

    f. $0.015 M$ boric acid ($pK_a = 9.2$);

    g. $0.005 M$ pyridine hydrochloride ($pK_a = 5.3$);

    h. $0.065 M$ methylamine hydrochloride ($pK_a = 10.6$);

    i. $0.065 M$ ammonium nitrate ($pK_a = 9.3$);

    j. $0.0065 M$ ammonium nitrate ($pK_a = 9.3$).

5. Calculate the dissociation constant of an acid for which a $1.0 \times 10^{-3} M$ solution has a pH equal to:

    a. 3.50;

    b. 4.00;

    c. 4.50;

    d. 5.00.

6. Calculate the hydrogen ion concentration and pH of the following solutions:

    a. $0.05M$ dichloroacetic acid $(K_a = 8.0 \times 10^{-2})$;

    b. $0.05M$ trichloroacetic acid $(K_a = 2.5 \times 10^{-1})$;

    c. $0.10M$ iodic acid $(K_a = 2.0 \times 10^{-1})$;

    d. $0.01M$ thiocyanic acid $(K_a = 1.6 \times 10^{-1})$;

    e. $0.1M$ sodium hydrogen sulphate $(K_a = 1.6 \times 10^{-2})$;

    f. $0.01M$ guanidine* $(K_b = 3.0 \times 10^{-1})$;

    g. $0.02M$ dimethylamine* $(K_b = 1.3 \times 10^{-3})$.

7. Calculate what the hydrogen ion concentration and pH would be in the following metal-ion solutions. Assume that the predominant reaction is the first protolysis step.

    a. $0.1M$ scandium chloride $(pK_a = 4.9)$;

    b. $0.01M$ lead perchlorate $(pK_a = 7.9)$;

    c. $0.01M$ beryllium nitrate $(pK_a = 5.7)$;

    d. $0.05M$ iron(II) perchlorate $(pK_a = 9.7)$.

8. Calculate the pH value of the following solutions:

    a. $0.01M$ aniline hydrochloride $(pK_a = 4.7, pK_w = 13.9)$;

    b. $0.1M$ sodium cyanide $(pK_b = 4.6, pK_w = 13.8)$;

    c. $0.02M$ sodium formate $(pK_b = 10.1, pK_w = 13.8)$;

    d. $0.001M$ ammonium chloride $(pK_a = 9.3, pK_w = 14.0)$;

    e. $0.10M$ sodium lactate $(K_a = 1.6 \times 10^{-4}, K_w = 1.7 \times 10^{-14})$;

    f. $0.20M$ sodium acetate $(K_a = 2.0 \times 10^{-5}, K_w = 2.0 \times 10^{-14})$;

    g. $0.02M$ sodium acetate $(K_a = 1.5 \times 10^{-5}, K_w = 1.3 \times 10^{-14})$;

    h. $0.002M$ sodium acetate $(K_a = 1.4 \times 10^{-5}, K_w = 1.1 \times 10^{-14})$.

9. Calculate the degree of dissociation of the acids or bases in the following solutions:

    a. $0.01M$ acetic acid $(K_a = 1.32 \times 10^{-5})$

    b. $0.005M$ acetic acid $(K_a = 1.32 \times 10^{-5})$

    c. $0.001M$ acetic acid $(K_a = 1.32 \times 10^{-5})$

    d. $0.0001M$ acetic acid $(K_a = 1.32 \times 10^{-5})$

    e. $0.1M$ aniline $(K_b = 4.0 \times 10^{-10})$

    f. $0.01M$ pyridine $(K_b = 2.0 \times 10^{-9})$

    g. $0.01M$ morpholine $(K_b = 2.5 \times 10^{-6})$

    h. $0.01M$ ethanolamine $(K_b = 2.5 \times 10^{-5})$

    i. $0.01M$ dimethylamine $(K_b = 1.0 \times 10^{-3})$

10. For the following electrolytes, calculate the concentration of a solution with a 5.0% degree of dissociation:

    a. hydrocyanic acid $(pK_a = 9.3)$;

    b. hypochlorous acid $(pK_a = 7.5)$;

    c. propionic acid $(pK_a = 4.9)$;

    d. lactic acid $(pK_a = 3.9)$;

    e. *o*-nitrobenzoic acid $(pK_a = 2.2)$;

    f. dichloroacetic acid $(pK_a = 1.3)$.

11. Calculate the pH in the following solutions:

    a. $2.0 \times 10^{-4}M$ aniline $(pK_b = 9.4)$;

---

  * Assume that the ion product of water is 1.0×10^{-14}.

$b.$ $5.0 \times 10^{-4} M$ aniline $(pK_b = 9.4)$;

$c.$ $1.0 \times 10^{-2} M$ dimethylamine hydrochloride $(pK_a = 11.0)$;

$d.$ $1.0 \times 10^{-3} M$ dimethylamine hydrochloride $(pK_a = 11.0)$;

$e.$ $1.0 \times 10^{-2} M$ hydrogen peroxide $(pK_a = 11.8)$;

$f.$ $2.0 \times 10^{-4} M$ hydrocyanic acid $(pK_a = 9.3)$;

$g.$ $5.0 \times 10^{-5} M$ acetic acid $(pK_a = 4.9)$;

$h.$ $4.0 \times 10^{-5} M$ trimethylamine $(pK_b = 4.2)$;

$i.$ $4.0 \times 10^{-4} M$ trimethylamine $(pK_b = 4.2)$.

12. Calculate the pH of the following buffer solutions:

$a.$ $0.10M$ HF $+ 0.10M$ NaF $(pK_a = 3.1)$;

$b.$ $0.10M$ HF $+ 0.05M$ NaF $(pK_a = 3.1)$;

$c.$ $0.05M$ HF $+ 0.10M$ NaF $(pK_a = 3.1)$;

$d.$ $0.10M$ NH_4Cl $+ 0.10M$ NH_3 $(pK_b = 4.6, pK_w = 13.8)$;

$e.$ $0.01M$ NH_4Cl $+ 0.01M$ NH_3 $(pK_b = 4.7, pK_w = 13.9)$;

$f.$ $0.10M$ benzoic acid $+ 0.01M$ sodium benzoate $(pK_a = 4.1)$;

$g.$ $0.05M$ acetic acid $+ 0.08M$ potassium acetate $(pK_a = 4.8)$;

$h.$ $0.10M$ trichloroacetic acid $+ 0.10M$ sodium trichloroacetate $(pK_a = 0.5)$;

$i.$ $0.10M$ dichloroacetic acid $+ 0.10M$ sodium dichloroacetate $(pK_a = 1.1)$;

$j.$ $0.10M$ monochloroacetic acid $+ 0.10M$ sodium monochloroacetate $(pK_a = 2.7)$;

$k.$ $0.10M$ dimethylamine $+ 0.10M$ dimethylamine hydrochloride $(pK_a = 11.1)$;

$l.$ $0.02M$ di-isopropylamine $+ 0.02M$ di-isopropylamine hydrochloride $(pK_a = 12.0, pK_w = 13.88)$;

$m.$ $0.02M$ di-isopropylamine $+ 0.05M$ di-isopropylamine hydrochloride $(pK_a = 12.0, pK_w = 13.82)$;

$n.$ $0.05M$ di-isopropylamine $+ 0.02M$ di-isopropylamine hydrochloride $(pK_a = 12.0, pK_w = 13.88)$.

13. For solutions containing a mixture of sodium acetate and acetic acid at pH values of

$a.$ 3.0;

$b.$ 4.0;

$c.$ 5.0

determine the ratio of the concentrations of the acid and its conjugate base $(pK_a = 4.8)$.

14. How much $1M$ sodium hydroxide should be added to 1 litre of

$a.$ $0.10M$ formic acid $(pK_a = 3.65)$;

$b.$ $0.10M$ acetic acid $(pK_a = 4.76)$;

$c.$ $0.10M$ benzoic acid $(pK_a = 4.12)$

to give a buffer solution with a pH of 4.00.

15. How much $1M$ hydrochloric acid should be added to 1 litre of $0.10M$ ammonia $(pK_a = 9.4)$ to give a buffer solution with a pH of:

$a.$ 8.50;

$b.$ 9.00;

$c.$ 9.4;

$d.$ 10.0.

16. Calculate the ionic strength of all the solutions in question 15.

17. Calculate the final pH of 100 ml of a buffer solution containing $0.10M$ formic acid $(pK_a = 3.7)$ and $0.10M$ sodium formate, to which has been added:

a. 1.00 ml of $1M$ perchloric acid;

b. 5.00 ml of $1M$ perchloric acid;

c. 1.00 ml of $1M$ potassium hydroxide;

d. 5.00 ml of $1M$ potassium hydroxide.

18. Calculate the concentration of the conjugate base which should be added to a solution of

a. $0.10M$ dichloroacetic acid ($pK_a = 1.1$);

b. $0.05M$ formic acid   ($pK_a = 3.7$);

c. $0.05M$ chloroacetic acid   ($pK_a = 2.7$);

d. $0.10M$ chloroacetic acid   ($pK_a = 2.7$)

to give a buffer solution of pH 2.5.

19. Calculate the buffer capacity of a solution containing:

a. $0.10M$ acetic acid + $0.10M$ sodium acetate ($pK_a = 4.8$);

b. $0.05M$ acetic acid + $0.05M$ sodium acetate ($pK_a = 4.8$);

c. $0.05M$ formic acid + $0.05M$ sodium formate ($pK_a = 3.7$);

d. $0.05M$ dichloroacetic acid + $0.05M$ sodium dichloroacetate ($pK_a = 1.1$);

e. $0.10M$ dichloroacetic acid + $0.05M$ sodium dichloroacetate ($pK_a = 1.1$).

20. By using the concept of buffer capacity, calculate the pH change when 5 millimoles of strong base are added to 100 ml of each of the buffer solutions in question 19.

21. Check the pH changes found in question 20 by using the usual formula for the pH of a buffer solution.

22. Calculate the concentrations of the two components of a buffer solution when buffers of capacity 0.2 and $pH = pK_a$ have to be prepared from:

a. ammonium chloride + ammonia ($pK_a = 9.4$);

b. formic acid + sodium formate ($pK_a = 6.4$);

c. tris(hydroxymethyl)aminomethane and its hydrochloride ($pK_a = 8.1$);

d. pyridine and its hydrochloride ($pK_a = 5.3$).

23. Calculate the concentrations of the two components of a buffer solution when the following buffers with capacity 0.15 have to be prepared:

a. ammonium chloride + ammonia pH = 9.0 ($pK_a = 9.4$);

b. formic acid + sodium formate pH = 4.0 ($pK_a = 3.7$);

c. benzoic acid + sodium benzoate pH = 4.0 ($pK_a = 4.1$);

d. dimethylamine + dimethylamine hydrochloride pH = 10.5 ($pK_a = 11.1$).

24. Calculate the pH values of the following solutions:

a. $0.10M$ succinic acid ($pK_{a1} = 4.4$, $pK_{a2} = 5.3$);

b. $0.02M$ selenious acid ($pK_{a1} = 2.6$, $pK_{a2} = 8.3$);

c. $0.01M$ telluric acid ($pK_{a1} = 7.6$, $pK_{a2} = 10.4$);

d. $0.01M$ o-phthalic acid ($pK_{a1} = 2.9$, $pK_{a2} = 5.4$);

e. $0.10M$ iminodiacetic acid ($pK_{a1} = 2.7$, $pK_{a2} = 9.5$);

f. $0.05M$ fumaric acid ($pK_{a1} = 3.0$, $pK_{a2} = 4.4$);

g. $0.05M$ maleic acid ($pK_{a1} = 1.9$, $pK_{a2} = 6.2$);

h. $0.05M$ oxalic acid ($pK_{a1} = 1.3$, $pK_{a2} = 4.3$);

i. $0.10M$ tartaric acid ($pK_{a1} = 3.0$, $pK_{a2} = 4.3$);

j. $0.01M$ tartaric acid ($pK_{a1} = 3.0$, $pK_{a2} = 4.3$);

$k.$ $0.002M$ tartaric acid ($pK_{a1} = 3.0$, $pK_{a2} = 4.3$);

$l.$ $0.001M$ sulphuric acid ($pK_{a2} = 1.9$);

$m.$ $0.05M$ hydrogen sulphide ($pK_{a1} = 7.0$, $pK_{a2} = 12.9$);

$n.$ $0.05M$ sulphurous acid ($pK_{a1} = 1.7$, $pK_{a2} = 7.1$);

$o.$ $0.10M$ phosphoric acid ($pK_{a1} = 2.0$, $pK_{a2} = 6.9$, $pK_{a3} = 11.7$).

25. Calculate the pH values of the following solutions:

$a.$ $1 \times 10^{-4}M$ sodium carbonate ($pK_{a1} = 6.4$, $pK_{a2} = 10.3$);

$b.$ $5 \times 10^{-2}M$ sodium arsenate ($pK_{a1} = 2.1$, $pK_{a2} = 6.7$, $pK_{a3} = 11.2$, $pK_w = 13.65$);

$c.$ $1 \times 10^{-2}M$ potassium oxalate ($pK_{b1} = 9.8$, $pK_{b2} = 12.7$, $pK_w = 13.8$);

$d.$ $1 \times 10^{-3}M$ potassium phosphate ($pK_{b1} = 2.1$, $pK_{b2} = 6.9$, $pK_{b3} = 11.9$, $pK_w = 13.9$);

$e.$ $2 \times 10^{-3}M$ tetrasodium ethylenediaminetetra-acetate acid ($pK_{b1} = 3.6$, $pK_{b2} = 7.7$, $pK_{b3} = 11.2$, $pK_{b4} = 11.8$, $pK_w = 13.9$);

$f.$ $2 \times 10^{-3}M$ disodium iminodiacetate ($pK_{b1} = 4.4$, $pK_{b2} = 11.2$, $pK_w = 13.9$).

26. Calculate the concentrations of all the species participating in protolytic equilibria in the following systems:

$a.$ $0.01M$ sulphur dioxide ($pK_{a1} = 1.8$, $pK_{a2} = 7.2$);

$b.$ $1 \times 10^{-4}M$ carbonic acid ($pK_{a1} = 6.4$, $pK_{a2} = 10.3$);

$c.$ $2 \times 10^{-3}M$ phosphoric acid ($pK_{a1} = 2.0$, $pK_{a2} = 7.0$, $pK_{a3} = 11.8$);

$d.$ $5 \times 10^{-3}M$ fumaric acid ($pK_{a1} = 3.0$, $pK_{a2} = 4.4$);

$e.$ $1.5 \times 10^{-3}M$ ethylenediamine ($pK_{b1} = 13.4$, $pK_{b2} = 6.5$, $pK_w = 14.0$);

$f.$ $1 \times 10^{-3}M$ disodium iminodiacetate ($pK_{b1} = 4.5$, $pK_{b2} = 11.3$, $pK_w = 14.0$).

27. Calculate the concentrations of all the species in the equilibria of question 26, after making $[H_3O^+]$ equal to $0.1M$ by addition of HCl.

28. Calculate the logarithms of the side-reaction coefficients at integral values of the pH, and plot a graph of $\log \alpha$ vs. pH for the protonation of:

$a.$ iminodiacetate anion ($pK_{a1} = 2.7$, $pK_{a2} = 9.5$);

$b.$ nitrilotriacetate anion ($pK_{a1} = 2.0$, $pK_{a2} = 2.6$, $pK_{a3} = 9.8$);

$c.$ salicylate anion ($pK_{a1} = 2.9$, $pK_{a2} = 13.1$);

$d.$ picolinate anion ($pK_{a1} = 1.5$, $pK_{a2} = 5.4$);

$e.$ diethylenetriamine ($pK_{a1} = 4.4$, $pK_{a2} = 9.2$, $pK_{a3} = 10.0$);

$f.$ triethylenetetramine ($pK_{a1} = 3.4$, $pK_{a2} = 6.8$, $pK_{a3} = 9.3$, $pK_{a4} = 10.0$);

$g.$ triethylenetetraminehexa-acetate anion ($pK_{a1} = 2.4$, $pK_{a2} = 3.0$, $pK_{a3} = 4.2$. $pK_{a4} = 6.2$, $pK_{a5} = 9.4$, $pK_{a6} = 10.2$).

29. From the stepwise stability constants of the hydroxo-complexes of the metal ions, calculate the stepwise acid dissociation constants:

$a.$ for Sc^{3+}: $k_1 = 1.26 \times 10^9$, $k_2 = 1.6 \times 10^8$, $k_3 = 2.5 \times 10^7$, $k_4 = 5.0 \times 10^4$;

$b.$ for Mg^{2+}: $k_1 = 1.0 \times 10^2$, $k_2 = 1.0 \times 10^1$;

$c.$ for Cu^{2+}: $k_1 = 6.3 \times 10^5$, $k_2 = 5.0 \times 10^4$, $k_3 = 5.0 \times 10^3$, $k_4 = 5.0 \times 10^2$;

$d.$ for Be^{2+}: $k_1 = 2.0 \times 10^8$, $k_2 = 6.3 \times 10^5$ $k_3 = 1.26 \times 10^4$, $K_4 = 1 \times 10^0$.

30. Calculate the logarithms of the side-reaction coefficients for metal ions in solutions with integral pH values. Neglect the formation of polynuclear complexes. Plot a graph of $\log \alpha$ vs. pH.

a. for Cd^{2+} $(p*\beta_1 = -10.3,\ p*\beta_2 = 20.6,\ p*\beta_3 = 33.8,\ p*\beta_4 = 46.9)$;
b. for Cu^{2+} $(p*\beta_1 = -8.2,\ p*\beta_2 = 17.5,\ p*\beta_3 = 27.8,\ p*\beta_4 = 39.1)$;
c. for Hg^{2+} $(p*\beta_1 = 3.8,\ p*\beta_2 = 6.2,\ p*\beta_3 = 21.1)$;
d. for Pb^{2+} $(p*\beta_1 = 7.9,\ p*\beta_2 = 17.3,\ p*\beta_3 = 28.0)$.

31. Calculate the minimum values of the solubility product of $M(OH)_2$ for which, at the following pH values, the hydroxide precipitate is not formed in a $10^{-3}M$ metal ion solution:
   a. pH $= 3$;
   b. pH $= 6$;
   c. pH $= 9$;
   d. pH $= 11$.

32. What is the lowest metal ion concentration at which the hydroxide precipitate will start to form for the following:
   a. $Pb(OH)_2$ $(K_{s0} = 1 \times 10^{-15},\ pH = 6)$;
   b. $Zr(OH)_4$ $(K_{s0} = 4 \times 10^{-57},\ pH = 3)$;
   c. $Fe(OH)_3$ $(K_{s0} = 3.2 \times 10^{-40},\ pH = 3)$.

33. Calculate the pH value of solutions containing these mixtures of acids:
   a. $0.01M$ $HClO_4$ + $0.10M$ acetic acid $(pK_a = 4.8)$;
   b. $0.01M$ HCl + $0.10M$ chloroacetic acid $(pK_a = 2.8)$;
   c. $0.005M$ HCl + $0.05M$ formic acid $(pK_a = 3.7)$.

34. Calculate the pH value of a solution containing $0.01M$ hydrogen sulphide $(pK_{a1} = 7.2,\ pK_{a2} = 13.2)$ and $0.01M$ carbon dioxide $(pK_{a1} = 6.4,\ pK_{a2} = 10.3)$

35. Calculate the pH of a solution containing $0.01M$ formic acid $(K_a = 2 \times 10^{-4})$ and $0.02M$ acetic acid $(K_a = 1.6 \times 10^{-5})$.

36. Check whether the pH value of $10^{-3}M$ acetic acid $(pK_a = 4.9)$ will change after it has been saturated with hydrogen sulphide to give a concentration of $0.1M$ $(pK_{a1} = 7.2,\ pK_{a2} = 13.2)$.

37. Calculate the pH of a solution containing $0.01M$ sulphuric acid $(pK_{a2} = 1.8)$ and $0.1M$ hydrogen sulphide $(pK_{a1} = 7.0,\ pK_{a2} = 12.9)$.

38. Calculate the pH of a solution containing $1 \times 10^{-2}M$ Na_2CO_3 $(pK_{b1} = 3.7,\ pK_{b2} = 7.5)$ and $1 \times 10^{-2}M$ Na_3PO_4 $(pK_{b1} = 2.1,\ pK_{b2} = 6.9,\ pK_{b3} = 11.8)$. Assume pK_w is 13.8.

39. Calculate the hydrogen ion concentration in solutions of the following salts:
   a. $0.05M$ methylamine formate $(pK_a = 3.7,\ pK_b = 3.1,\ pK_w = 13.8)$;
   b. $0.10M$ ammonium formate $(pK_a = 3.7,\ pK_b = 4.4,\ pK_w = 13.8)$;
   c. $0.10M$ ammonium chloroacetate $(pK_a = 2.7,\ pK_b = 4.4,\ pK_w = 13.8)$;
   d. $0.01M$ hexamethylenetetramine acetate $(pK_a = 4.8,\ pK_b = 8.7,\ pK_w = 13.9)$;
   e. $0.02M$ ammonium benzoate $(pK_a = 4.1,\ pK_b = 4.5,\ pK_w = 13.9)$;
   f. $0.01M$ cadmium acetate $(pK_a = 4.8,\ pK_b = 3.6,\ pK_w = 13.9)$.

40. Calculate the dissociation constants of the bases which give $0.01M$ solutions of acetates with the following pH values. The pK_a of acetic acid is 4.8, $pK_w = 13.9$.
   a. anilinium acetate (pH 4.70);

   b. diethylammonium acetate (pH 7.90);
   c. nickel acetate (pH 7.50);
   d. pyridine acetate (pH 6.70).

41. Calculate the pH values of solutions obtained by mixing equal volumes of:
   a. $0.02M$ ammonia ($pK_b = 4.8$) and $0.04M$ methylamine hydrochloride ($pK_a = 10.6$);
   b. $0.02M$ benzoic acid ($pK_a = 4.2$) and $0.01M$ ammonia ($pK_b = 4.8$);
   c. $0.01M$ formic acid ($pK_a = 3.7$) and $0.005M$ methylamine ($pK_b = 3.4$);
   d. $0.01M$ chloroacetic acid ($pK_a = 2.9$) and $0.015M$ pyridine ($pK_b = 8.7$).

42. Calculate the pH value of mixtures of 50 ml of $0.10M$ acetic acid ($pK_a = 4.8$) with 50 ml of the following solutions of ethylenediamine ($pK_{b1} = 3.8$, $pK_{b2} = 6.6$ $pK_w = 13.9$):
   a. $0.025M$;
   b. $0.050M$;
   c. $0.075M$;
   d. $0.10M$;
   e. $0.125M$.

43. Calculate the pH values of the following solutions:
   a. $0.005M$ aminoacetic acid ($pK_{a1} = 2.35$, $pK_{a2} = 9.77$);
   b. $0.01M$ anthranilic acid ($pK_{a1} = 2.1$, $pK_{a2} = 5.0$);
   c. $0.01M$ sodium hydrogen tartrate ($pK_{a1} = 3.0$, $pK_{a2} = 4.3$);
   d. $0.03M$ potassium hydrogen sulphite ($pK_{a1} = 1.7$, $pK_{a2} = 6.9$);
   e. $0.05M$ disodium hydrogen phosphate ($pK_{a1} = 2.0$, $pK_{a2} = 6.9$, $pK_{a3} = 11.7$);
   f. $0.01M$ potassium hydrogen oxalate ($pK_{a1} = 1.1$, $pK_{a2} = 4.0$);
   g. $0.001M$ potassium hydrogen oxalate ($pK_{a1} = 1.1$, $pK_{a2} = 4.0$);
   h. $0.005M$ potassium hydrogen o-phthalate ($pK_{a1} = 2.9$, $pK_{a2} = 5.3$);
   i. $0.01M$ disodium ethylenediaminetetra-acetate ($pK_{a1} = 2.1$, $pK_{a2} = 2.7$, $pK_{a3} = 6.2$, $pK_{a4} = 10.3$).

44. How much $1M$ hydrochloric acid can be added to 100 ml of a buffer solution containing $0.1M$ NaH$_2$PO$_4$ and $0.1M$ NaH$_2$PO$_4$, to change the pH by not more than:
   a. 0.2 pH units;
   b. 0.4 pH units;
   c. 1.0 pH units.

CHAPTER 4

Theory of acid-base titrations

4.1 GENERAL REMARKS

The aim of an acid–base titration is to determine the concentration of a solution by measuring the volume of titrant (of known concentration) that reacts with it according to a stoichiometric proton-transfer reaction.

We can distinguish the following variants of acid–base titrations:

(1) strong acid with strong base,
(2) strong base with strong acid,
(3) weak acid with strong base,
(4) weak base with strong acid,
(5) weak acid with weak base or vice versa.

The first two cases will be discussed together because they consist of the same basic reaction of protolysis, in which the H_3O^+ and OH^- ions are the only substrates. The next two cases are also similar to each other. The last case hardly ever occurs in analytical practice because the analyst can choose the titrant, and will always use a strong acid (or a strong base) to obtain better results.

An additional variant is formed by titrations of multiple systems, such as systems of polyprotic acids and bases, and mixtures of acids and mixtures of bases of different concentrations and strengths. These will be considered jointly because in these systems, which are of great practical and theoretical importance, many common features can be observed.

The volume of titrant required for stoichiometric proton transfer is determined by means of an indicator which allows location of the end-point of the titration. In principle, the end-point ought to agree with the theoretical equivalence point of the reaction (which corresponds to the stoichiometric ratios of the reagents), but seldom does in practice. Nevertheless, when the difference between the two points is small, the titration error remains within acceptable limits.

The graphical representation of a titration is the *titration curve*. This term can be used with two different meanings. In a general sense the titration curve is a functional relation between the logarithm of the concentration of one of the reagents and the volume of the titrant added or the *titration fraction f* (which is proportional to the volume of titrant added, which has a value between 0 and 1 at the equivalence point, and then takes values higher than 1). In a practical sense, the term titration curve is often applied to a graph of the measured quantity (which is a function of concentration) vs. the volume of titrant added or the titration fraction. If the measured quantity is exactly proportional to the logarithm of concentration, as in potentiometric titrations, the practical titration curve will be the same as the theoretical one. In other cases the practical titration curve, such as a graph of absorbance changes (spectrophotometric titration), of conductivity (conductometric titration), or of current intensity (amperometric titration), is different in shape from the theoretical titration curve.

In this chapter we shall consider titration curves primarily in the general sense. Knowledge of these curves allows us to predict the course of the reaction and to choose the most advantageous method of end-point location. In acid–base titrations the dependent variable in titration curves is, as a rule, the pH of the solution.

There are two methods for calculating titration curves. The simpler method in acid–base titration is the calculation of pH values corresponding to the changing compositions of the solution during the titration. Such calculations can be done by using the equations discussed in Chapter 3. The other method is to determine the equation of the titration curve for the whole process of titration. In this case we start from a general assumption of equality of the sums of positive and negative charges, i.e. from the charge-balance equation. This method is mathematically more difficult and seldom used, unless the titration curve is to be calculated with the aid of a computer. This procedure is unquestionably more general, and it allows calculations to be done in the vicinity of the equivalence point, for which the usual simplifications are not valid. In this chapter we shall first discuss the calculation of titration curves from the solution composition for given points, and then give the general equations for titration curves.

4.2 TITRATION OF A STRONG ACID WITH A STRONG BASE

In the titration of a strong acid with a strong base in aqueous medium the product is water, since the reaction is between the H_3O^+ and OH^- ions.

Thus, the pH value of the titration solution at any stage nearly up to the equivalence point can be calculated from the concentration of untitrated strong acid. The expression

$$\text{pH} = -\log C_{\text{HA}} \tag{4.1}$$

is satisfied for pH values sufficiently far below 7 for the concentration of H_3O^+ resulting from the dissociation of water to be neglected. In calculation of the pH, an essential factor is the change in the volume resulting from dilution of the solution as titrant is added.

If the initial volume is V_0 and the volume of titrant added is V, the initial number of moles of acid (or mmoles if the volume is expressed in ml) will be $C_{\text{HA}}V_0$, and the amount of acid already reacted at a point in the titration will be $C_B V$, where C_B is the molar concentration of the strong base. Accordingly, the concentration of H_3O^+ ions at each point of the titration curve before the equivalence point is

$$[H_3O^+] = \frac{C_{\text{HA}}V_0 - C_B V}{V_0 + V} \tag{4.2}$$

The volume of water produced in the reaction is negligible.

At the equivalence point the concentration of H_3O^+ ions is equal to the concentration of OH^- ions; consequently for a solution of a salt of a strong acid and a strong base, with no excess of acid or excess of base, this concentration is $10^{-7}M$ (at 25°C).

To calculate the concentration of H_3O^+ beyond the equivalence point we assume that we have a solution of a salt and a strong base. The salt does not affect the pH of the solution, so we can calculate the concentration of OH^- ions from the amount of added base and the total volume of solution $(V_0 + V)$. If the total number of moles (mmoles) of the base is $C_B V$, the concentration of OH^- ions after the equivalence point is

$$[OH^-] = \frac{C_B V - C_{\text{HA}}V_0}{V_0 + V} \tag{4.3}$$

or

$$[H_3O^+] = \frac{K_w(V_0 + V)}{C_B V - C_{\text{HA}}V_0} \tag{4.4}$$

Let us now introduce the concept of the titration fraction f, which is the ratio of the number of moles (mmoles) of base added as titrant to the number of moles (mmoles) of acid being titrated

$$f = \frac{C_B V}{C_{\text{HA}}V_0} \tag{4.5}$$

The values of f range from 0 at the start of the titration, through 1 at the equivalence point, to numbers greater than 1 beyond the equivalence point. Sometimes, in calculations, instead of the titration fraction we use 'titration percentage', which has a value of 100% at the equivalence point (and is equal to $100f$).

From Eqs. (4.2) and (4.5) we obtain

$$[H_3O^+] = \frac{(1-f)C_{HA}V_0}{V_0 + V} \tag{4.6}$$

and similarly from equation (4.4) we obtain

$$[H_3O^+] = \frac{K_w(V_0 + V)}{(f-1)C_{HA}V_0} \tag{4.7}$$

Table 4.1 shows the use of these equations. From a series of individual

Table 4.1

Calculation of points on the titration curve of 50 ml of $0.10M$ hydrochloric acid with $0.10M$ sodium hydroxide

V ml	f	$C_{HA}V_0 - C_B V$ mmole	$V_0 + V$ ml	$[H_3O^+]$ M	pH
0	0	5.00	50	0.10	1.00
10	0.20	4.00	60	0.0667	1.18
20	0.40	3.00	70	0.0429	1.37
30	0.60	2.00	80	0.0250	1.60
40	0.80	1.00	90	0.0111	1.95
45.0	0.90	0.50	95.0	0.00526	2.28
48.0	0.96	0.20	98.0	0.00204	2.69
49.0	0.98	0.10	99.0	0.00101	3.00
49.5	0.99	0.05	99.5	0.00050	3.30
49.8	0.996	0.02	99.8	0.00020	3.70
49.9	0.998	0.01	99.9	0.00010	4.00
50.0	1.00	—	100.0	1.00×10^{-7}	7.00
50.1	1.002	0.01	100.1	1.00×10^{-10}	10.00
50.2	1.004	0.02	100.2	5.01×10^{-11}	10.30
50.5	1.01	0.05	100.5	2.01×10^{-11}	10.70
51.0	1.02	0.10	101.0	1.01×10^{-11}	11.00
52.0	1.04	0.20	102.0	5.10×10^{-12}	11.29
55.0	1.10	0.50	105.0	2.10×10^{-12}	11.68
60	1.20	1.00	110	1.10×10^{-12}	11.96
70	1.40	2.00	120	6.00×10^{-13}	12.22
80	1.60	3.00	130	4.23×10^{-13}	12.36
90	1.80	4.00	140	3.50×10^{-13}	12.46
100	2.00	5.00	150	3.00×10^{-13}	12.52

titration points calculated in this manner the titration curve can be plotted as pH vs. volume of titrant or as pH vs. titration fraction, f. Because of the increase in the volume of the solution, the curve is not symmetrical with respect to the equivalence point. A more symmetrical curve is obtained if the titrant is significantly more concentrated than the solution being titrated, i.e. when the change in volume, in relation to the initial volume, is small (in practice, this would increase the error of titrant volume measurement). However, close to the equivalence point, in the range $0.98 \leqslant f \leqslant 1.02$, the curve is sufficiently symmetrical to allow the use, for determination of the end-point, of methods which assume that the titration

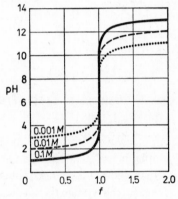

Fig. 4.1. Titration curves of a strong acid with 0.1, 0.01 and 0.001M strong base.

curve is symmetrical. Figure 4.1 shows titration curves of strong acids with initial concentrations 10^{-1}, 10^{-2} and $10^{-3}M$. The titrant used was a solution of a strong base with a concentration equal to the initial acid concentration; thus the $10^{-1}M$ acid was titrated by a base with the same concentration ($10^{-1}M$), etc.

A similar method is used to calculate the titration curves of a strong base with a strong acid. Before the equivalence point we have

$$[OH^-] = \frac{C_B V_0 - C_{HA} V}{V_0 + V} \qquad (4.8)$$

so

$$[H_3O^+] = \frac{K_w(V_0 + V)}{C_B V_0 - C_{HA} V} \qquad (4.9)$$

At the equivalence point we have

$$[OH^-] = [H_3O^+] = \sqrt{K_w} \qquad (4.10)$$

and beyond the equivalence point

$$[H_3O^+] = \frac{C_{HA}V - C_BV_0}{V_0 + V} \qquad (4.11)$$

If we introduce, instead of the symbols V_0 and V, the symbols V_B and V_{HA}, which denote the volume of the base solution and the volume of the acid solution respectively, we would obtain identical expressions in both cases.

If we replace $C_{HA}V$ in Eqs. (4.9) and (4.11) by the titration fraction (in this case of course $f = C_{HA}V/C_BV_0$), we obtain

$$[H_3O^+] = \frac{K_w(V_0 + V)}{(1 - f)C_BV_0} \qquad (4.12)$$

and

$$[H_3O^+] = \frac{(f - 1)C_BV_0}{V_0 + V} \qquad (4.13)$$

4.3 GENERAL EQUATION FOR THE TITRATION OF A STRONG ACID WITH A STRONG BASE

Although the method of calculating the titration curve which we have given is comparatively simple, it has a number of defects. First, in different regions of the titration curve we make different theoretical assumptions; this is because we first regard the system as an acid solution, next as a mixture of an acid and a salt, at the equivalence point as a salt solution and beyond this point as a solution containing a strong base and a salt. Particular equations can be derived for each of these cases; but it is better to formulate a general equation to give unified description of the whole titration curve.

Such an equation can be derived if we remember that the total number of positive charges in a solution must be equal to the total number of negative charges. Thus the corresponding sums of ions will be equal, if we introduce suitable coefficients when the charges of the ions are different from unity. Thus, if a strong acid HA with concentration C_{HA} and initial volume V_0 is titrated with a strong base BOH with concentration C_B and volume V, then at any moment we have

$$[H_3O^+] + [B^+] = [OH^-] + [A^-] \qquad (4.14)$$

If the concentration of B^+ and A^- are expressed in terms of the initial concentrations of the solution being titrated and the titrant, i.e.

$$[A^-] = \frac{C_{HA}V_0}{V_0 + V}, \qquad [B^+] = \frac{C_BV}{V_0 + V} \qquad (4.15)$$

then, on substitution of these expressions in Eq. (4.14), we obtain

$$[H_3O^+] + \frac{C_B V}{V_0 + V} = [OH^-] + \frac{C_{HA} V}{V_0 + V} \tag{4.16}$$

and after rearrangement and elimination of $[OH^-]$

$$\frac{C_B V - C_{HA} V_0}{V_0 + V} = \frac{K_w}{[H_3O^+]} - [H_3O^+] \tag{4.17}$$

A more compact form of this equation is obtained if we use the titration fraction

$$f = \frac{C_B V}{C_{HA} V_0} \tag{4.18}$$

We then obtain

$$\frac{C_{HA} V_0}{V_0 + V} (f - 1) = \frac{K_w}{[H_3O^+]} - [H_3O^+] \tag{4.19}$$

which is the final form of the equation of the titration curve.

This equation may be simplified in particular ranges of the titration curve. Before the equivalence point, when $[H_3O^+] \gg [OH^-]$, it takes the form

$$\frac{C_{HA} V_0}{V_0 + V} (1 - f) = [H_3O^+] \tag{4.20}$$

and beyond the equivalence point, when $[H_3O^+] \ll [OH^-]$, it is

$$\frac{C_{HA} V_0}{V_0 + V} (f - 1) = \frac{K_w}{[H_3O^+]} \tag{4.21}$$

These relations are identical to those given before, which shows that they represent particular cases of the general expression (4.19). The general form is the most useful for very dilute solutions. Figure 4.2 shows titra-

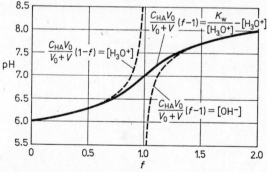

Fig. 4.2. Exact and approximate titration curves for $1.0 \times 10^{-6} M$ hydrochloric acid with $1.0 \times 10^{-4} M$ sodium hydroxide.

tion curves calculated from simplified forms of Eqs. (4.20) and (4.21) and the titration curve calculated from the general Eq. (4.19). As can be seen, when the concentration of the solution being titrated is $1.0 \times 10^{-6} M$ and the titrant concentration is $1.0 \times 10^{-4} M$, there are already differences between the pH values at titration fractions $f = 0.5$ and $f = 1.5$ and these differences increase as the equivalence point is approached. However, since we very seldom use such extremely dilute solutions (which require special working techniques); in ordinary titrations, the simplified equations are the most useful.

To describe the titration curve for a strong base titrated with a strong acid we make the same basic assumption (Eq. (4.14)) and change only the symbols of Eq. (4.15). Thus we obtain

$$[A^-] = \frac{C_{HA}V}{V_0 + V}, \quad [B^+] = \frac{C_B V_0}{V_0 + V} \tag{4.22}$$

and

$$\frac{C_B V_0 - C_{HA}V}{V_0 + V} = \frac{K_w}{[H_3O^+]} - [H_3O^+] \tag{4.23}$$

and on substitution of the titration fraction, we get the general equation of the titration curve of a strong base

$$\frac{C_B V_0}{V_0 + V}(1 - f) = \frac{K_w}{[H_3O^+]} - [H_3O^+] \tag{4.24}$$

For calculation of titration curves for highly dilute solutions, provided that the pH is more than 0.7 units away from 7, the simplified equations may be used because $K_w/[H_3O^+] \gg [H_3O^+]$ or $K_w/[H_3O^+] \ll [H_3O^+]$. However, for $6.3 \leqslant$ pH $\leqslant 7.7$ it is advisable to use the general equations (4.19) or (4.24), which are second-degree equations.

Example 4.1. Calculate the titration curve for titration of 100 ml of $1.0 \times \times 10^{-6} M$ NaOH with $2.0 \times 10^{-4} M$ HCl.

In this case, which is rather hypothetical, the volume of the solution titrated up to the equivalence point changes only very slightly—the volume increase is 0.5% of the initial volume, and this can be neglected. Thus Eq. (4.24) takes the form

$$C_B(1 - f) = \frac{K_w}{[H_3O^+]} - [H_3O^+]$$

We will calculate $[H_3O^+]$ for $f = 0.50$, 0.75 and 0.90. We use first an approximation in which we neglect $[H_3O^+]$ in relation to $K_w/[H_3O^+]$ because the solution before the equivalence point is alkaline. If $f = 0.50$, then

$$1.0 \times 10^{-6}(1 - 0.50) = 10^{-14}/[H_3O^+]$$

and

$$[H_3O^+] = 2.0 \times 10^{-8}M, \quad pH = 7.70$$

The pH value indicates that in this case the approximation is valid. However, when $f = 0.75$, we obtain

$$[H_3O^+] = \frac{10^{-14}}{(1 - 0.75)10^{-6}} = 4 \times 10^{-8}M, \quad pH = 7.40$$

Since this pH value is inside the range 6.3–7.7, we must use the quadratic equation. Then

$$1.0 \times 10^{-6}(1 - 0.75) = \frac{10^{-14}}{[H_3O^+]} - [H_3O^+]$$

and on rearrangement

$$[H_3O^+]^2 + 1.0 \times 10^{-6}(1 - 0.75)[H_3O^+] - 10^{-14} = 0$$

Hence

$$[H_3O^+] = 3.5 \times 10^{-8}M, \quad pH = 7.46$$

An analogous calculation for $f = 0.90$ gives

$$[H_3O^+] = 6.2 \times 10^{-8}M, \quad pH = 7.21$$

If we used the simplified formula in this case, we would again obtain a false result

$$[H_3O^+] = 1.0 \times 10^{-7}M, \quad pH = 7.00$$

4.4 TITRATION OF A WEAK ACID WITH A STRONG BASE

To calculate the pH values at a series of points on the curve of a weak acid for titration with a strong base, various assumptions can be made.

At the start of the titration, $f = 0$, the solution is essentially a solution of a weak acid, so one of the equations given in Section 3.3 or 3.5 can be used. Most often it is

$$[H_3O^+] = \sqrt{K_a C_{HA}} \tag{4.25}$$

which is satisfied if the acid is not very dilute and not too strong.

On the addition of small quantities of a strong base to a solution of a weak acid the reaction

$$HA + OH^- \rightleftharpoons H_2O + A^- \tag{4.26}$$

proceeds to the right practically stoichiometrically, and consequently a weak base A⁻ (conjugate to the titrant acid) is formed in the solution in concentration equivalent to the amount of strong base added. This amount corresponds to the decrease in amount of acid present; consequently in the solution

$$[\text{HA}] = \frac{C_{\text{HA}}V_0 - C_\text{B}V}{V_0 + V} \tag{4.27}$$

and

$$[\text{A}^-] = \frac{C_\text{B}V}{V_0 + V} \tag{4.28}$$

If we substitute these values in the equation for the pH of a solution containing a conjugate acid–base pair, i.e. in the formula for the calculation of the pH of a buffer solution, we obtain

$$[\text{H}_3\text{O}^+] = \frac{K_\text{a}[\text{HA}]}{[\text{A}^-]} = \frac{K_\text{a}(C_{\text{HA}}V_0 - C_\text{B}V)}{C_\text{B}V} \tag{4.29}$$

In practice this equation may be used almost throughout the range of titration fractions $0 < f < 1$; it fails very seldom, and only near the limits of this range. It should be noted that in this equation neither the volume nor the concentration of the solution being titrated is used directly, so the change in pH during the titration of a weak acid with a strong base, and indeed the course of the titration curve, does not depend on the initial concentration or the dilution of the solution during titration. This becomes more obvious if we realize that the ratio $[\text{HA}]/[\text{A}^-]$ can be replaced by the ratio of the number of moles of acid to the number of moles of base, this ratio being independent of dilution, and if we use the reciprocal of the titration fraction

$$\frac{1}{f} = \frac{C_{\text{HA}}V_0}{C_\text{B}V} \tag{4.30}$$

From Eq. (4.29) we thus obtain

$$[\text{H}_3\text{O}^+] = K_\text{a} \frac{(1 - f)}{f} \tag{4.31}$$

which confirms the lack of dependence of pH on C_{HA} and $V_0 + V$.

At the equivalence point, when $f = 1$, the solution being titrated is a solution of a salt of a strong base and a weak acid, so according to the Brønsted–Lowry theory, it is a solution of a weak anionic base with concentration

$$C_\text{A} = C_{\text{HA}} \frac{V_0}{V_0 + V} \tag{4.32}$$

Thus, in order to calculate pH, we first determine $[OH^-]$, which is equal to

$$[OH^-] = \sqrt{C_A K_b} \tag{4.33}$$

where K_b is the basic dissociation constant of base A^-. Since $K_b = K_w/K_a$, and C_A is determined by Eq. (4.32), we obtain

$$[OH^-] = \sqrt{\frac{C_{HA} K_w V_0}{K_a(V_0 + V)}} \tag{4.34}$$

or, on converting into $[H_3O^+]$,

$$[H_3O^+] = \sqrt{\frac{K_w K_a(V_0 + V)}{C_{HA} V_0}} \tag{4.35}$$

According to the equation, both the initial concentration and an increase in solution volume affect the pH value.

The titration process beyond the equivalence point, when $f > 1.0$, is identical to that discussed in Section 4.2 and corresponds to a mixture of a strong base and a weak anionic base. Since in most cases the difference between the dissociation constants of these bases is quite large, we can regard the system as a solution of a strong base, and neglect the weak electrolyte. Thus, we calculate pOH as if there were $C_B V - C_{HA} V_0$ moles of OH^- in the volume $V_0 + V$. Hence

$$[OH^-] = \frac{C_B V - C_{HA} V_0}{V_0 + V} \tag{4.36}$$

$$[H_3O^+] = \frac{K_w(V_0 + V)}{C_B V - C_{HA} V_0} \tag{4.37}$$

If we introduce f we obtain

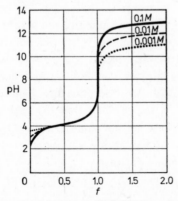

Fig. 4.3. Titration curves for 0.1, 0.01 and 0.001M benzoic acid ($K_a = 6.3 \times 10^{-5}$) with 0.2, 0.02 and 0.002M sodium hydroxide, respectively.

Table 4.2

Calculation of points on the titration curve of 50 ml of 0.10M benzoic acid ($K_a = 6.3 \times$
$\times 10^{-5}$) with 0.20M sodium hydroxide

V ml	f	$\dfrac{C_{HA}V_0 - C_B V}{}$ mmole	$\dfrac{C_B V}{} = C_{HA}V_0 f$ mmole	$V_0 + V$ ml	$[H_3O^+]$ M	pH
0	0	5.0	0	50	2.51×10^{-3}	2.60
5	0.20	4.0	1.0	55	2.52×10^{-4}	3.60
10	0.40	3.0	2.0	60	9.45×10^{-5}	4.02
15	0.60	2.0	3.0	65	4.20×10^{-5}	4.38
20	0.80	1.0	4.0	70	1.58×10^{-5}	4.80
23	0.92	0.40	4.60	73	5.48×10^{-6}	5.26
24	0.96	0.20	4.80	74	2.62×10^{-6}	5.58
25	1.00	—	5.00	75	3.08×10^{-9}	8.51
26	1.04	0.20	5.20	76	3.80×10^{-12}	11.43
27	1.08	0.40	5.40	77	1.93×10^{-12}	11.71
30	1.20	1.00	6.00	80	8.00×10^{-13}	12.10
35	1.40	2.00	7.00	85	4.25×10^{-13}	12.37
40	1.60	3.00	8.00	90	3.00×10^{-13}	12.52
45	1.80	4.00	9.00	95	2.38×10^{-13}	12.62
50	2.00	5.00	10.00	100	2.00×10^{-13}	12.70

$$[H_3O^+] = \frac{K_w(V_0 + V)}{(f - 1)C_{HA}V_0} \qquad (4.38)$$

Table 4.2 shows the method of calculating the titration curve for
a weak acid, from Eqs. (4.25), (4.31) and (4.38). Figure 4.3 shows the
titration curve of three different solutions of benzoic acid, namely 0.1M,
0.01M and 0.001M. Differences in the graphs of these curves appear for
f values corresponding to the beginning of the titration, to the equivalence
point and to the region beyond it. It should be stressed, however, that
no differences dependent on dilution will be seen in the region close to
$f = 0$ and $f = 1$ if we use the simplest formula for the pH of a buffer
solution. In precise calculations we must use a more accurate expression
in these ranges to allow for the fact that the real concentrations of the
acid and the conjugate base are not equal to their stoichiometric concentrations if none of the components is in large excess. The following is
such an expression used close to $f = 0$ and $f = 1$:

$$K_a = \frac{[H_3O^+]\{C_B V/(V_0 + V) + [H_3O^+] - [OH^-]\}}{\left\{\dfrac{C_{HA}V_0 - C_B V}{V_0 + V} - [H_3O^+] + [OH^-]\right\}} \qquad (4.39)$$

This equation, however, is used only in exceptional cases.

Comparison of the titration curves of weak acids having different dissociation constant values (Fig. 4.4) indicates that the weaker the acid the smaller the break in the titration curve at the equivalence point. The titration of an acid with a dissociation constant smaller than about 10^{-8},

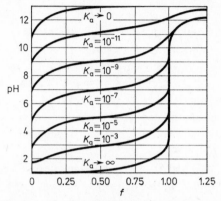

Fig. 4.4. Titration curves of $0.10M$ solutions of acid with various values of the dissociation constant, with $0.10M$ sodium hydroxide solution.

even in $0.10M$ solutions, does not give good results; the break in the curve is so small that it is practically impossible to determine the end-point.

The expressions we have derived, which are the basis for calculation of titration curves, can be used both for molecular acids and for cationic acids. The titration curve of the anilinium ion (a solution of aniline hydro-chloride) ($pK_a = 4.7$) will be very similar to the titration curve of acetic acid ($pK_a = 4.76$). However, if we titrate the ammonium ion ($pK_a = 9.37$) then, as in the case of boric acid ($pK_a = 9.10$) we do not observe a clear break in the curve at the equivalence point.

In calculations of the titration curve usually a single value of each equilibrium constant will be assumed to apply throughout the titration. This is not always quite correct because the ionic strength may vary during the titration. At the beginning of titration, in the solution of a molecular acid the ionic strength is small, but increases up to the end where the solution contains a dissociated salt. In very exact calculations ionic strength corrections to equilibrium constant values may be applied at individual points.

Finally, we list the basic differences between the titration curves of strong and weak acids.

1. Titration of a weak acid in the region $0 < f < 1$ gives an S-shaped curve, identical with the dissociation curve of a weak acid (apart from the strongly acidic regions). At the start, the titration curve of a strong acid is flatter, which means that a strong acid solution is a better buffer than a weak acid solution containing very small quantities of the conjugate base.

2. The titration curve of a weak acid in the region $0 < f < 1$, unlike the curve for a strong acid, does not depend on the concentration of the acid. In the case of a strong acid a change of concentration by one order of magnitude moves the first part of the curve by one pH unit.

3. The equivalence point of the titration of a weak acid with a strong base always lies above pH $= 7.0$. The weaker the acid titrated and the higher its concentration, the higher the pH of that point. In the titration of a strong acid, the equivalence point, independently of the concentration, is always at pH $= 7.0$.

4. The weaker the acid titrated and the lower its concentration, the smaller is the pH break at the equivalence point.

5. Beyond the equivalence point, when $f > 1$, the titration curve depends only on the dilution of the solution irrespective of the kind of acid titrated.

4.5 TITRATION OF A WEAK BASE WITH A STRONG ACID

The calculation of the curve for titration of a weak base with a strong acid follows a procedure similar to that used for weak acid titration with strong base. The only difference is the replacement of the characteristic data of the acid systems by the corresponding data of the base systems.

Thus we calculate the initial point, $f = 0$, in the same way as we calculated the pH of a weak base, i.e. mostly by use of the approximate equation

$$[OH^-] = \sqrt{C_B K_b} \tag{4.40}$$

or

$$[H_3O^+] = \frac{K_w}{C_B K_b} = \sqrt{\frac{K_w K_a}{C_B}} \tag{4.41}$$

where K_a denotes the acid dissociation constant of the conjugate acid of the titrated base.

In the buffer region, where $0 < f < 1$, we use the equation for calculation of the pH of the buffer; the equation is based either on the value of the basic dissociation constant

$$[OH^-] = K_b \frac{[B]}{[HB^+]}, \quad [H_3O^+] = \frac{K_w[HB^+]}{K_b[B]} \tag{4.42}$$

or on the value of the acid dissociation constant

$$[H_3O^+] = K_a \frac{[HB^+]}{[B]} \tag{4.43}$$

When we substitute into these equations

$$[HB^+] = \frac{C_{HA}V}{V_0 + V} \tag{4.44}$$

$$[B] = \frac{C_B V_0 - C_{HA}V}{V_0 + V} \tag{4.45}$$

we obtain a form which is applicable except in the regions close to $f = 0$ and $f = 1$

$$[H_3O^+] = K_a \frac{C_{HA}V}{(C_B V_0 - C_{HA}V)} \tag{4.46}$$

If the titration fraction $f = C_{HA}V/C_B V_0$ is introduced into the equation, we obtain

$$[H_3O^+] = K_a \frac{f}{(1 - f)} \tag{4.47}$$

To calculate the pH at the equivalence point we consider our system as a solution of a weak acid conjugate to the base being titrated. The number of mmoles of this acid is equal to the number of mmoles of the base present at the beginning of titration, but the volume of the solution has changed from V to $V_0 + V$. Hence

$$[H_3O^+] = \sqrt{K_a \frac{C_B V_0}{V_0 + V}} \tag{4.48}$$

or, if the value of the basic dissociation constant is used,

$$[H_3O^+] = \sqrt{\frac{K_w C_B V_0}{K_b(V_0 + V)}} \tag{4.49}$$

Beyond the equivalence point, where $f > 1$, the pH value is determined by the amount of excess of acid added, and so

$$[H_3O^+] = \frac{C_{HA}V - C_B V_0}{V_0 + V} \tag{4.50}$$

or, if we introduce the titration fraction f,

$$[H_3O^+] = \frac{(f - 1)C_B V_0}{(V_0 + V)} \tag{4.51}$$

The titration curves for several $0.10M$ solutions of weak bases with a $0.1M$ solution of strong acid (such as hydrochloric acid) are shown in Fig. 4.5. Similar titration curves are obtained when anionic bases are used. For example the titration curve of ammonia ($pK_b = 4.4$) is nearly identical with the titration curve of borate ion as a base ($pK_b = 4.7$). This has the following practical consequences if an acid is too weak to be titrated with a strong base, then its conjugate base is sufficiently strong

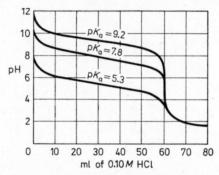

Fig. 4.5. Titration curves of $0.10M$ weak base solutions: ammonia ($pK_a = 9.2$), triethanolamine ($pK_a = 7.8$), pyridine ($pK_a = 5.3$) with $0.10M$ strong acid (e.g. HCl).

to give a distinct end-point on the titration curve with a strong acid. For example, boric acid is so weak that it cannot be titrated accurately with sodium hydroxide, but its sodium salt (borax, $Na_2B_4O_7$) is a sufficiently strong base in aqueous solution to be used as a standard for acid standardization.

In the case of aniline ($pK_b = 9.1$) we have the reverse situation. The aniline is too weak a base to be titrated with an acid, but the anilinium ion is a sufficiently strong acid for titration with a strong base.

4.6 GENERAL EQUATION FOR THE TITRATION OF A WEAK ACID OR A WEAK BASE

The titration curve of a weak acid HA with a strong base BOH is treated in a way similar to that for strong acid titration. The starting point is a charge-balance equation, i.e. a statement that the total number of positive charges must be equal to the total number of negative charges. Here, the positive charges are on H_3O^+ ions and B^+ ions corresponding to the strong base; the negative charges are on OH^- ions A^- ions from the weak acid being titrated. Consequently

$$[B^+] + [H_3O^+] = [OH^-] + [A^-] \tag{4.52}$$

Since the strong base (the titrant) is fully dissociated, the concentration of B^+, after addition of V ml of base, concentration C_B, to the solution being titrated, with volume V_0 and initial concentration C_{HA}, is

$$[B^+] = \frac{C_B V}{V_0 + V} \tag{4.53}$$

The concentration of the anion A^- can be found from the dissociation constant K_a of the acid HA

$$[A^-] = \frac{K_a[HA]}{[H_3O^+]} \tag{4.54}$$

and if we remember that $[A^-] + [HA]$ is equal to the total initial concentration, lowered as a result of dilution by the titrant added,

$$[A^-] + [HA] = \frac{C_{HA} V_0}{V_0 + V} \tag{4.55}$$

we can determine from Eqs. (4.54) and (4.55) that

$$[A^-] = \frac{K_a}{([H_3O^+] + K_a)} \frac{C_{HA} V_0}{(V_0 + V)} \tag{4.56}$$

If we substitute Eqs. (4.53) and (4.56) into (4.52), we obtain

$$\frac{C_B V}{V_0 + V} + [H_3O^+] = [OH^-] + \frac{K_a}{([H_3O^+] + K_a)} \frac{C_{HA} V_0}{(V_0 + V)} \tag{4.57}$$

On rearrangement, we obtain

$$\frac{C_B V}{V_0 + V} - \frac{K_a}{([H_3O^+] + K_a)} \frac{C_{HA} V_0}{(V_0 + V)} = [OH^-] - [H_3O^+] \tag{4.58}$$

Substitution of the titration fraction $f = C_B V / C_{HA} V_0$ and $[OH^-] = K_w / [H_3O^+]$ finally gives the expression

$$\frac{C_{HA} V_0}{V_0 + V} \left(f - \frac{K_a}{[H_3O^+] + K_a} \right) = \frac{K_w}{[H_3O^+]} - [H_3O^+] \tag{4.59}$$

This general equation for the titration curve of a weak acid is a cubic equation in $[H_3O^+]$ it is difficult to solve but in most cases it can be reduced to equations that are at most quadratic.

Comparison of Eq. (4.59) with Eq. (4.19) shows extensive analogy. The only difference is the replacement of unity in the equation for a strong acid by the fraction $K_a/([H_3O^+] + K_a)$. This fraction is equal to the degree of dissociation of a weak acid. Since strong acids (or bases) are fully dissociated, the value of their degree of dissociation is 1. Therefore Eq. (4.19) can be regarded as a particular case of the more general Eq. (4.59).

Let us now consider in what way and on what assumptions Eq. (4.59) can be transformed into the expressions given previously (Section 4.4) for particular parts of the titration curve. If $f = 0$ and thus also $V = 0$, we have

$$C_{HA} \frac{K_a}{[H_3O^+] + K_a} = [H_3O^+] \tag{4.60}$$

$K_w/[H_3O^+] = [OH^-]$ is omitted here because at the start of a titration it is considerably smaller than $[H_3O^+]$. On transformation we obtain

$$C_{HA}K_a = [H_3O^+]^2 + K_a[H_3O^+] \tag{4.61}$$

which is one of the forms used to calculate the pH of a moderately weak acid. If we consider a very weak acid for which $[H_3O^+] \gg K_a$, then

$$[H_3O^+]^2 = K_a C_{HA} \tag{4.62}$$

i.e. we obtain the simplest equation for calculating the pH of a weak acid.

If we take values of the titration fraction inside the range $0 < f < 1$, we most often obtain

$$([OH^-] - [H_3O^+]) \ll \frac{C_{HA}V_0}{V_0 + V} \tag{4.63}$$

The right-hand side of Eq. (4.59) is thus practically equal to zero, and we obtain

$$\frac{C_{HA}V_0}{V_0 + V} \left(f - \frac{K_a}{[H_3O^+] + K_a} \right) = 0 \tag{4.64}$$

since the factor outside the brackets is always different from zero

$$f = \frac{K_a}{[H_3O^+] + K_a} \tag{4.65}$$

and we obtain an expression identical to Eq. (4.31)

$$[H_3O^+] = K_a \frac{(1 - f)}{f} \tag{4.66}$$

At the equivalence point of the titration, when $f = 1$, the pH of the solution is higher than 7. Consequently, $[H_3O^+] \ll [OH^-] = K_w/[H_3O^+]$ and we obtain an equation which is a good approximation, namely

$$\frac{C_{HA}V_0}{V_0 + V}\left(1 - \frac{K_a}{[H_3O^+] + K_a}\right) = \frac{K_w}{[H_3O^+]} \tag{4.67}$$

or

$$\frac{C_{HA}V_0}{V_0 + V}\left(\frac{[H_3O^+]}{[H_3O^+] + K_a}\right) = \frac{K_w}{[H_3O^+]} \tag{4.68}$$

Since at the equivalence point $[H_3O^+] \ll K_a$, we can introduce another approximation that results in the expression

$$\frac{C_{HA}V_0}{V_0 + V}\frac{[H_3O^+]}{K_a} = \frac{K_w}{[H_3O^+]} \tag{4.69}$$

which can be transformed into

$$[H_3O^+] = \sqrt{\frac{K_w K_a(V_0 + V)}{C_{HA}V_0}} \tag{4.70}$$

and is thus identical with Eq. (4.35).

If we continue the titration, then for the range beyond the equivalence point, when $f > 1$, $[H_3O^+] \ll K_a$ and $[OH^-] \gg [H_3O^+]$, we obtain

$$\frac{C_{HA}V_0}{V_0 + V}(f - 1) = \frac{K_w}{[H_3O^+]} \tag{4.71}$$

which is identical with Eqs. (4.38) and (4.21), corresponding to the addition of excess of a strong base to the acid being titrated.

The general equation for the titration curve for a weak acid is used both for molecular acids and for cationic acids in spite of the fact that the derivation of the equation for a cationic acid, which is based on the charge balance, is somewhat different from that for a molecular acid. If we titrate acid HA^+ which is present in solution in an amount equivalent to the negative ion X^-, for example aniline hydrochloride titrated with a strong base B^+OH^-, the charge balance is of the form

$$[HA^+] + [B^+] + [H_3O^+] = [OH^-] + [X^-] \tag{4.72}$$

If we make the appropriate substitutions

$$[HA^+] = \frac{C_{HA}V_0}{V_0 + V}\left(\frac{[H_3O^+]}{[H_3O^+] + K_a}\right) \tag{4.73}$$

$$[X^-] = \frac{C_{HA}V_0}{V_0 + V}, \qquad [B^+] = \frac{C_B V}{V_0 + V} \tag{4.74}$$

(where K_a is the dissociation constant of the cationic acid HA^+), we obtain

$$\frac{C_{HA}V_0}{V_0 + V}\left(\frac{[H_3O^+]}{[H_3O^+] + K_a}\right) + \frac{C_B V}{V_0 + V} + [H_3O^+]$$

$$= [OH^-] + \frac{C_{HA}V_0}{V_0 + V} \tag{4.75}$$

and on rearrangement and multiplication of both sides by $(V_0+V)/C_{HA}V_0$, we have

$$\frac{[H_3O^+]}{[H_3O^+] + K_a} - 1 + \frac{C_B V}{C_{HA}V_0} = ([OH^-] - [H_3O^+])\frac{V_0 + V}{C_{HA}V_0} \tag{4.76}$$

or

$$\frac{C_B V}{C_{HA}V_0} - \frac{K_a}{[H_3O^+] + K_a} = ([OH^-] - [H_3O^+])\frac{V_0 + V}{C_{HA}V_0} \tag{4.77}$$

If we introduce the titration fraction f into this equation and replace $[OH^-]$ by $K_w/[H_3O^+]$, we obtain

$$\frac{C_{HA}V_0}{V_0 + V}\left(f - \frac{K_a}{[H_3O^+] + K_a}\right) = \frac{K_w}{[H_3O^+]} - [H_3O^+] \tag{4.78}$$

which is identical with Eq. (4.59) for the titration of a molecular acid.

As we mentioned before, the fraction which is subtracted from f is equal to the degree of dissociation of the acid (x). Consequently Eq. (4.78) can be represented in the form

$$f = x + \frac{V_0 + V}{C_{HA}V_0}\left(\frac{K_w}{[H_3O^+]} - [H_3O^+]\right) \tag{4.79}$$

Thus, if the second term on the right-hand side of the equation is small compared to the degree of dissociation, as occurs in moderately alkaline or acidic solutions, then the graph of the titration fraction vs. pH, (the titration curve) is identical with the graph of degree of dissociation vs. pH. Figure 4.6 shows both curves. However, it should be remembered that in the case of a titration curve the pH axis is usually the x-axis, whereas for a dissociation curve the pH axis is the y-axis.

In a similar way we can derive the equation for the titration of a weak base with a strong acid. If the dissociation constant of the weak base B is

$$K_b = \frac{[BH^+][OH^-]}{[B]} \tag{4.80}$$

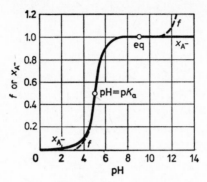

Fig. 4.6. Titration curve (f) and dissociation curve (x_{A^-}) of a weak acid.

then the equation of the titration curve has the form

$$\frac{C_B V_0}{V_0 + V}\left(\frac{K_b}{[OH^-] - K_b} - f\right) = [OH^-] - [H_3O^+] \tag{4.81}$$

and when we eliminate $[OH^-]$ by substituting K_w and the dissociation constant K_a of the conjugate acid of the base titrated, we obtain

$$\frac{C_B V_0}{V_0 + V}\left(\frac{[H_3O^+]}{[H_3O^+] + K_a} - f\right) = \frac{K_w}{[H_3O^+]} - [H_3O^+] \tag{4.82}$$

In practice, we almost always introduce some simplifying assumptions. This makes it possible to solve equations which in unsimplified form are cubic. In certain limiting cases, however, such simplifying assumptions may be a source of errors.

We can also obtain the equations for titration curves for more complicated cases, e.g. for polyprotic acids. In such cases, however, the mathematical form of the equations is very complicated, and it is much easier to introduce simplifying assumptions first and then calculate the titration curve point by point as is also often done for monoprotic systems.

4.7 TITRATION CURVES FOR POLYPROTIC ACIDS AND MIXTURES OF ACIDS

The titration curves of polyprotic acids show separate breaks if there are considerable differences between the successive dissociation constants of the given acid. In such cases the calculation of pH at particular points of the titration curve is relatively simple because we can introduce a number of simplifying assumptions. The first segment of the titration curve corresponds to the reaction

$$H_nA + OH^- \rightleftharpoons H_2O + H_{n-1}A^- \tag{4.83}$$

The points of this segment, except the points situated very close to $f = 1$, can be calculated in the same way as for a weak monoprotic acid. Of course, this is justified only if the difference between the first and the second degree of dissociation is so large that the concentration of anions with a charge greater than one can be omitted.

For the starting point of the curve we thus use the equation

$$[H_3O^+] = \sqrt{K_{a1}C_{H_nA}} \tag{4.84}$$

or a better approximation for this equation, e.g.

$$[H_3O^+] = \frac{-K_{a1} + \sqrt{K_{a1}^2 + 4K_{a1}C_{H_nA}}}{2} \tag{4.85}$$

We are sometimes forced to use Eq. (4.85), because the first degree of dissociation is usually rather large.

In the buffer region, between $f = 0$ and $f = 1$, we use the usual expression

$$[H_3O^+] = K_{a1} \frac{[H_nA]}{[H_{n-1}A^-]} = K_{a1} \frac{C_{H_nA}V_0 - C_BV}{C_BV} \tag{4.86}$$

At the first equivalence point the system should be treated as a solution of ampholyte $H_{n-1}A^-$, which can both accept a proton and lose the next one, and not as a solution of the anionic base. In this case we calculate $[H_3O^+]$ from the more general equation

$$[H_3O^+] = \sqrt{\frac{K_{a1}K_{a2}C_{H_nA} + K_{a1}K_w}{K_{a1} + C_{H_nA}}} \tag{4.87}$$

or, if $C_{H_nA} \gg K_{a1}$ and $C_{H_nA} \gg K_w/K_{a2}$, from the simplest equation

$$[H_3O^+] = \sqrt{K_{a1}K_{a2}} \tag{4.88}$$

Addition of the base beyond the first equivalence point causes the reaction

$$H_{n-1}A^- + OH^- \rightleftharpoons H_2O + H_{n-2}A^{2-} \tag{4.89}$$

consequently the solution is a buffer consisting of an acid $H_{n-1}A^-$ and a base $H_{n-2}A^{2-}$, and so

$$[H_3O^+] = K_{a2} \frac{[H_{n-1}A^-]}{[H_{n-2}A^{2-}]} \tag{4.90}$$

The concentrations of $H_{n-1}A^-$ and $H_{n-2}A^{2-}$ occurring in this equation can be estimated from the quantity and concentration of the solution being titrated and the titrant. Thus the concentration of $H_{n-1}A^-$ is equal to twice the number of moles of polyprotic acid initially present, minus the total number of moles of base added since the beginning of the titration, all divided by the total volume of the solution

$$[H_{n-1}A^-] = \frac{2C_{H_nA}V_0 - C_BV}{V_0 + V} \tag{4.91}$$

The concentration of $H_{n-2}A^{2-}$, on the other hand, is equal to the difference between the number of moles of base added since the beginning of the titration and the number of moles of base added to the point $f = 1$ (equal to $C_{H_nA}V_0$), divided as before by the total volume of the solution

$$[H_{n-2}A^{2-}] = \frac{C_BV - C_{H_nA}V_0}{V_0 + V} \tag{4.92}$$

Substituting Eqs. (4.91) and (4.92) in Eq. (4.90), we obtain

$$[H_3O^+] = K_{a2} \frac{2C_{H_nA}V_0 - C_BV}{C_BV - C_{H_nA}V_0} = K_{a2} \frac{2-f}{f-1} \tag{4.93}$$

In the general case, in the nth buffer region we have

$$[H_3O^+] = K_{an} \frac{n-f}{f-(n-1)} \tag{4.94}$$

The last equivalence point of the titration of a polyprotic acid corresponds to a solution of a polyprotic base A^{n-}. In calculating the concentration of OH^- and then of H_3O^+ we use the equation for the $[OH^-]$ of a weak base solution, remembering that a polyprotic base (e.g. CO_3^{2-}, PO_4^{3-}, AsO_4^{3-}) is usually a fairly strong base. Consequently, to calculate $[OH^-]$ we should use the equation

$$[OH^-] = \frac{-K_{b1} + \sqrt{K_{b1}^2 + 4K_{b1}C_{H_nA}\left(\frac{V_0}{V_0 + V}\right)}}{2} \tag{4.95}$$

where $C_{H_nA}V_0/(V_0+V)$ is the total concentration of the base A^{n-} in the final volume of the solution, and K_{b1} denotes the first basic dissociation constant of the base A^{n-}, equal to K_w/K_{an}, K_{an} being the last acid dissociation constant of H_nA.

If the base A^{n-} is fairly strong, so that the last equivalence point of the acid titration lies in a strongly alkaline range, then the calculation of the points of the titration curve in the last buffer region becomes less and less exact as we approach $f = n$, if we use the simplest equation for a buffer solution. Therefore, we should use a more exact formula in this range, namely

$$K_{an} = \frac{[H_3O^+]\left\{\dfrac{C_BV - (n-1)C_{H_nA}V_0}{V_0 + V} + [H_3O^+] - [OH^-]\right\}}{\left\{\dfrac{nC_{H_nA}V_0 - C_BV}{V_0 + V} - [H_3O^+] + [OH^-]\right\}}$$

(4.96)

The introduction of the value $f = C_BV/C_{H_nA}V_0$ in this equation does not give a simpler expression. We should note, however, that in the region in question, where the values of pH are high, we have $[H_3O^+] \ll [OH^-]$ and consequently

$$K_{an} = \frac{[H_3O^+]\left\{\dfrac{C_BV - (n-1)C_{H_nA}V_0}{V_0 + V} - [OH^-]\right\}}{\left\{\dfrac{nC_{H_nA}V_0 - C_BV}{V_0 + V} + [OH^-]\right\}}$$

(4.97)

Example 4.2. Compare the pH values of the curve for titration of 100 ml of $0.10M$ orthophosphoric acid with $0.20M$ sodium hydroxide, calculated by use of (1) the simplified and (2) the exact equations, for the following points: $f = 2.50$, $f = 2.75$, $f = 3.00$. The third dissociation constant of orthophosphoric acid is $K_{a3} = 10^{-11.7}$ and the ion-product of water $K_w = 10^{-13.7}$.

We first use the simplified equations.

For $f = 2.50$

$$[H_3O^+] = K_{a3}\frac{n - f}{f - (n-1)} = 10^{-11.7}\frac{(3 - 2.5)}{(2.5 - 2)} = 10^{-11.7}M$$

i.e. pH = 11.70.

The same equation used for $f = 2.75$ gives

$$[H_3O^+] = 10^{-11.7}\frac{3 - 2.75}{2.75 - 2} = 10^{-11.7}\frac{1}{3} = 10^{-12.18}M$$

i.e. pH = 12.18

At the point $f = 3.00$ we calculate pH by use of the equation for a weak base with a negligible degree of dissociation

$$[OH^-] = \sqrt{K_{b1}C_{H_nA}\frac{V_0}{V_0 + V}} = \sqrt{10^{-2.0}\times 0.10\frac{100}{250}}$$

$$= \sqrt{4.0 \times 10^{-4}} = 2 \times 10^{-2} M$$

i.e. pOH = 1.70 and pH = 13.7 − 1.7 = 12.00

The comparison of these three results shows that the procedure was wrong since on passing from $f = 2.75$ to $f = 3.00$ we get a decrease in the pH, which is contrary to the normal course of a titration!

We now use the exact Eq. (4.97) to calculate pH when $f = 2.50$. On substituting the appropriate numbers, we obtain

$$10^{-11.7} = [H_3O^+] \frac{\left(\dfrac{0.20 \times 125 - 2 \times 0.10 \times 100}{225} - [OH^-] \right)}{\left(\dfrac{3 \times 0.10 \times 100 - 0.20 \times 125}{225} + [OH^-] \right)}$$

We replace $[H_3O^+]$ by $10^{-13.7}/[OH^-]$ and obtain

$$\frac{10^{-11.7}}{10^{-13.7}} [OH^-] = \frac{\dfrac{5}{225} - [OH^-]}{\dfrac{5}{225} + [OH^-]}$$

and on rearrangement the quadratic equation

$$[OH^-]^2 + 3.22 \times 10^{-2} [OH^-] - 2.22 \times 10^{-4} = 0$$

from which

$$[OH^-] = 5.85 \times 10^{-3} M$$

and pH = 13.70 − 2.23 = 11.47.

Proceeding in the same way for $f = 2.75$, we obtain

$$[OH^-]^2 + 2.05 \times 10^{-2} [OH^-] - 3.16 \times 10^{-4} = 0$$

so that

$$[OH^-] = 1.03 \times 10^{-2} M \quad \text{and} \quad pH = 13.70 - 1.99 = 11.71$$

To calculate pH for $f = 3.0$ we use Eq. (4.95); substituting the appropriate numbers, we obtain

$$[OH^-] = \frac{-1.0 \times 10^{-2} + \sqrt{1 \times 10^{-4} + 4 \times 10^{-2} \times 10^{-1} \left(\dfrac{100}{250} \right)}}{2}$$

$$= 1.56 \times 10^{-2} M$$

and pH = 13.70 − 1.81 = 11.89.

The results obtained show the size of the errors (sometimes leading to chemically absurd conclusions) which can arise if unjustified approximations are made. Our last set of results, obtained by means of exact calculations, indicate that during this titration process no break-point is seen in the titration curve at $f = 3$.

If the polyprotic base obtained in the last stage of titration is weak and the concentration of the titrant large, we calculate the titration curve beyond the last equivalence point directly from the excess of the strong base

$$[OH^-] = \frac{C_B V - n C_{H_n A} V_0}{V_0 + V} = \frac{(f - n) C_{H_n A} V_0}{V_0 + V} \tag{4.98}$$

However, if the polyprotic base is relatively strong, the curve beyond the last equivalence point should be calculated as in the case of a mixture of a strong base and a weak base (Section 3.10). Thus, if K_b denotes the basic dissociation constant, we have

$$K_b = \frac{(C_1 + Y) Y}{(C_2 - Y)} \tag{4.99}$$

where C_1 and C_2 denote the concentrations of the strong and the weak base, respectively, and Y is the concentration of OH^- ions arising from the dissociation of the weak base, and equal to the concentration of the conjugate acid. The total concentration of OH^- ions, which is the basis for the calculations of pH, is equal to $C_1 + Y$. Equation (4.99) can be solved for Y. Then

$$K_b C_2 - K_b Y = C_1 Y + Y^2 \tag{4.100}$$

and on rearrangement

$$Y^2 + (C_1 + K_b) Y - K_b C_2 = 0 \tag{4.101}$$

The concentration of the strong base in this equation is

$$C_1 = \frac{(f - n) C_{H_n A} V_0}{V_0 + V} \tag{4.102}$$

as in Eq. (4.98), according to which the presence of strong base is the only factor which determines the pH beyond the equivalence point. The concentration of the weak base is given by the equation

$$C_2 = \frac{C_{H_n A} V_0}{V_0 + V} \tag{4.103}$$

which can also be readily calculated for a particular titration. Solution of the quadratic Eq. (4.101) gives values for Y and hence the total concentration of hydroxide ions

$$[OH^-] = C_1 + Y \tag{4.104}$$

or

$$[H_3O^+] = \frac{K_w}{C_1 + Y} \tag{4.105}$$

These expressions allow us to calculate the titration curve in a strongly alkaline region. However, it should be remembered that, for a very weak acid (i.e. where a comparatively strong conjugated base is formed) we do not obtain in practice a clear break-point on the titration curve at the equivalence point, so the titration has no practical application.

Example 4.3. For the case given above (titration of 100 ml of $0.1M$ H_3PO_4 with $0.20M$ NaOH) calculate the pH at the following points of the titration curve: $f = 3.10$ and $f = 3.40$. The basic dissociation constant of the anion is $K_{b1} = 0.010$.

To start with, let us neglect the basic dissociation of the PO_4^{3-} ion and use Eq. (4.98), which enables us to calculate the concentration of OH^- ions from the excess of strong base added. Then, if $V = 155$ ml, we obtain

$$[OH^-] = \frac{0.10 \times 0.10 \times 100}{100 + 155} = \frac{1}{255} = 0.0039M$$

and so

$$pOH = 2.41 \quad \text{and} \quad pH = 13.70 - 2.41 = 11.29$$

Similarly we calculate pH at the point $f = 3.40$

$$[OH^-] = \frac{0.40 \times 0.10 \times 100}{100 + 170} = \frac{4}{270} = 0.0148M$$

$$pOH = 1.83 \quad \text{and} \quad pH = 13.70 - 1.83 = 11.87.$$

These calculations are obviously incorrect because in Example 4.2 we obtained for an earlier point of the curve ($f = 3.00$) the pH value 11.89.

A more exact expression, viz. Eq. (4.99), should give correct results. For the point $f = 3.10$ we calculate successively

$$V_0 + V = 255 \text{ ml}$$

$$C_1 = \frac{0.10 \times 0.10 \times 100}{255} = 0.0039M$$

$$C_2 = \frac{0.10 \times 100}{255} = 0.0392M$$

We substitute these values into Eq. (4.101)

$$Y^2 + (0.0039 + 0.010)Y - 0.010 \times 0.0392 = 0$$

and therefore

$$Y = \frac{-0.0139 + \sqrt{1.93 \times 10^{-4} + 4 \times 3.92 \times 10^{-4}}}{2} = 0.0140M$$

The total concentration of OH^- in the solution is (Eq. (4.104))

$$[OH^-] = 0.0039 + 0.0140 = 0.0179M$$

and so

$$pOH = 1.75 \quad \text{and} \quad pH = 11.95$$

Similar calculations for $f = 3.40$ lead to the following results:

$$V_0 + V = 270 \text{ ml}$$

$$C_1 = \frac{0.40 \times 0.10 \times 100}{270} = 0.0148M$$

$$C_2 = \frac{0.10 \times 100}{270} = 0.0370M$$

$$Y^2 + (0.0148 + 0.010)Y - 0.010 \times 0.0370 = 0$$

$$Y = \frac{-0.0248 + \sqrt{6.15 \times 10^{-4} + 4 \times 1.48 \times 10^{-3}}}{2} = 0.0105M$$

$$[OH^-] = 0.0148 + 0.0105 = 0.0253M$$

$$pOH = 1.60 \quad \text{and} \quad pH = 12.10$$

These values enable us to draw the titration curve. Along with the data obtained in Example 4.2, they show the absence of a pH break at the third equivalence point.

Figure 4.7 shows calculated titration curves for tartaric acid and Fig. 4.8 those for orthophosphoric acid. To show the effect of concentration, the curves have been drawn for two different initial concentrations of each acid, $0.10M$ and $0.010M$. The curves show that an analytical

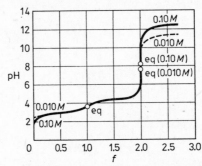

Fig. 4.7. Titration curves of 0.10 and 0.010M tartaric acid with a strong base.

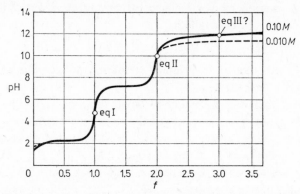

Fig. 4.8. Titration curves of 0.10 and 0.010M orthophosphoric acid with a strong base.

titration for tartaric acid should be continued up to the total neutralization of the acid, whereas for orthophosphoric acid, analytically useful
breaks on the titration curve are obtained only at $f = 1.0$ and $f = 2.0$.

The position of the buffer region does not depend on the initial acid
concentration in either of these cases and differences in the graphs appear
only in the initial and final segments of the titration curves.

Titration of a mixture of acids or a mixture of bases follows the same
basic rules. If we titrate with a strong base a mixture of a strong acid
and a weak acid with a sufficiently small dissociation constant, we obtain
two separate breaks, almost independent of each other, which are clearly
marked on the curve. In such cases, however, the quantity plotted along
the x-axis should be the volume of the titrant added and not the titration
fraction. The correct end-point for titration of a strong acid is obtained
if the strong acid is already 99.9% titrated at the pH calculated for the
solution of the weak acid (before titration).

For example, a strong acid with concentration 0.10M will be 99.9%
titrated when, in accordance with Eq. (4.6)

$$[H_3O^+] = \frac{(1-f)C_{HA}V_0}{V_0 + V} = \frac{0.001 \times 0.10 \times 100}{200}$$

$$= \frac{0.0001}{2} = 5 \times 10^{-5}M$$

i.e. when pH $= 4.30$.

For a weak acid the pH at the initial point of titration is given by

$$pH = (pK_a - \log C)/2$$

Hence for a solution of a weak acid of concentration about $0.10M$ and pH 4.30, the pK_a value would be

$$pK_a = 2 \times 4.30 + 1.00 = 9.43, \quad K_a = 2.5 \times 10^{-8}$$

For an acid with such a dissociation constant, however, the end-point of titration is not very clear.

In practice, it is not necessary to be so rigorous about the sharpness of either end-point. The end-point of the strong acid titration can be read from the graph with sufficient accuracy even when the dissociation constant of the weak acid is of the order of 10^{-7}.

The titration curve for a mixture of two weak acids is similar to the titration curve for a diprotic acid. The buffer segments are in regions dependent on the pK_a of the individual acids. The pH at the end-point of titration of the stronger acid is calculated, just as for a diprotic acid, from the expression

$$[H_3O^+] = \sqrt{K_{a1}K_{a2}} \tag{4.106}$$

Often, if the values of the dissociation constants differ from each other by 6–7 orders of magnitude, it is possible to obtain good accuracy in estimating the first end-point, but when the difference between the dissociation constants is smaller, it is more common to observe merely an inflexion on the titration curve. However, this can be used as a basis for the graphical determination of the end-point.

In many practical cases it may be helpful to add other solvents to increase the ratio of the dissociation constants. It may also be possible to add some reagent to react with one of the acids, for example to form a stable complex, which shifts the equilibrium. A well-known example is the change in the apparent dissociation constant of boric acid from 6.3×10^{-10} to 5×10^{-6} on addition of a polyhydroxy alcohol (e.g. mannitol) to the solution being titrated.*

Improvements in the techniques for potentiometric determination of hydrogen ion activity in acid–base titrations allow titration curves to be plotted very accurately. Separate potential breaks that are clearly marked on a titration curve can readily be used for quantitative determination of the acids (or bases) titrated, with the aid of graphical or numerical methods. The situation is more complicated if the titrant contains acid–base systems with similar pK_a values. In Chapter 3 we discussed methods of calculating the $[H_3O^+]$ of different systems containing polyprotic acids

---

* See, for example, R. Belcher, G. W. Tully and G. Svehla, *Anal. Chim. Acta*, 1970, **50**, 261.

(Section 3.4) or mixtures of several acids (Section 3.11). If there are n simultaneous acid–base equilibria, the equation will be of degree $n+1$ with respect to $[H_3O^+]$. For two weak acids, a quartic equation is obtained, because in an exact solution it is necessary to consider also the dissociation of water. For a triprotic acid, there are four equilibria, so we obtain an equation of the fifth degree. In these equations, the parameters are the appropriate concentrations of acids and equilibrium constants of the reaction.

Similarly, the general equation for the titration curve of a single weak acid is a cubic equation in $[H_3O^+]$ (Section 4.6), and each additional equilibrium caused by the presence of an additional acid raises the degree of the equation obtained. As we stated above, the solution of such an equation in the general case is complicated and difficult to accomplish in an exact manner. The situation changes if we reverse the situation, as happens in the calculation of titration results, i.e. when $[H_3O^+]$ is known and the unknowns are the concentrations of the acids. The number of unknowns is then equal to the number of acids present in the solution. If we measure $[H_3O^+]$ for at least as many titration points as there are unknowns, we obtain a set of equations which can be solved, particularly with the use of a computer.

It is of great importance to select suitable experimental points, placed not only close to the expected equivalence points but also for pH values corresponding to the entire course of the titration. If the solution titrated contains a strong acid and weak acids, addition of titrant base causes neutralization to occur at each point of the titration curve to a different extent for each of the acids to be determined. This is the basis of the method for calculating the results, and it can be applied even for rather complicated mixture of acids. The same reasoning can of course be applied to the titration of a mixture of bases.

One of the methods that has been described* is based on the following consideration: at each point k of the titration curve of a mixture containing n weak acids and possibly also a strong acid, the total concentration, $T_{H,k}$, of protons capable of titration at point k, can be expressed as the sum of the concentrations of protons derived from weak acids (taking into account the number of protons N_i of each of the acids) with initial concentration $C_{i,0}$ and from the strong acid with initial concentration C_A. For point k the concentrations of these acids must be multiplied by a factor to allow for the dilution of the initial volume V_0 of solution by

---

* A. Ivaska and I. Nagypál, *Talanta*, 1980, **27**, 721.

addition of volume V of titrant. The sum of the acid concentrations is diminished by the concentration of the added base, given by the concentration of the titrant solution C_B and the term defining its dilution $V/(V_0 + V)$. Thus we obtain the equation

$$T_{H,k} = \sum_{i=1}^{n} N_i C_{i,0} \frac{V_0}{V_0 + V} + C_A \frac{V_0}{V_0 + V} - C_B \frac{V}{V_0 + V}$$

$$(4.107)$$

in which protons potentially capable of titration are described by the parameters of the initial state and by the corresponding volumes of the solutions.

The same quantity $T_{H,k}$ may be described by means of quantities characterizing the equilibrium state at point k, which results in a definite and experimentally determined value of $[H_3O^+]_k$. This quantity is $\bar{n}_{i,k}$, the formation constant of acid i at point k, which, when multiplied by the initial concentration of the weak acid, gives the concentration (after volume changes are considered) of the protons of this acid that are still available for titration. To this concentration we add the existing concentration of $[H_3O^+]_k$ and then subtract the existing concentration of $[OH^-]_k$ = $K_w/[H_3O^+]_k$.

Thus the second equation which describes $T_{H,k}$ is

$$T_{H,k} = \sum_{i=1}^{n} \bar{n}_{i,k} C_{i,0} \frac{V_0}{V_0 + V} + [H_3O^+]_k - K_w/[H_3O^+]_k$$

$$(4.108)$$

In this equation it is convenient to replace $\bar{n}_{i,k}$ by an expression containing the appropriate overall constants of the proton reaction $\beta_{j,i}$ where the indices j denote the successive stages of proton attachment

$$\bar{n}_{i,k} = \frac{\sum\limits_{j=0}^{N_i} j\beta_{j,i}[H_3O^+]_k^j}{\sum\limits_{j=0}^{N_i} \beta_{j,i}[H_3O^+]_k^j} \tag{4.109}$$

with $\beta_{0,i} = 1$. From Eqs. (4.107)–(4.109) we obtain after simple transformations the expression

$$\sum_{i=1}^{n} \frac{\sum\limits_{j=0}^{N_i} (N_i - j)\beta_{j,i}[H_3O^+]_k^j}{\sum\limits_{j=0}^{N_i} \beta_{j,i}[H_3O^+]_k^j} C_{i,0} + C_A$$

$$= C_B \frac{V}{V_0} + \frac{V_0 + V}{V_0} \left([H_3O^+] - \frac{K_w}{[H_3O^+]} \right) \qquad (4.110)$$

which is written for each experimental point k. There should be at least as many such points as the number of acids (weak and strong), i.e. $(n+1)$. We then obtain a system of $(n+1)$ linear equations in the unknowns $C_{i,0}$ and C_A; this system may be solved with the aid of a computer.

It is clear that the values used for the equilibrium constants should have been determined under exactly the same conditions as those in the titration. The accuracy of determination of these constants has a direct influence on the accuracy of the titration results. The accuracy of this method depends also on the relative concentrations of the acids and on the degree of superposition of the appropriate ionic equilibria. The precision of determination may be improved by increasing the number of points measured. For multicomponent systems, this method has produced results with an error not greater than 2%.

4.8 TITRATION ERROR

The accuracy of a titration result depends on the exactness of determination of the equivalence point. In analytical practice, the end-point of a titration should be as close as possible to the theoretical equivalence point. The difference between the end-point and the equivalence point indicates the absolute value of the *titration error*. This difference can be expressed in terms of, for example, volumes of titrant solution, i.e. as the difference between the volume of the titrant added up to the end-point, V_{ep}, and the volume which should have been added to reach the equivalence point, V_{eq}. If this difference is positive ($V_{ep} > V_{eq}$) the end-point lies beyond the equivalence point, and the titration error is said to be positive. If $V_{ep} < V_{eq}$, the titration error is negative.

Most often the titration error (TE) is expressed in relative terms, as a fraction of the volume of titrant added up to the moment of reaching the equivalence point, i.e.

$$TE = \frac{V_{ep} - V_{eq}}{V_{eq}} \qquad (4.111)$$

Since the volume at the equivalence point is almost the same as the volume at the end-point, the following expression is essentially the same:

$$TE = \frac{V_{ep} - V_{eq}}{V_{ep}} \qquad (4.112)$$

If the titration error is expressed as a percentage, then

$$\text{TE}(\%) = \frac{V_{ep} - V_{eq}}{V_{ep}} \times 100\% \qquad (4.113)$$

In practice, this form of the equation is inconvenient because we usually do not know the volume of the titrant solution necessary to reach the equivalence point. The pH value at the titration end-point, however, can be found either by means of colour indicators or by the potentiometric method. If we know the concentration of H_3O^+ ions at the titration end-point, we can calculate the concentration of the strong acid titrated. If, in addition, we know the change in the solution volume during titration, we can calculate the titration error of the strong acid by means of the equation

$$\text{TE}(\%) = \frac{V_0 + V}{C_{HA}V_0} X \times 100 \qquad (4.114)$$

where X denotes the concentration of the acid titrated (with a minus sign) or the concentration of base added in excess, in relation to the equivalence point (with a plus sign).

Example 4.4. Fifty ml of $0.1M$ hydrochloric acid were titrated to pH = 4.4 with $0.10M$ sodium hydroxide. Calculate the titration error made because the end-point was not at pH 7.0.

In a solution of pH 4.4 $[H_3O^+]$ is $4.0 \times 10^{-5}M$, and we can assume that this is equal to the concentration of untitrated acid. To reach the end-point, about 50 ml of the titrant should be used, so we may assume that $V_0 + V = 100$ ml. On substituting the appropriate numerical values in Eq. (4.114) we obtain

$$\text{TE}(\%) = \frac{100}{0.10 \times 50} (-4.0 \times 10^{-5}) \times 100\% = -0.08\% \simeq -0.1\%$$

Because the error calculated is so small, pH = 4.4 is an acceptable end-point for the titration.

Example 4.5. What is the highest permissible end-point pH for a titration of $0.1M$ hydrochloric acid with $0.05M$ sodium hydroxide, if the titration error is not to exceed 0.1%?

We calculate the concentration of base (OH^- ions), which, when added to the acid solution in excess (a positive titration error), will cause an error of 0.1%. Since the sodium hydroxide solution has half the concentration of the hydrochloric acid, the volume at the end-point will be three times the initial volume. Thus

$$0.1\% = \frac{3}{0.10} X \times 100\%, \quad X = 3.3 \times 10^{-5} M$$

This concentration of OH^- ions corresponds to $pOH = 4.5$ or $pH = 14 - -4.5 = 9.5$. Thus, to avoid an error greater than 0.1% we should titrate to an end-point pH not greater than 9.5.

Calculation of the titration error as above is really an approximation because the concentration of H_3O^+ at the end-point results not only from the presence of untitrated acid but also from the dissociation of water, to form equal quantities of H_3O^+ and OH^- ions. The titration error is not affected by the ions obtained from the dissociation of the product of neutralization. Thus, in exact calculations we should substitute for X the difference between the total concentration of H_3O^+ at the end-point and the concentration of H_3O^+ from the dissociation of water being equal to the concentration of OH^- ions at the end-point. The formula for the titration error is thus

$$TE(\%) = -\frac{V_0 + V}{C_{HA}V_0}\left([H_3O^+] - \frac{K_w}{[H_3O^+]}\right) \times 100 \qquad (4.115)$$

The minus sign results from the fact that, in the titration of a strong acid, a negative error corresponds to a deficiency of base, which makes the solution acid and so $[H_3O^+] \gg K_w/[H_3O^+]$. In the case of an excess of the base (titrant) with respect to the equivalence amount, the solution is alkaline, $[H_3O^+] \ll K_w/[H_3O^+]$, which makes the titration error positive. In most real titrations, one of the terms in the brackets is much smaller than the other and can be omitted. Such simplifying assumptions cannot be used in the titration of very dilute solutions close to the equivalence point.

If we titrate a strong base instead of an acid, then an excess of OH^- ions causes a negative error, so the minus sign after the equality sign should be eliminated from Eq. (4.115). We then obtain

$$TE(\%) = \frac{V_0 + V}{C_B V_0}\left([H_3O^+] - \frac{K_w}{[H_3O^+]}\right) \times 100 \qquad (4.116)$$

In Eqs. (4.115) and (4.116) we have the solution volumes V_0 and V and the initial concentration of titrant, C_{HA} or C_B. If in the expression $(V_0 + V)/V_0$ we replace the end-point volume of titrant, of concentration C, by the equivalence point volume $V_{eq} = C_{HA}V_0/C$, we obtain

$$\frac{V_0 + V}{V_0} \simeq \frac{V_0 + V_{eq}}{V_0} = \frac{V_0 + C_{HA}V_0/C}{V_0} = \frac{C + C_{HA}}{C} \qquad (4.117)$$

The final form of Eq. (4.115) for titration of an acid is

$$\text{TE}(\%) = \frac{C + C_{HA}}{CC_{HA}} \left(\frac{K_w}{[H_3O^+]} - [H_3O^+] \right) \times 100 \qquad (4.118)$$

and the final form of Eq. (4.116) for titration of a base, is

$$\text{TE}(\%) = \frac{C + C_B}{CC_B} \left([H_3O^+] - \frac{K_w}{[H_3O^+]} \right) \times 100 \qquad (4.119)$$

Example 4.6. Calculate the range of end-point pH for which the titration error in the titration of $1.0 \times 10^{-4} M$ hydrochloric acid with $1.0 \times 10^{-3} M$ sodium hydroxide does not exceed 0.1%.

We first calculate

$$\frac{\text{TE}(\%)}{100} \frac{CC_{HA}}{C + C_{HA}} = \frac{0.1}{100} \frac{1 \times 10^{-3} \times 10^{-4}}{1.1 \times 10^{-3}} = 9.1 \times 10^{-8} M$$

This value is equal to the expression in brackets in Eq. (4.118). Then, solution of the quadratic equation

$$\frac{K_w}{[H_3O^+]} - [H_3O^+] = 9.1 \times 10^{-8} M$$

gives

$$[H_3O^+] = 6.5 \times 10^{-8} M \quad \text{and} \quad \text{pH} = 7.19$$

When the negative error is calculated similarly, we obtain pH = 6.81. The end-point thus must be located with an accuracy of better than ± 0.2 pH unit, which requires the use of a pH-meter as the end-point detector. If we had omitted the influence of the dissociation of water, the result obtained would indicate, incorrectly, that pH_{ep} should be within the limits 7.00 ± 0.04.

Equations identical with those given above can be obtained from the general equation for the titration curve of a strong acid (Eq. (4.19)), because $(f-1)$ is just the value of the titration error, which can readily be shown by substituting the value $V_{eq} = C_{HA} V_0 / C_B$ into Eq. (4.111).

In the titration of a weak acid HA, the titration error is calculated in a similar way. From the definition of titration error we get

$$\text{TE}(\%) = - \frac{[HA]}{[HA] + [A^-]} \times 100 \qquad (4.120)$$

and on substituting the concentration of untitrated acid, calculated from the dissociation constant of the acid,

$$[HA] = \frac{[H_3O^+][A^-]}{K_a} \qquad (4.121)$$

we obtain

$$TE(\%) = -\frac{\dfrac{[H_3O^+][A^-]}{K_a}}{\dfrac{[H_3O^+][A^-]}{K_a} + [A^-]} \times 100$$

$$= -\frac{[H_3O^+]}{[H_3O^+] + K_a} \times 100 \qquad (4.122)$$

where $[H_3O^+]$ is the concentration at the end-point. This error is negative because the titration end-point lies before the equivalence point. It should be noted that the titration error calculated according to this simplest Eq. (4.122) does not depend either on the initial concentration of the acid or on its dilution during the titration. This is related to the concentration-independence of the buffer region of the titration curve of a weak acid.

In many practical calculations when the titration error is small and the acid is not very weak, Eq. (4.122) becomes still simpler, namely

$$TE(\%) = \frac{[H_3O^+]}{K_a} \times 100 \qquad (4.123)$$

Example 4.7. Calculate the titration error for titration of formic acid with sodium hydroxide to pH 6 or 7. The dissociation constant of formic acid is $10^{-3.65}$.

We calculate the titration error for both pH values according to Eq. (4.122).

$$\text{If } pH_{ep} = 6 \quad TE(\%) = -\frac{10^{-6}}{10^{-6} + 10^{-3.65}} \times 100 = -0.44\%$$

$$\text{If } pH_{ep} = 7 \quad TE(\%) = -\frac{10^{-7}}{10^{-7} + 10^{-3.65}} \times 100 = -0.045\%$$

These calculations indicate that the titration error for pH 7 is small and can be neglected, even though formic acid is a weak acid, and the equivalence point for titration of a $0.1M$ solution with the same concentration of strong base occurs at

$$[H_3O^+] = \sqrt{\frac{10^{-14} \times 10^{-3.65} \times 2}{10^{-1}}} = 10^{-8.17}M$$

in accordance with Eq. (4.35) and thus $pH_{ep} = 8.17$.

Example 4.8. What is the smallest end-point pH for the error of titrating a tartaric acid solution with sodium hydroxide not to exceed -0.2%?

The dissociation constants of tartaric acid are $K_{a1} = 10^{-2.9}$ and $K_{a2} = 10^{-4.1}$.

We continue the titration of tartaric acid until both acid protons are neutralized, because the small difference between the dissociation constants makes it impossible to obtain a clear pH break after the first proton is neutralized. Thus we treat the system as a monoprotic acid with $K_a = K_{a2}$, remembering, however, that the relative value of the titration error calculated in the usual way should be divided by two.

According to Eq. (4.122)

$$TE(\%) = - \frac{[H_3O^+]}{([H_3O^+] + K_a)2} \times 100$$

i.e.

$$[H_3O^+] = \frac{-2 \times TE \times K_a}{2 \times TE + 100}$$

On substitution of numerical values, we obtain

$$[H_3O^+] = 10^{-6.8} M \quad \text{and} \quad pH = 6.8$$

This indicates that the end-point pH should not be smaller than 6.8, but it does not define its upper limit.

If the end-point pH is larger than the equivalence-point pH, the titration error should be calculated in the same way as for a strong acid titration, i.e. by use of Eq. (4.114).

Example 4.9. Calculate the error in the titration of $0.1M$ formic acid with $0.1M$ strong base, if the end-point pH is 9.5.

We assume that as a result of dilution during titration the initial acid concentration is halved, i.e. $(V_0 + V)/V_0 = 2$. At pH = 9.5, the excess of added base will be equal to the concentration of OH^-, i.e. $10^{-14}/10^{-9.5} = 10^{-4.5}$. By use of Eq. (4.114) we obtain

$$TE(\%) = \frac{2}{10^{-1}} \times 10^{-4.5} \times 100\% = 10^{-1.2}\% = 0.063\%$$

Assuming the ion-product of water $K_w = 10^{-13.8}$ the result is slightly larger, namely 0.1%. The results obtained in this example and in Example 4.7 indicate that taking the end-point in the pH range 7–9.5 should not cause an error larger than 0.1%.

This method of calculating the titration error is valid in most practical cases. However, if the solutions are very dilute or if the end-point pH differs greatly from 7.0, more exact calculations should be used, taking account of the dissociation of the untitrated part of the acid and of the influence of H_3O^+ and OH^- ions arising from dissociation of the solvent. In the numerator of Eq. (4.120) the concentration of HA should thus be decreased by subtracting the untitrated part which has dissociated. The concentration of this dissociated part is equal to the total concentration of H_3O^+ ions, decreased by subtracting the concentration of H_3O^+ ions which arise from water dissociation, i.e. equal to the concentration of OH^- ions. Hence the concentration of the untitrated acid is

$$[HA] - ([H_3O^+] - [OH^-]) \tag{4.124}$$

and the titration error

$$TE(\%) = - \frac{[HA] - ([H_3O^+] - [OH^-])}{[HA] + [A^-]} \times 100 \tag{4.125}$$

By substituting for [HA] the expression obtained from the dissociation constant of the acid, and transformation, we obtain

$$TE(\%) = - \frac{[H_3O^+] - \dfrac{K_a}{[A^-]}([H_3O^+] - [OH^-])}{[H_3O^+] + K_a} \times 100 \tag{4.126}$$

If we multiply the numerator and the denominator of the fraction $K_a/[A^-]$ by the total concentration of the titrated acid, $[HA] + [A^-]$, then after simple transformation, the fraction will assume the form

$$
\begin{aligned}
\frac{K_a}{[A^-]} &= \frac{K_a([A^-] + [HA])}{\dfrac{[HA]K_a}{[H_3O^+]}([A^-] + [HA])} \\
&= \frac{K_a + [H_3O^+]}{[A^-] + [HA]} = \frac{K_a + [H_3O^+]}{\dfrac{C_{HA}V_0}{V_0 + V}}
\end{aligned}
\tag{4.127}
$$

On substituting for the volume ratio the appropriate ratio of the concentrations, in accordance with Eq. (4.117), we obtain from Eqs. (4.127) and (4.126) the final form of the expression for the error of titrating a weak acid with a strong base

$$TE(\%)$$

$$= \frac{\dfrac{C_{HA} + C_B}{C_{HA}C_B}([OH^-] - [H_3O^+])([H_3O^+] + K_a) - [H_3O^+]}{[H_3O^+] + K_a} \times 100 \tag{4.128}$$

This equation reduces to much simpler forms if it is possible to omit terms which are much smaller than the largest term of the sum.

Example 4.10. Calculate the titration error in the titration of $1 \times 10^{-3} M$ acetic acid with $1 \times 10^{-3} M$ sodium hydroxide to pH 7.5. The dissociation constant of acetic acid is 1.6×10^{-5}.

In this example the titration conditions are very unfavourable because the solutions are very dilute. The pH at the equivalence point is 7.8, so it differs from the end-point pH by only 0.3 units. If we use the full Eq. (4.128) for our calculations, then on substituting numerical values we obtain

$$TE = \frac{+2 \times 10^3 (3.2 \times 10^{-8} - 3.2 \times 10^{-7})(3.2 \times 10^{-8} + 1.6 \times 10^{-5}) - 3.2 \times 10^{-8}}{3.2 \times 10^{-8} + 1.6 \times 10^{-5}}$$

$$= \frac{+2 \times 10^3 (-2.9 \times 10^{-7})(1.6 \times 10^{-5}) - 3.2 \times 10^{-8}}{1.6 \times 10^{-5}}$$

$$= -5.8 \times 10^{-4} - 2 \times 10^{-3} = -2.6 \times 10^{-3}$$

i.e. $TE(\%) = -0.26\%$

The contribution of the second term of the sum is small, however, and if it were neglected, the calculated error would be -0.20%. Thus, even in this case, the simplified Eq. (4.122) can be used. The result obtained in this example indicates that the titration error for very dilute weak acid solutions is rather large, even if the difference between the end-point pH and the equivalence-point pH is small.

Example 4.11. Calculate the titration error when $1.0 \times 10^{-3} M$ dichloro-acetic acid ($K_a = 5 \times 10^{-2}$) is titrated with sodium hydroxide of similar concentration, if the end-point pH is 6.

In this case the concentration of undissociated and untitrated acid close to the equivalence point is small because dichloroacetic acid is a moderately weak acid. We thus must use the most general Eq. (4.128). We obtain

$$TE(\%) = \frac{2 \times 10^3 (10^{-8} - 10^{-6})(10^{-6} + 5 \times 10^{-2}) - 10^{-6}}{10^{-6} + 5 \times 10^{-2}} \times 100\%$$

and, on omitting small terms in the sums, we have

$$TE(\%) = \frac{2 \times 10^3 (-10^{-6})(5 \times 10^{-2}) - 10^{-6}}{5 \times 10^{-2}} \times 100\%$$

$$= (-2 \times 10^{-3} - 2 \times 10^{-5}) \times 100\% = -0.2\%$$

The first term of this sum is considerably larger than the second and therefore it is of primary importance in calculating the titration error. If we used here the simplest Eq. (4.122) we would obtain the completely false result $TE(\%) = 10^{-4}\%$.

In this example, the calculation of the equivalence-point pH by use of the simple formula (4.35) gives the completely wrong result, $pH = 6.2$. This is because the dichloroacetate ion is a very weak base, and thus water dissociation cannot be neglected. If we use Eq. (3.47) we obtain an equivalence-point pH of 7.0, which is correct.

The titration of a weak base with a strong acid is analogous to the titration of a weak acid with a strong base. The derivation of the appropriate equations is very similar: we merely replace $[H_3O^+]$, $[OH^-]$, K_a and C_{HA} by $[OH^-]$, $[H_3O^+]$, K_b and C_B, respectively. In place of Eq. (4.122) we obtain

$$TE(\%) = \frac{-[OH^-]}{[OH^-] + K_b} \times 100 \tag{4.129}$$

and in place of Eq. (4.128), without any simplifying assumptions, we obtain

$$TE(\%)$$

$$= \frac{\dfrac{C_B + C_{HA}}{C_B C_{HA}}([H_3O^+] - [OH^-])([OH^-] + K_b) - [OH^-]}{[OH^-] + K_b} \times 100 \tag{4.130}$$

Example 4.12. Calculate the error on titrating $0.02M$ hydrazine solution with $0.02M$ hydrochloric acid to an end-point pH of 6.0. The hydrazine dissociation constant is $K_b = 10^{-5.7}$.

First, the pH at the equivalence point should be found. Because of the small value of the dissociation constant and the comparatively large base concentration, we have

$$[H_3O^+] = \sqrt{\frac{K_w C_B}{K_b} \frac{V_0}{V_0 + V}} = \sqrt{\frac{10^{-14} \times 2 \times 10^{-2}}{10^{-5.7} \times 2}}$$

$$= 10^{-5.15}M$$

$$pH = 5.15$$

Consequently, to calculate the titration error we use Eq. (4.129) and obtain

$$TE(\%) = \frac{-10^{-8}}{10^{-8} + 10^{-5.7}} \times 100\% = -0.5\%$$

The examples given above show that errors in the titration of the more concentrated solutions of strong acids and bases are small, even if the end-point pH is quite distant from the equivalence point pH, whereas in very dilute solutions of weak acids or bases it is easy to make a significant systematic error. The equations we have given permit selection of an appropriate end-point pH value, and also of the most advantageous method of end-point detection.

The titration error resulting from the difference between the pH at the theoretical equivalence point (pH_{eq}) and the pH at the end-point (pH_{ep}) can be estimated readily with the aid of logarithmic diagrams.* If we titrate an acid HA of initial concentration C_{HA}, with a strong base OH^- with concentration C_{OH}^0, then the concentration of the added base at the end-point of the titration is $C_{OH} = C_{OH}^0 V/V_0$, if V and V_0 have the same meaning as before, and if the volume of base added is sufficiently smaller than the volume of acid for the dilution correction to be disregarded.

The percentage titration error TE% is defined by the difference between the acid and added base concentrations at the end-point

$$TE(\%) = 100 \times \frac{C_{OH} - C_{HA}}{C_{HA}} \tag{4.131}$$

If we express the concentrations of the reagents in terms of the actual concentrations of the species formed in the solution

$$C_{HA} = [HA] + [A^-] \tag{4.132}$$

$$C_{OH} = [A^-] + [OH^-] - [H_3O^+] \tag{4.133}$$

then the titration error is

$$TE(\%) = 100 \times \frac{[OH^-] - [H_3O^+] - [HA]}{C_{HA}} \tag{4.134}$$

or, if $[HA] \gg [H_3O^+]$, which is true for most weak acids, we obtain

$$TE(\%) = 100 \times \frac{[OH^-] - [HA]}{C_{HA}} \tag{4.135}$$

The concentration terms appearing in this equation can be read from a logarithmic diagram. Figure 4.9 shows the equivalence point (2), and the points (3) and (3′) corresponding to [HA] and $[OH^-]$ at the end-point, for which pH_{ep} differs from pH_{eq} by ΔpH.

It follows from the diagram that

$$\log[OH^-]_{ep} = \log[HA]_{eq} + \Delta pH \tag{4.136}$$

$$\log[HA]_{ep} = \log[HA]_{eq} - \Delta pH \tag{4.137}$$

---

* E. Wänninen, *Talanta*, 1980, **27**, 29.

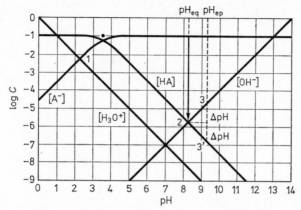

Fig. 4.9. Logarithmic diagram for a weak acid $(K_a = 10^{-3.5}, C_{HA} = 10^{-1}M)$. The thick line $(\log[HA]_{eq}/C_{HA})$ is the basis for titration error calculation. ΔpH denotes the difference between the equivalence point and the end-point of the titration.

When the end-point concentrations resulting from these equations are introduced into Eq. (4.135) we obtain

$$TE(\%) = 100 \times \frac{[HA]_{eq}}{C_{HA}}(10^{\Delta pH} - 10^{-\Delta pH}) = 100 \times \frac{[HA]_{eq}}{C_{HA}}\Delta F$$

$$(4.138)$$

This equation shows that the titration error is determined by two factors. The first, $[HA]_{eq}/C_{HA}$, depends on the properties of the acid titrated, and thus—as we shall prove later—on the dissociation constant K_a; the second factor (ΔF) is a function only of ΔpH, and thus results from inexact determination of the titration end-point. ΔF may be calculated for each case from the value of ΔpH. Alternatively, values of ΔF or $\log(100\Delta F)$ may be read from tables (Table 4.3). In this case it is sufficient to read from the logarithmic diagram (Fig. 4.9) the value of $\log([HA]_{eq}/C_{HA})$ (marked by a length of thick line with an arrow) and add to it the value of $\log(100\Delta F)$ corresponding to the error of reading the pH at the end-point.

If a weak base is titrated with a strong acid for which the logarithm of the protolytic equilibrium constant is marked on the logarithmic diagram for the system, the equivalence point corresponds to the point of intersection of the curves for $\log[A^-]$ and $\log[H_3O^+]$. For this point we calculate $\log[A^-]/C_{A^-}$ and, by use of the value of $\log(100\Delta F)$ (Table 4.3), can calculate the titration error.

Table 4.3
Values of the term $\Delta F = 10^{\Delta pH} - 10^{-\Delta pH}$
for various values of ΔpH

ΔpH	ΔF	$\log(100\Delta F)$
0	0	—
0.1	0.5	1.67
0.2	1.0	1.98
0.3	1.5	2.17
0.4	2.1	2.33
0.5	2.8	2.45
0.6	3.7	2.57
0.7	4.8	2.68
0.8	6.2	2.79
0.9	7.8	2.89
1.0	9.9	3.00
1.5	32	3.50
2.0	100	4.00

Titration errors can be evaluated rapidly with the aid of logarithmic diagrams. If ΔF is constant (and corresponds to a definite degree of precision of end-point determination), then the greater the value of $\log[HA]/C_{HA}$ (i.e. the shorter the distance between the equivalence point and the abscissa of the value of $\log C_{HA}$), the larger the titration error.

It is particularly advantageous to use logarithmic diagrams for error evaluation in titrations of polyprotic acids or mixtures of two acids with different concentrations. On the appropriate diagrams, the distances which are the basis for error calculations are marked by lines with arrows.

In the example shown in Fig. 4.10 it is more useful, from the point of view of the titration error, to titrate up to the first equivalence point, because the error will then be much smaller than for a titration to complete neutralization of the acid.

In the example shown in Fig. 4.11 the error in the titration of the first, stronger, base is much smaller than that of the titration of the sum of the bases. If we also titrate the protonated forms of the two bases (with the same concentrations as before) with a strong base, the error for titration of the acid $B'H^+$, corresponding to the weaker base, to the equivalence point at pH 7.3 (the segment between the point $[B'H^+] = [B]$ and $\log C = -4$) would be smaller than the error of titrating to the second equivalence point (at pH 11.6) of the neutralization of the two bases ($[BH^+] = [OH^-]$).

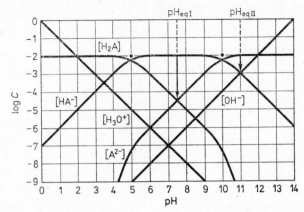

Fig. 4.10. Logarithmic diagram for a diprotic acid (iminodiacetic acid, K_{a1} = $10^{-5.0}$, K_{a2} = $10^{-10.0}$, C_{H_2A} = $10^{-2}M$. The lines with arrows are equal to $\log[HA]_{eqII}/C_{H_2A}$ and $\log[H_2A]_{eqI}/C_{H_2A}$ respectively, and their lengths are the basis for calculation of titration error.

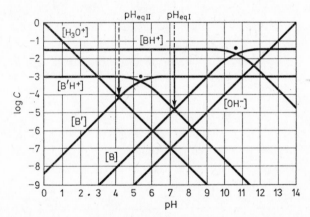

Fig. 4.11. Logarithmic diagram for a mixture of two bases (e.g. methylamine K_{BH^+} = $10^{-10.6}$, C_B = $10^{-1.5}M$, and pyridine $K_{B'H^+}$ = $10^{-5.2}$, $C_{B'}$ = $10^{-3.0}M$) titrated with a strong acid. The lines marked by arrows are the basis for calculating the titration error of the stronger base (eq I) and the sum of the bases (eq II).

A certain complication in the use of logarithmic diagrams arises when $[H_3O^+]$ cannot be omitted in Eq. (4.134) because it is not small enough in relation to [HA]. We must then determine an auxiliary intersection point of the line $\log[OH^-]$ with the segment corresponding to $\log([HA] + [H_3O^+])$. The situation is discussed in Example 3.49.

4.9 PRECISION OF TITRATIONS

The difference between the end-point and the equivalence point of a titration is the source of the systematic error of determination. The selection of an ideal indicator that changes colour exactly at the equivalence point, or the measurement of pH by the potentiometric method, might actually reduce the systematic error to zero. However, even then the results of individual titrations would differ from one another, although for a sufficient number of measurements the mean result would be very close to the true value. The scatter in the results is caused by random errors, which may result, for example, from imperfect measurement technique, the subjective estimation of a colour change, or accidental fluctuations of temperature. The scatter in a series of measurements is a consequence of insufficient precision of the method.

In acid–base titrations the most significant source of error is the imprecision of end-point location. The colour change of the indicator does not occur suddenly at a definite pH, but progressively over a range of one or two pH units. Even if a pH-meter is used to observe pH changes, each reading has a small random error connected with the rate of mixing of the solution and of the attainment of the equilibrium potential by the electrodes, the stability of readings, the fluctuations of temperature and so on.

Since the location of the end-point is related to the uncertainty of reading the pH, the steeper the titration curve close to the equivalence point the higher the precision of the result. The slope of the titration curve is affected by the value of the equilibrium constant of the neutralization reaction and by the concentrations of the reagents.

One method of estimating the slope of a titration curve close to the equivalence point is by the 'break' in the titration curve, which can be defined as the change in pH caused by a given increment of titrant. Instead of measuring the $\Delta pH/\Delta V$, it is convenient to calculate the pH change corresponding to a particular change of the titration fraction f (Fig. 4.12). The break in the titration curve is then equal to the ratio $\Delta pH/\Delta f$. The value to use for Δf is a matter of personal choice. A convenient value is $\Delta f = \pm 0.001$, i.e. we calculate the difference in pH between points $f = 0.999$ and $f = 1.001$.

Example 4.13. Calculate the break in the curves for titration of $0.10M$ NaOH and $0.10M$ NH_3 with $0.10M$ HCl, taking $\Delta f = \pm 0.001$.

We calculate the pH at the point $f = 0.999$. In the titration of NaOH

$$[H_3O^+] = \frac{K_w(V_0 + V)}{(1 - f)C_BV_0} = \frac{10^{-14} \times 2}{10^{-3} \times 10^{-1}} = 2 \times 10^{-10} M$$

and pH = 9.7.

In the titration of NH_3

$$[H_3O^+] = \frac{K_a f}{(1 - f)} = 10^{-6.2} M$$

and pH = 6.2.

Beyond the equivalence point the titration curve is the same for both bases and if $f = 1.001$

$$[H_3O^+] = \frac{(f - 1)C_BV_0}{(V_0 + V)} = 10^{-4.3} M$$

and pH = 4.3.

Thus, for sodium hydroxide, the break in the titration curve is 9.7 −4.3 = 5.4 pH units, and for NH_3 it is 6.2−4.3 = 1.9 pH units, i.e. it is considerably smaller.

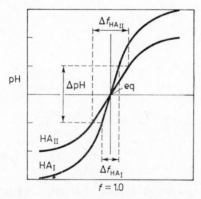

Fig. 4.12. Effect of the error of pH determination (ΔpH) on the error of concentration determination expressed by the change of the titration fraction (Δf) for different slopes of the titration curve at the equivalent point.

The break in the titration curve at the end-point is a good measure of titration precision, which is dependent on the nature of the titration curve, but since the choice of Δf is fairly arbitrary, it is not totally satisfactory.

A better and more objective measure is provided by the *titration sharpness index*, which can be defined as

$$\eta = \left| \frac{dpH}{df} \right| \tag{4.139}$$

The sharpness index can be calculated for each point of a titration curve—it is the slope of the titration curve at a given point. In the case of 'symmetrical' acid–base titrations it reaches its maximum at the equivalence point.

To calculate the value of the sharpness index in the titration of an acid with a base, we must take into consideration that

$$f = \frac{C_{\text{base}}}{C_{\text{acid}}} \tag{4.140}$$

where C_{base} and C_{acid} are the total concentrations of base and acid close to the equivalence point, whatever the extent of reactions. From Eqs. (4.139) and (4.140) we obtain

$$\eta = \frac{dpH}{dC_{\text{base}}} C_{\text{acid}} \tag{4.141}$$

If we express C_{acid} as $C_{\text{HA}} C_{\text{B}}/(C_{\text{HA}} + C_{\text{B}})$, in accordance with Eq. (4.118) and C_{base}/dpH as the buffer capacity β of the system, we have

$$\eta = \frac{C_{\text{HA}} C_{\text{B}}}{C_{\text{HA}} + C_{\text{B}}} \frac{1}{\beta} \tag{4.142}$$

On substituting the expression for the buffer capacity of a system (Section 3.7) and assuming $K_a \gg [\text{H}_3\text{O}^+]$ we obtain

$$\eta = \frac{0.434}{\dfrac{C_{\text{HA}} + C_{\text{B}}}{C_{\text{HA}} C_{\text{B}}} \left(\dfrac{K_w}{[\text{H}_3\text{O}^+]} + [\text{H}_3\text{O}^+] \right) + \dfrac{[\text{H}_3\text{O}^+]}{K_a}} \tag{4.143}$$

On substitution of the value for $[\text{H}_3\text{O}^+]$ at the equivalence point, this expression allows calculation of the sharpness index of a titration. In certain cases those terms in the sum which are small in comparison with the largest may be omitted.

Example 4.14. Calculate the sharpness index for titration of $0.01M$ hydrochloric acid with $0.01M$ sodium hydroxide.

At the equivalence point, $[\text{H}_3\text{O}^+] = 10^{-7}M$, and the term $[\text{H}_3\text{O}^+]/K_a$ is very small compared with other terms. Hence

$$\eta = \frac{0.434}{\dfrac{2 \times 10^{-1}}{10^{-2}} \times 2 \times 10^{-7}} = 1.08 \times 10^5$$

This value is very large and indicates that determinations should have high precision.

Example 4.15. Calculate the sharpness index for the titration of $0.10M$ acetic acid with $0.10M$ NaOH.

We substitute in Eq. (4.143) the value $[H_3O^+]/K_a$ obtained for the equivalence point from the equation of the titration curve

$$\frac{[H_3O^+]}{K_a} = \frac{C_{HA} + C_B}{C_{HA}C_B}\left(\frac{K_w}{[H_3O^+]} - [H_3O^+]\right)$$

and obtain

$$\eta = 0.217 \times \frac{C_{HA}C_B}{C_{HA} + C_B} \times \frac{[H_3O^+]}{K_w}$$

and, on substituting the numerical values for the concentrations, K_w and $[H_3O^+]$ (which is $10^{-8.7}$ for the equivalence point)

$$\eta = 0.217 \times \frac{10^{-2}}{2 \times 10^{-1}} \times \frac{10^{-8.7}}{10^{-14}} = 1.08 \times 10^{3.3} = 2.17 \times 10^3$$

In order to understand the relationship between the numerical values of the sharpness indices and the precision of locating the titration end-point, we must realize that a sharpness index of 10^3 corresponds (by definition) to an uncertainty of 0.1% in the ratio C_{base}/C_{acid} for a pH change of 1 unit. A precision of 0.1% is usually attainable, the limiting factor being the precision of volume reading. Thus, if we use a colour indicator for which ΔpH (the uncertainty of pH determination) is less than one pH unit, satisfactory precision will be obtained if $\eta \geqslant 10^3$.

The quality of a titration can be characterized in a somewhat different way by means of direct calculation of the precision error for a given value of ΔpH. From the slope of the titration curve we can determine values for the finite differences, ΔpH and ΔC,

$$\frac{dpH}{dC} = \frac{\Delta pH}{\Delta C} \tag{4.144}$$

where ΔpH denotes the uncertainty in pH measurement at the end-point (the error of precision of pH measurement), and ΔC is the absolute precision error of concentration determination. To calculate the relative error of titration precision (relative precision, RP) we substitute ΔC calculated from Eq. (4.144) in the expression for the relative error and obtain

$$RP(\%) = \pm\frac{\Delta C}{C} \times 100 = \frac{dC}{dpH}\frac{\Delta pH}{C} \times 100 \tag{4.145}$$

On introducing the buffer capacity β, we have

$$\text{RP}(\%) = \pm \frac{\Delta\text{pH}}{C} \times \beta \times 100 \tag{4.146}$$

which on introduction of the expression for β becomes

$$\text{RP}(\%) = \pm 2.3 \times \frac{\Delta\text{pH}}{C} \left(\frac{K_w}{[\text{H}_3\text{O}^+]} + [\text{H}_3\text{O}^+] \right.$$

$$\left. + \frac{CK_a[\text{H}_3\text{O}^+]}{(K_a + [\text{H}_3\text{O}^+])^2} \right) \times 100 \tag{4.147}$$

In these equations the symbol C denotes the total concentration of acid in a solution, i.e. the initial concentration of acid, C_{HA}, multiplied by the appropriate dilution. Close to the equivalence point this factor is equal to $C_B/(C_{\text{HA}} + C_B)$. The concentration of H_3O^+ refers to the equivalence point, and depends on the concentrations of the solutions used for titration and on the dissociation constant of the acid.

In the titration of acids by a strong base, Eq. (4.147) can be simplified, since $[\text{H}_3\text{O}^+] \ll K_a$ at or near the equivalence point. We then obtain

$$\text{RP}(\%) = \pm 2.3 \frac{\Delta\text{pH}}{C} \left(\frac{K_w}{[\text{H}_3\text{O}^+]} + [\text{H}_3\text{O}^+] + \frac{C[\text{H}_3\text{O}^+]}{K_a} \right) \times 100 \tag{4.148}$$

This equation enables us to consider the following simplifications.

1. In the titration of strong acids, $C \ll K_a$ and $[\text{H}_3\text{O}^+] = K_w/[\text{H}_3\text{O}^+]$ $= \sqrt{K_w}$; hence the third term in brackets can be omitted. On substituting for K_w we obtain

$$\text{RP}(\%) = \pm 2.3 \times \frac{\Delta\text{pH}}{C} \times 2\sqrt{K_w} \times 100 = \pm 4.6 \times 10^{-5} \times \frac{\Delta\text{pH}}{C} \tag{4.149}$$

2. In the titration of weak acids, if it can be assumed that $[\text{OH}^-] = \sqrt{K_w C/K_a}$ and $[\text{H}_3\text{O}^+] \ll [\text{OH}^-]$ at the equivalence point, then

$$\text{RP}(\%) = \pm 2.3 \times \frac{\Delta\text{pH}}{C} \times 2 \sqrt{\frac{K_w C}{K_a}} \times 100$$

$$= \pm 4.6 \times 10^{-5} \times \frac{\Delta\text{pH}}{\sqrt{CK_a/K_w}} \tag{4.150}$$

Example 4.16. Calculate the relative precision of the titration of $0.01M$ weak acid (dissociation constant $K_a = 10^{-6}$) with $0.10M$ NaOH, if the precision of pH measurement is 0.4 unit.

Since the titrant solution is much more concentrated than the titrant, the dilution during the titration can be neglected. The pH at the equivalence point is calculated as follows:

$$[H_3O^+] = \sqrt{\frac{K_w K_a}{C}} = \sqrt{\frac{10^{-14} \times 10^{-6}}{10^{-2}}} = 10^{-9} M$$

so pH = 9. This means that we can omit $[H_3O^+]$, in comparison with $[OH^-]$. On substitution of numerical data, we obtain

$$RP(\%) = 4.6 \times 10^2 \times \frac{0.4}{\sqrt{10^6}} \% = 0.18\%$$

The precision of weak base titration is calculated analogously. Instead of the equilibrium constant of acid neutralization

$$HA + OH^- \rightleftharpoons A^- + H_2O, \quad K = \frac{K_a}{K_w} \tag{4.151}$$

we require the equilibrium constant of base neutralization

$$B + H_3O^+ \rightleftharpoons BH^+ + H_2O, \quad K = \frac{1}{K_a} = \frac{K_b}{K_w} \tag{4.152}$$

In Eq. (4.150) the value of ΔpH remains unchanged because it is equal to the value of ΔpOH with the opposite sign, which is of no importance when absolute values are considered. Thus, in the case of a weak base we obtain

$$RP(\%) = \pm\, 4.6 \times 10^2 \frac{\Delta pH}{\sqrt{CK_b/K_w}} \tag{4.153}$$

The equation for the relative precision of titrating a dibasic acid to the first equivalence point has a slightly different form. Since in a system of this kind the buffer capacity is (approximately)

$$\beta = 2.3 \times 2C\sqrt{K_{a2}/K_{a1}} \tag{4.154}$$

the relative precision is

$$RP(\%) = \pm 2.3 \times 2 \times \Delta pH \sqrt{(K_{a2}/K_{a1})} \times 100 \tag{4.155}$$

and thus does not depend on the concentration of the titrant. This is not surprising if we note that the pH at the equivalence point and over the two buffer regions is independent of the concentration of the solution.

Example 4.17. Calculate the relative precision for the titration of a solution of orthophosphoric acid to the first equivalence point. $K_{a1} = 10^{-2.1}$, $K_{a2} = 10^{-7.2}$ and $\Delta pH = 0.4$.

From Eq. (4.155) we find

$$RP(\%) = 4.6 \times 10^2 \times 0.4\sqrt{(10^{-7.2}/10^{-2.1})}$$
$$= 5.2 \times 10^{-1}\% \cong 0.5\%$$

The precision of this determination is thus poor. Since the ratio K_{a2}/K_{a1} is almost the same as the ratio K_{a3}/K_{a2}, titration up to the second equivalence point will not improve the result.

If a polyprotic acid is titrated to the last equivalence point, the system can be treated as if it were a monoprotic acid with dissociation constant equal to the last dissociation constant of the polyprotic acid. A typical example is the titration of tartaric acid to the second equivalence point. Since the dissociation constants are $K_{a1} = 10^{-3.0}$ and $K_{a2} = 10^{-4.3}$, the system is regarded as a monoprotic acid with acid dissociation constant $K_a = 10^{-4.3}$.

Example 4.18. Calculate the relative precision for the titration $0.10M$ tartaric acid with $0.10M$ NaOH, to each of the two end-points if ΔpH $= 0.2$, which corresponds to the precision of potentiometric measurement. $K_{a1} = 10^{-3.0}$ and $K_{a2} = 10^{-4.3}$.

The precision for use of the first end-point is calculated from Eq. (4.155), according to which

$$\text{RP}(\% =) \pm 4.6 \times 0.2 \times 10^{-0.7\cdot} \times 100\% = \pm 18.4\%$$

This result indicates that the first equivalence point is useless as a titration end-point.

For the second equivalence point, we treat the system as a monoprotic acid, remembering that the relative error will be halved. Thus

$$\text{RP}(\%) = \pm \frac{4.6 \times 0.2 \times 10^{-7}}{2 \times 10^{-2.9}} \times 100\% = \pm 0.004\%$$

which shows the high precision of determination. There is therefore a large break in the potential at the end-point.

4.10 LOCATION OF TITRATION END-POINTS BY USE OF INDICATORS

In Section 2.14 we discussed some properties of the acid–base indicators used for pH measurement. However, the most common use for pH indicators is in acid–base titrations. The indicator changes colour as the pH changes during the titration: if it has been chosen correctly, the colour change occurs at the equivalence point.

However, the use of colour indicators for end-point location is another source of titration errors. One such error arises from the difference between the pH at the titration equivalence point and the pH range for the indicator colour change. In Section 4.8 we discussed this kind of titration error, on the assumption that the pH at the end-point can be freely chosen by the analyst. In practice, with indicator titrations this is only approximately correct. There are only a limited number of indicators, and it is not always possible to choose one for which the pH of the colour change is close to the theoretical equivalence-point pH.

One way to achieve better matching of end-point to equivalence point is to use for comparison a reference solution with a pH equal to or close to the pH of the equivalence point. The concentration ratio of the two coloured forms of the indicator can then be fairly arbitrary. The only restriction is that close to the pH chosen, the colour change due to a slight change in the pH should be distinctly noticeable. The use of mixed indicators is also helpful, and by a suitable selection of component concentrations we can optimize the colour change so that very small changes in pH can be seen. In the case of one-colour indicators, a change in the total indicator concentration can affect the pH of colour appearance or disappearance, and thus be helpful in reducing the error.

Another source of titration error is the fact that the indicator colour change does not occur sharply at some specified pH value, but gradually over a range of about 1.5 pH units, and the human eye is often not sensitive enough to allow precise identification of the colour corresponding to the equivalence point. More precise determination of the end-point is possible if, instead of visual observation of the solution, we use spectrophotometric measurement, in a spectrophotometric titration.

In order to obtain maximum precision in titrations with an indicator, the colour of the solution is often compared with the colour of a solution with a pH corresponding exactly to the equivalence point of the titration, and containing the same quantity of indicator.

The properties of an indicator and a titrated system can conveniently be plotted together on one diagram. Figure 4.13 is a diagram of this kind for the titration of strong acids with strong bases of different concentrations. The ranges of colour change for Methyl Orange, Bromothymol Blue and phenolphthalein are marked. Even this simple small-scale drawing clearly indicates that Methyl Orange is at best only usable for titrating solutions with a concentration of about $0.1M$; phenolphthalein, with its colour change on the other side of the equivalence point, is a slightly better indicator, suitable for titrations at a concentration down to about $0.01M$, and Bromothymol Blue is suitable for the titration of all the marked concentrations of strong acids (and strong bases).

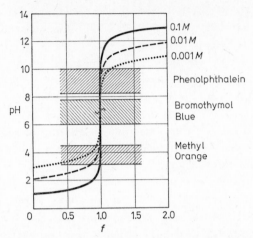

Fig. 4.13. Titration curves of 0.1, 0.01 and 0.001M solutions of a strong acid with strong base solutions. The ranges of colour change of the indicators are indicated.

The information contained in such a diagram is of course approximate. To obtain more exact data, for example when the equilibrium constants are not big enough to ensure a large pH change at the equivalence point, we must do calculations similar to those in Sections 4.8 and 4.9.

Example 4.19. Calculate, on the basis of the sharpness index, the probable precision of titration of $10^{-1}M$ and $10^{-3}M$ acetic acid with a solution of sodium hydroxide of the same concentration, if the range of colour change is about one pH unit. State which of the indicators listed in Table 2.3 is suitable for these titrations. Will the titration have a systematic titration error? If so, calculate the size of the error, taking as the end-point the middle of the colour-change range of a two-colour indicator or the initial point of the colour-change range of a one-colour indicator.

First, we must determine the pH at the equivalence point for both titrations. In calculations we use the ion product of water $K_w = 10^{-14}$, and the dissociation constant of acetic acid $K_a = 10^{-4.8}$ independently of the changing ionic strength.

$$[H_3O^+] = \sqrt{\frac{K_w K_a (V_0 + V)}{C_{HA} V_0}} = \sqrt{\frac{10^{-14} \times 10^{-4.8}}{10^{-1}} \times 2}$$
$$= 10^{-8.8}M$$

For the 0.001M acid

$$[H_3O^+] = \sqrt{\frac{10^{-14} \times 10^{-4.8}}{10^{-3}} \times 2} = 10^{-7.8}M$$

The pH values at the two equivalence points are thus 8.8 and 7.8, respectively.

To calculate the error resulting from the uncertainty in the determination of pH at the end-point, we must estimate the sharpness index at the end-point. The end-point will depend on the indicator used, and therefore the next stage is to select the indicator. The table of indicators contains no indicator suitable for both titrations. For the titration of $0.10M$ acetic acid, phenolphthalein seems to be the most suitable although colour begins to appear at pH = 8.2. If we take this pH as that of the end-point, we can calculate the sharpness index

$$\eta = \frac{0.434}{\dfrac{2 \times 10^{-1}}{10^{-2}}\left(\dfrac{10^{-14}}{10^{-8.2}} + 10^{-8.2}\right) + \dfrac{10^{-8.2}}{10^{-4.8}}}$$

$$= \frac{0.434}{2 \times 10^{1} \times 10^{-5.8} + 10^{-3.4}} = 1.0 \times 10^{3}$$

From the calculated value of the sharpness index we can estimate the error connected with the uncertainty in the pH determination. It is $1/1.0 \times \times 10^{3} = 1 \times 10^{-3}$, i.e. 0.1%, which is within the usual limits (0.1%).

Next we can find the titration error resulting from the difference between the end-point and equivalence points

$$TE(\%) = \frac{\left(\dfrac{2 \times 10^{-1}}{10^{-2}} \times 10^{-5.8} \times 10^{-4.8}\right) - 10^{-8.2}}{10^{-8.2} + 10^{-4.8}} \times 100$$

$$= \frac{2 \times 10^{1} \times 2.0 \times 10^{-11} - 6.4 \times 10^{-9}}{1.6 \times 10^{-5}} \times 100 = -3.8 \times 10^{-2}\%$$

This systematic error, equal to about 0.04%, can be disregarded, particularly since in practice the pH at the end-point will be somewhat higher than the expected value of 8.2, since we are more likely to overtitrate the solution than to estimate the end-point too soon.

If we consider in the same way the titration of the $0.001M$ acid, phenolphthalein is inconvenient because positive errors would result. The titration error calculated for an end-point pH of 8.2 would be +0.27%, but since the real end-point usually occurs at pH > 8.2, the error will increase. Increasing the indicator concentration would cause some improvement. In any case, however, the titration should be based on a comparison with a buffer solution with pH corresponding to the equivalence point and containing the same quantity of indicator. If we take pH 8.2 as the end-point, the sharpness index will be

$$\eta = \cfrac{0.434}{\cfrac{2 \times 10^{-3}}{10^{-6}}\left(\cfrac{10^{-14}}{10^{-8.2}} + 10^{-8.2}\right) + \cfrac{10^{-8.2}}{10^{-4.8}}} = 1.22 \times 10^2$$

This value signifies that $\Delta pH = 1$ can cause a precision error of $1/1.22 \times \times 10^2 = 0.008$, i.e. 0.8%. This error is so large that the titration is practically unacceptable. However, if we take as the end-point pH $=7.7$, equal to the equivalence point, then for

$$\eta = \frac{0.434}{2 \times 10^3(10^{-6.3} + 10^{-7.7}) + 10^{-3.0}} = 2.1 \times 10^2$$

the precision error will be 0.005. This calculation shows the great importance of correct location of the end-point, since otherwise not only the systematic titration error but also the precision error is increased.

Another indicator which could be used for our example is Phenol Red, for which the centre of the range of colour change is at about 7.4. By using a suitable control buffer with pH 7.7, we can obtain an improved result; however, the colour change from yellow to red is usually not very distinct, so the use of such an indicator requires some practice.

Example 4.20. Which of the indicators, Methyl Red or Bromothymol Blue, should be used in the titration of $0.01M$ ethanolamine with $0.1M$ hydrochloric acid? To a first approximation, the colour-change pH can be taken as the middle of the range given in Table 2.3.

First we need to calculate the equivalence-point pH. We can use the formula for calculation of the pH of a weak acid solution, in this case the ethanolammonium ion, $pK_a = 9.6$. If we disregard dilution (which is only 10%) then

$$[H_3O^+] = \sqrt{K_a C_{HA}} = \sqrt{10^{-9.6} \times 10^{-2}} = 10^{-5.8}M$$

i.e. $pH_{eq} = 5.8$.

Let us now assume that if Methyl Red is used, the end-point is at pH $(6.2+4.4)/2 = 5.3$ and if Bromothymol Blue is used it is at pH $(7.6 +6.0)/2 = 6.6$. We can now calculate the titration error in each case, by use of Eq. (4.128) modified by substituting $[OH^-]$ for $[H_3O^+]$ and vice versa, and by substituting $K_b = K_w/K_a = 10^{-4.4}$ for K_a. Then we obtain for Methyl Red

$$TE(\%) = \frac{10^2(10^{-5.3} - 10^{-8.7})(10^{-8.7} + 10^{-4.4}) - 10^{-8.7}}{10^{-8.7} + 10^{-4.4}} \times 100$$

$$= 0.04\%$$

and likewise for Bromothymol Blue

$$TE(\%) = \frac{10^2(10^{-6.6} - 10^{-7.4})(10^{-7.4} + 10^{-4.4})}{10^{-7.4} + 10^{-4.4}} \times 100$$

$$= -0.1\%$$

If we now compare, for the two indicators, the uncertainty of pH determination resulting from the sharpness indices, we obtain, on the basis of Eq. (4.143) modified as before, for Methyl Red

$$\eta = \frac{0.434}{10^2(10^{-5.3} + 10^{-8.7})\dfrac{10^{-8.7}}{10^{-4.4}}} = 1.1 \times 10^3$$

and for Bromothymol Blue

$$\eta = \frac{0.434}{10^2(10^{-6.6} + 10^{-7.4})\dfrac{10^{-7.4}}{10^{-4.4}}} = 4.3 \times 10^2$$

Thus the uncertainty values are 0.09% and 0.23% respectively. This result shows that Methyl Red is superior to Bromothymol Blue as an indicator for the titration of ethanolamine with a strong acid.

PROBLEMS

1. Calculate the pH values at titration fractions of 0.50, 0.70, and 0.90 in the following titrations:
  a. $1 \times 10^{-2} M$ NaOH with $1 \times 10^{-2} M$ HCl;
  b. $1 \times 10^{-2} M$ NaOH with $1 \times 10^{-1} M$ HCl;
  c. $1 \times 10^{-3} M$ NaOH with $1 \times 10^{-1} M$ HCl;
  d. $1 \times 10^{-4} M$ HCl with $1 \times 10^{-2} M$ NaOH.

2. Calculate the pH values at titration fractions of 0.25, 0.50, 0.75, 0.90 and 1.50 in the following titrations:
  a. $1 \times 10^{-2} M$ acetic acid ($pK_a = 4.8$) with $1 \times 10^{-2} M$ NaOH;
  b. $1 \times 10^{-4} M$ acetic acid ($pK_a = 4.8$) with $1 \times 10^{-2} M$ NaOH;
  c. $2 \times 10^{-3} M$ dichloroacetic acid ($pK_a = 1.3$) with $0.1 M$ NaOH;
  d. $2 \times 10^{-2} M$ pyridine hydrochloride ($pK_a = 5.2$) with $0.2 M$ KOH.

3. Calculate the pH values at titration fractions of 0.50, 0.70, 0.90 and 1.50 in the following titrations:
  a. $2 \times 10^{-2} M$ pyridine ($pK_b = 8.8$) with $0.2 M$ HCl;
  b. $2 \times 10^{-3} M$ trimethylamine ($pK_b = 4.2$) with $0.2 M$ HCl;
  c. $1 \times 10^{-2} M$ sodium borate ($pK_b = 4.8$) with $0.2 M$ HCl;
  d. $5 \times 10^{-3} M$ hydroxylamine ($pK_b = 7.9$) with $0.01 M$ HCl.

4. Calculate the pH values at titration fractions of 0, 0.50, 0.75, 1.0, 1.50, 1.75, 2.0, and 2.50 in the following titrations:

    a. $1 \times 10^{-2} M$ oxalic acid ($pK_{a1} = 1.3$, $pK_{a2} = 4.3$) with $0.1M$ NaOH;

    b. $5 \times 10^{-3} M$ ethylenediamine ($pK_{b1} = 3.4$, $pK_{b2} = 6.5$) with $0.1M$ HCl;

    c. $5 \times 10^{-4} M$ ethylenediamine ($pK_{b1} = 3.4$, $pK_{b2} = 6.5$) with $0.1M$ HCl;

    d. $2 \times 10^{-3} M$ sodium sulphide ($pK_{b1} = 0.8$, $pK_{b2} = 6.8$) with $0.02M$ HCl;

    e. $1 \times 10^{-2} M$ tartaric acid ($pK_{a1} = 2.9$, $pK_{a2} = 4.3$) with $0.1M$ KOH;

    f. $1 \times 10^{-3} M$ iminodiacetic acid ($pK_{a1} = 2.7$, $pK_{a2} = 9.5$) with $0.05M$ KOH;

    g. $5 \times 10^{-3} M$ ascorbic acid ($pK_{a1} = 4.2$, $pK_{a2} = 11.6$) with $0.05M$ KOH.

5. Give the pH range for which the titration error is below 0.1% in:

    a. titration of 100 ml of $0.01M$ HCl with $0.01M$ NaOH;

    b. titration of 100 ml of $0.001M$ HCl with $0.001M$ NaOH;

    c. titration of 100 ml of $0.001M$ HCl with $0.1M$ NaOH.

6. Calculate the titration error when the end point was taken at pH 5.0:

    a. in titration of $0.1M$ $HClO_4$ with $0.1M$ NaOH;

    b. in titration of $0.01M$ $HClO_4$ with $0.01M$ NaOH;

    c. in titration of $0.001M$ $HClO_4$ with $0.001M$ NaOH;

    d. in titration of $0.001M$ $HClO_4$ with $0.1M$ NaOH.

7. Calculate the titration error when the end point was taken at pH 7.0:

    a. in titration of $0.1M$ acetic acid ($pK_a = 4.8$) with $0.1M$ KOH;

    b. in titration of $0.01M$ acetic acid ($pK_a = 4.8$) with $0.1M$ KOH;

    c. in titration of $0.01M$ dichloroacetic acid ($pK_a = 1.3$) with $0.1M$ KOH;

    d. in titration of $0.01M$ phenol ($pK_a = 10.0$) with $0.1M$ KOH;

    e. in titration of $0.01M$ pyridine hydrochloride ($pK_a = 5.2$) with $0.01M$ KOH;

    f. in titration of $0.001M$ oxalic acid ($pK_{a1} = 1.3$, $pK_{a2} = 4.3$) with $0.01M$ KOH.

8. Calculate the titration error for the titrations of question 7, but with the end-point taken at pH 9.0.

9. Give the pH range for which the titration error is below 0.2% in:

    a. titration of $0.01M$ acetic acid ($pK_a = 4.8$) with $0.01M$ KOH;

    b. titration of $0.01M$ ethanolamine ($pK_a = 4.6$) with $0.1M$ HCl;

    c. titration of $0.01M$ citric acid ($pK_{a1} = 3.1$, $pK_{a2} = 4.8$, $pK_{a3} = 6.4$) with $0.1M$ NaOH;

    d. titration of $0.05M$ sodium carbonate ($pK_{b1} = 3.7$, $pK_{b2} = 7.5$, $pK_w = 13.8$) with $0.1M$ HCl.

10. Calculate the pH change which corresponds to $\Delta f = \pm 0.001$ in the titrations of:

    a. $0.10M$ hydrochloric acid;

    b. $0.10M$ acetic acid;

    c. $0.01M$ hydrochloric acid;

    d. $0.01M$ acetic acid

with NaOH at a concentration 10 times larger than the solution titrated.

11. Calculate the sharpness index in titrations of strong acids with strong bases of the following concentrations:

    a. $0.01M$ acid vs. $0.01M$ base;

b. 0.01M acid vs. 0.1M base;

c. 0.001M acid vs. 0.01M base;

d. 0.001M acid vs. 0.10M base.

12. Calculate the sharpness index in the following titrations. Take $pK_w = 13.8$.
   a. 0.10M formic acid ($pK_a = 3.7$) with 0.1M NaOH;
   b. 0.10M hydroxylammonium chloride ($pK_a = 6.2$) with 0.1M NaOH;
   c. 0.01M formic acid ($pK_a = 3.7$) with 0.1M NaOH;
   d. 0.01M hydroxylammonium chloride ($pK_a = 6.2$) with 0.1M NaOH.

13. Calculate the relative precision of the following titrations, when the pH value at the end point can be determined with a precision of 0.5 pH units. Take $pK_w = 13.8$.
   a. 0.01M HCl with 0.1M NaOH;
   b. 0.001M HCl with 0.1M NaOH;
   c. 0.001M HCl with 0.01M NaOH;
   d. 0.01M acetic acid ($pK_a = 4.8$) with 0.1M NaOH;
   e. 0.01M acetic acid ($pK_a = 4.8$) with 0.01M NaOH;
   f. 0.1M hydrocyanic acid ($pK_a = 9.2$) with 0.1M NaOH.

14. Calculate the relative precision of titration for all the neutralization steps of the following polyprotic acids. At the end point Δ pH = 0.2. Take $pK_w = 13.8$.
   a. 0.10M oxalic acid ($pK_{a1} = 1.1$, $pK_{a2} = 4.0$) with 0.1M NaOH;
   b. 0.10M iminodiacetic acid ($pK_{a1} = 2.7$, $pK_{a2} = 9.5$) with 0.1M NaOH;
   c. 0.10M citric acid ($pK_{a1} = 3.0$, $pK_{a2} = 4.4$, $pK_{a3} = 6.1$) with 0.1M NaOH.

CHAPTER 5

Acids and bases and the structure of molecules

5.1 EVALUATION OF ACID–BASE PROPERTIES FROM THERMO-DYNAMIC DATA

In Section 1.4 we considered the dependence of the equilibrium constant of a protolytic reaction on changes of the Gibbs free energy and the enthalpy. The discussion concerned only effects appearing in a definite medium of reaction, for example in water as a solvent, and so they referred to the properties of the hydrated acid, base or proton. Acid–base properties are extremely dependent on the solvent, and correlation of the properties observed in a particular medium with the nature of the molecules themselves is a difficult task. It is necessary to consider the energy changes that occur on transfer from the gas phase (in which reactions with the medium do not occur) to the solution phase. Only a small number of such systems can be studied experimentally, so usually the energy changes must be calculated from theoretical data.

A measure of the acid properties of a molecule HA in the gas phase is the proton affinity (P) which is equal to the enthalpy change when a proton is lost to yield the gaseous ion $H^+_{(g)}$ and the conjugate base $A^-_{(g)}$

$$HA_{(g)} \overset{P}{\rightleftharpoons} H^+_{(g)} + A^-_{(g)}$$

There are three factors which contribute to the value of the proton affinity. The first is the enthalpy of (D) dissociation of the acid molecule into $H_{(g)}$ and $A_{(g)}$ in the gas phase. This depends on the energy of the H–A bond and can be determined experimentally. The second factor is the ionization potential (I) corresponding to the energy of the loss of an electron from the atom or a molecule

$$H_{(g)} \overset{I}{\rightleftharpoons} H^+_{(g)} + e^-$$

The last factor is the electron affinity (E), which is the energy liberated on acceptance of an electron by an atom or a molecule

$$A_{(g)} + e^- \overset{E}{\rightleftharpoons} A_{(g)}^-$$

We can now calculate the proton affinity from the expression

$$P = D + I - E$$

Values for the proton affinity for some bases are listed in Table 5.1. A comparison of proton-affinity values with the strengths of bases in aqueous solution does not show any meaningful correlation, because the

Table 5.1

Proton affinities for certain bases (at 25°C)

Base	P(kJ/mole)	Base	P(kJ/mole)
CH_3OH	754	F^-	1541
H_2O	762	OH^-	1612
NH_3	875	NH_2^-	1645
I^-	1302	NH^{2-}	2720
Br^-	1340	N^{3-}	3433
Cl^-	1386		

Table 5.2

Contribution of dissociation energy, electron affinity and ionization potential to proton affinity for acids (kJ/mole)

	D	I	E	P
HF/F^-	565	1319	343	1541
HCl/Cl^-	431	1319	364	1386
HBr/Br^-	364	1319	343	1340
HI/I^-	297	1319	314	1302

proton affinity does not include the reaction with a solvent. However, for some groups of bases (e.g. H_2O, NH_3 or NH_3, NH_2^-, NH^{2-}, N^{3-}) the sequence of values is the same.

The contribution of particular factors to proton affinity values is illustrated by the data for hydrogen halide acids, which are given in Table 5.2.

In order to correlate proton-affinity values with acid–base properties in solutions, i.e. with the change of the reaction enthalpy

$$HA_{(sol)} \overset{\Delta H}{\rightleftharpoons} H_{(sol)}^+ + A_{(sol)}^-$$

we must consider the solvation of HA in the solvent and the corresponding value of ΔH_{HA}, and also the solvation of each of the ions H^+ and A^- (solvation enthalpies ΔH_{H^+} and ΔH_{A^-}). For univalent cations, the heats of hydration are in the range from -200 to -500 kJ/mole, but for the proton, because of its small size, the value is considerably larger, -1093 kJ/mole.

If we consider a cycle of reactions taking into account all the effects mentioned, we can calculate the enthalpy change in the protolytic reaction of the acid HA

$$
\begin{array}{ccc}
HA_{(g)} & \xrightarrow{\;\;D\;\;} & H_{(g)} + A_{(g)} \\
 & & \downarrow I \quad\quad \downarrow -E \\
\Delta H_{HA}\downarrow & & H^+_{(g)} + A^-_{(g)} \\
 & & \downarrow \Delta H_{H^+} \quad \downarrow \Delta H_{A^-} \\
HA_{(sol)} & \xrightarrow{\;\Delta H^0\;} & H^+_{(sol)} + A^-_{(sol)}
\end{array}
$$

This value is

$$\Delta H^0 = D + I - E - \Delta H_{HA} + \Delta H_{H^+} + \Delta H_{A^-}$$

or

$$\Delta H^0 = P - \Delta H_{HA} + \Delta H_{H^+} + \Delta H_{A^-}$$

From the value of ΔH^0 and the corresponding values of $T\Delta S^0$, we can calculate, as in Section 1.4, ΔG^0 and then the equilibrium constant of the protolytic reaction. For simple molecules such as hydrogen halides, such calculations give results which are in fairly good agreement with the measured equilibrium-constant values (Table 5.3). This agreement is rea-

Table 5.3

Comparison of the calculated and measured equilibrium constants of the protolysis reaction of hydrogen halide acids (kJ/mole)

	P	ΔH_{HA}	ΔH_{H^+}	ΔH_{A^-}	ΔH_0	$T\Delta S_0$	ΔG_0	$\log K_{calc}$	$\log K_{exp}$
HF/F$^-$	1541	-46	-1093	-507	-13	-25	12	-2.1	-3.14
HCl/Cl$^-$	1386	-17	-1093	-368	-58	-17	-41	7.2	6
HBr/Br$^-$	1340	-21	-1093	-331	-63	-13	-50	8.8	9
HI/I$^-$	1302	-25	-1093	-293	-59	-4	-55	9.6	10

sonable, provided that we remember that the uncertainty in ΔH^0 or ΔG^0, which is 4 kJ/mole, corresponds to an uncertainty in pK of 0.73, i.e. to a change of over 500% in K at 25°C. The influence of the entropy change on the reactions taking place in the gas phase, and also on the solvation reaction, is relatively small. However, in view of the small value of ΔH^0, the entropy change can have a significant effect on the values for a reaction in a solvent.

Characterization of acid–base properties independently of the reaction medium is of little practical importance in analytical chemistry, for under any given conditions it is always necessary to account for the solvation effects and for specific reactions. Such parameters are generally best determined empirically.

5.2 DEPENDENCE OF ACID–BASE PROPERTIES ON MOLECULAR STRUCTURE

In molecules which have acidic properties, from the point of view of the proton theory of acids there is an ionic bond or a polarized covalent bond between the hydrogen atom and the rest of the molecule. The electron pair forming the bond is placed asymmetrically, farther from hydrogen and closer to an atom more electronegative than hydrogen. Such polarization of the bond in a molecule with acid properties readily leads to a proton loss if a molecule which is a strong proton acceptor is nearby. The negative ion, or suitable dipole molecule, is a conjugate base. The more strongly the bond with the hydrogen atom is polarized, i.e. the more strongly the electron pair forming the bond is withdrawn from the hydrogen atom, the stronger the acid and the weaker the conjugate base. A base which corresponds to a weak conjugate acid attracts an electron pair less strongly and has a more distinct tendency to accept a proton, i.e. it is stronger.

The acid–base properties of particles depend fundamentally on the value of their charge. It follows from simple electrostatic theory that the more negative the charge on an acid, the more difficult is the loss of a proton from the molecule. Thus, an increase in the negative charge decreases the strength of the acid, but increases the strength of the conjugate base. Numerous observations confirm this statement. For example, hydrated ferric ions $Fe(H_2O)_n^{3+}$ with positive charge 3 are a stronger acid than ferrous ions $Fe(H_2O)_n^{2+}$ with positive charge 2. We have evidence of this in the lower pH of a solution of an iron(III) salt in comparison with the corresponding iron(II) salt, and also in that ferric hydroxide is precipitated in more acidic solutions than ferrous hydroxide. Solutions of salts of such species as Sn(IV), Bi(III), Ti(IV), Sb(III) and Sb(V) will produce precipitates of the sparingly soluble hydroxides, unless they are adequately acidified.

Further confirmation is available from the increase in the strengths of bases with similar structures and increasing negative charge. Manganate MnO_4^{2-} is a stronger base than permanganate MnO_4^-, and $Fe(CN)_6^{4-}$ ions attach protons more readily than do $Fe(CN)_6^{3-}$ ions; this can be seen in the values for the dissociation constants: $K_b = 10^{-9.8}$ for $Fe(CN)_6^{4-}$, and $K_b < 10^{-13}$ for $Fe(CN)_6^{3-}$.

The most frequently quoted proof that acid strength decreases as the negative charge increases is the fact that the values of the successive acid dissociation constants of polyprotic acids greatly decrease. In the series $H_3PO_4-H_2PO_4^--HPO_4^{2-}-PO_4^{3-}$, the strength of the acids decreases, while the strength of the conjugate bases increases. The successive values of the acid dissociation constants are: $K_{a1} = 7 \times 10^{-3}$, $K_{a2} = 6 \times 10^{-8}$, $K_{a3} = 1.3 \times 10^{-12}$. If we look at the series consisting of water and the ions derived from it, we observe the same regularity. The hydronium ion is a strong acid, water has weaker acid properties and hydroxide and oxide have increasing base strength

$$\xleftarrow{\text{increase of acid strength}}$$
$$H_3O^+\!-\!H_2O\!-\!OH^-\!-\!O^{2-}$$
$$\xrightarrow[\text{increase of base strength}]{}$$

The next factor which should be taken into consideration in the comparison of the acid–base properties of particles is the electronegativity of the atom binding the proton. This enables us to draw conclusions regarding the properties of the compounds of different elements with hydrogen. In each period of the periodic system, the electronegativity of the elements considerably increases from left to right, and there is a corresponding growth of acidity of the hydrogen compounds. Let us consider the second period. Electronegativity, defined in accordance with Pauling's definition, increases from 1.0 for lithium to 4.0 for fluorine. Lithium hydride has no acid character, and the hydride ion H^- has very strong basic properties. Beryllium hydride, next in the series, has a similar character but not so strongly marked. We disregard BH_3, which is unknown as a monomer, and pass to methane (CH_4), which, not having any free electron pair, does not show basic properties; however, the CH_3^- ion has very strong basic properties. The next compound, ammonia (NH_3), has a free electron pair and thus can react as a base. The acid properties of ammonia, though not observable in aqueous solutions, can be found in other solvents, such as liquid ammonia, in which the NH_2^- ion is a strong base. We are familiar with the properties of water (H_2O) as a stronger acid than NH_3, and also with the properties of the OH^- ion as a strong base, though weaker than the NH_2^- ion. Hydrogen fluoride HF, the last in the series, is definitely acidic in character, and its conjugate base is, in an aqueous solution, a rather weak base

← increase of basic properties

$LiH–BeH_2–CH_4–NH_3–H_2O–HF$

increase of acidic properties →

These considerations can be extended to acids conjugated to these hydrogen compounds. However, this relates only to those combinations which have free electron pairs and thus can combine with a proton (i.e. fluorine, oxygen and nitrogen compounds). Hydrogen fluoride, which is the strongest of the acids mentioned above, has at the same time the weakest base properties; so the conjugate acid H_2F^+ is a strong acid. The conjugate acid H_3O^+—corresponding to water—is also a strong acid, whereas the acid NH_4^+ is considerably weaker. These relations are presented in the following diagram:

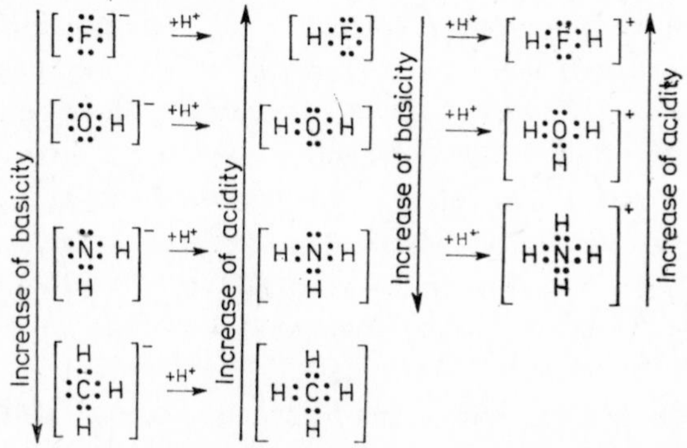

If a molecule of water or ammonia has one of its hydrogen atoms replaced by a more electronegative atom or group of atoms, the effective electronegativity of the oxygen or nitrogen atoms will increase—they will attract the electron pair more powerfully. As a result the acidic nature of the compound will be increased. That is why, for example, hypochlorous acid HOCl or nitric acid $HONO_2$, in which hydrogen is replaced by Cl or the electronegative group NO_2, are stronger acids than water; their basic properties, on the other hand, are considerably weaker. Comparison of the properties of the halogen oxyacids distinctly shows that the more electronegative the atom substituted for hydrogen the stronger the acidic properties of the compound

$$\xrightarrow{\text{increase of acid strength}}$$

$$\text{HOI—HOBr—HOCl}$$

A similar relation can be predicted for other halogen acids such as iodic acid, bromic acid and chloric acid.

Prediction of the acidic character on the basis of the electronegativity of the elements permits us also to estimate the properties of oxygen compounds for elements of other groups of the periodic system. The relevant acids of the fifth group of elements, i.e. nitric acid HNO_3, orthophosphoric acid H_3PO_4, orthoarsenic acid H_3AsO_4, orthoantimonic acid H_3SbO_4 and bismuthic acid $HBiO_3$ are successively weaker; the acid dissociation constants in aqueous solution decrease in value in the direction towards $HBiO_3$.

If we replace the hydrogen atoms in the ammonia molecule by electronegative substituents, a similar effect is observed. Compounds such as NH_2Cl (chloramine), NH_2NO_2 (nitroamine), NH_2NH_2 (hydrazine), and NH_2OH (hydroxylamine) have less basic (more acidic) properties than ammonia. A special position among these compounds is held by hydroxylamine NH_2OH. In fact, it could be considered as a derivative of water in which one hydrogen atom is replaced by the more negative group NH_2. This would suggest that the basic properties of hydroxylamine should be weaker than those of water, which contradicts the observed facts. Indeed, in an aqueous solution, hydroxylamine reacts as a base, forming the NH_3OH^+ ion, and its base dissociation constant is $K_b = 1.3 \times 10^{-8}$. This behaviour is caused by the fact that the nitrogen atom in an NH_2OH molecule is the stronger proton acceptor, and its properties determine the nature of the whole molecule, dominating over the properties of the oxygen atom.

Substitution of less electronegative atoms, such as alkali metals for hydrogen atoms, results in a compound with weaker acidic properties and more pronounced basic properties. Thus such substitution causes a decrease in the electronegativity of oxygen or nitrogen atoms. Indeed, the compounds KOH and KNH_2 (potassium hydroxide and potassium amide) have an ionic structure and are stronger bases than water and ammonia, respectively. It is worth noting that, just as ammonia is more basic than water, all its derivatives are more basic than the corresponding derivatives of water. These relations can be shown schematically in the following way:

$$\begin{array}{c} \xrightarrow{\text{increase of acid strength}} \\ \begin{array}{ccccc} KOH & \to & H_2O & \to & HOCl & \to & HONO_2 \\ \uparrow & & \uparrow & & \uparrow & & \uparrow \\ KNH_2 & \to & NH_3 & \to & NH_2Cl & \to & NH_2NO_2 \end{array} \end{array} \quad \uparrow \begin{array}{l} \text{increase of acid} \\ \text{strength} \end{array}$$

The dependence of the acidic properties of compounds on the electronegativity of atoms bound to oxygen can be observed on a series of derivatives containing the successive elements of one period of the periodic system. If, in compounds of the XOH type, both the electronegativity of element X and its valence increase, we shall obtain a series of compounds with properties that change from strongly basic to strongly acidic

$$[Na]OH—[HOMg]OH—[(HO)_2Al]OH—[(HO)OSi]OH—$$

$$—[(HO)_2OP]OH—[(HO)O_2S]OH—[O_3Cl]OH$$

The first two members of this series are ionic compounds which dissociate with formation of OH^- ions. Aluminium hydroxide $Al(OH)_3$ has an intermediate position—it is an amphiprotic compound, that dissociates as an acid in aqueous solution with the liberation of protons, or as a base, in which case OH^- ions are formed. The next compounds are acidic in nature, from the very weak silicic acid to the very strong perchloric acid.

As a factor in the acidic properties of the series from NaOH to $HClO_4$, not only the nature of the atom bound to oxygen, but also the number of electronegative groups bound to that atom is of importance. Starting with magnesium, there is an increase in the number of OH^- or O^{2-} groups, which influence the electronegativity of the oxygen actually bound to the dissociable hydrogen. This dependence is shown more clearly in the increase of acidity in the oxyacids of chlorine

$$\xrightarrow{\text{increase of acid strength}}$$

$$ClOH—OClOH—O_2ClOH—O_3ClOH$$

Similar relationships are shown by the transition metal elements. If we list the oxygen compounds of one element we observe an increase of acidity, which can be attributed, first, to the influence of negative groups in the molecule and, second, to a reduction in the size of the atom bound to oxygen. The smaller the radius of this atom and the larger the positive charge, the more powerfully it influences the electron distribution of oxygen. Thus a large density of positive charge causes stronger electron attraction to the central atom, which facilitates proton dissociation. As an illustration, the compounds of chromium can be used. In oxidation state (II), $Cr(OH)_2$, the chromium is distinctly basic, in state (III), $Cr(OH)_3$, it is amphoteric, and in state (VI), CrO_3, it is strongly acidic. The ionic radii of chromium in these oxidation state are: $r_{Cr^{2+}} = 0.080$ nm, $r_{Cr^{3+}} = 0.064$ nm, $r_{Cr^{6+}} = 0.052$ nm. Manganese compounds show similar changes in properties

$$Mn(OH)_2—Mn(OH)_3—Mn(OH)_4—H_3MnO_4—H_2MnO_4—$$

| basic | weakly basic | ampho-teric | weakly acidic | acidic |

$$—HMnO_4$$

strongly acidic

A quantitative measure of this gradual change in properties is given by the values of the corresponding equilibrium constants (the solubility product K_{s0} or the dissociation constant K_a)

$$Mn(OH)_2 \rightleftharpoons Mn^{2+} + 2OH^- \qquad K_{s0} = 10^{-14}$$
$$Mn(OH)_3 \rightleftharpoons Mn^{3+} + 3OH^- \qquad K_{s0} = 10^{-36}$$
$$Mn(OH)_4 \rightleftharpoons Mn^{4+} + 4OH^- \qquad K_{s0} = 10^{-56}$$
$$H_3MnO_4 \rightleftharpoons H^+ \;\;\; + H_2MnO_4^- \qquad K_{a1} = 10^{-8}$$
$$H_2MnO_4 \rightleftharpoons H^+ \;\;\; + HMnO_4 \qquad K_{a1} = 10^{-1}$$
$$HMnO_4 \rightleftharpoons H^+ \;\;\; + MnO_4^- \qquad \text{strong acid}$$

An increase in acidity is observed even when the electronegative atoms in a molecule are located at a distance from the oxygen–hydrogen bond. This is illustrated by comparison of the strength of acetic acid with its mono-, di- and trichloroderivatives. Acetic acid (CH_3COOH) is a weak acid with $K_a = 1.3 \times 10^{-5}$, monochloroacetic acid $CH_2ClCOOH$ is a somewhat stronger acid ($K_a = 1.3 \times 10^{-3}$), dichloroacetic acid $CHCl_2COOH$ is still stronger ($K_a = 5 \times 10^{-2}$), and trichloroacetic acid CCl_3COOH can almost be regarded as a strong acid ($K_a = 2 \times 10^{-1}$). A similar comparison made for other aliphatic chloroacids shows that the greater the distance from the carboxyl group of the electronegative atoms in a molecule, the smaller their effect on acid dissociation. However, in this case the increase in acidity on substitution is *not* caused by the inductive effect changing the strength of the O–H bond, but by the inductive effect causing more uniform charge distribution on the *anion*, and hence a lower ordering effect on the solvating water molecules, and thus a less negative entropy change.

Other factors also increase the ease of proton loss. Their influence usually either agrees with the electronegativity changes, or does not play a decisive role. Sometimes, however, when the changes in electronegativity are comparatively small, acidity trends may be determined by other factors including the size of the ions and atoms. If we compare the hydrogen compounds of the elements in the same group of the periodic table, it will be found that the acidic properties change in the opposite direction to that indicated by the electronegativity

increase of acidity
$$\xrightarrow{\hspace{4cm}}$$

$$HF \ll HCl < HBr < HI$$

$$H_2O < H_2S < H_2Se < H_2Te$$

$$NH_3 < PH_3 < AsH_3 < SbH_3$$

$$\xleftarrow{\hspace{4cm}}$$
increase of electronegativity
$$\xrightarrow{\hspace{4cm}}$$
increase of ionic radii

In all these cases the radius of the negative ion (the conjugate base) increases when moving downwards in the group.

Consequently, the unit of negative charge is spread over a larger volume so its density is smaller. This factor is important here because it counteracts the effect of electronegativity, and actually determines the direction of the changes in the acidic properties.

Predictions based on theoretical data are simple and in agreement with experimental facts only if there are no contradictory factors. In general, it is only the knowledge of the real behaviour of compounds that permits judgements as to the importance of particular factors for particular groups of compounds.

5.3 SOME TRENDS IN THE CHANGES OF ACID STRENGTH IN SOLUTIONS

In the previous section we discussed some qualitative rules for prediction of acid–base properties of molecules and inorganic ions. However, in aqueous solutions the evaluation of acid properties is more complicated, because of hydration of ions and hydrogen bonding. Nevertheless, for some types of acids, relationships have been established which permit us to predict at least approximately, changes in the value of the dissociation constant.

Typical inorganic acids are the oxyacids, in which protons are connected through oxygen atoms with the central atoms of various elements, usually in fairly high oxidation states. The general formula of these acids is $XO_m(OH)_n$. If all the oxygen atoms in an acid molecule have hydrogen atoms, attached, i.e. if $m = 0$, the first acid dissociation constant is very small, ($< 10^{-7}$) and the acid is very weak. Some examples of such acids are boric acid $B(OH)_3$, $K_{a1} = 6.3 \times 10^{-10}$, arsenious acid H_3AsO_3, $K_{a1} = 6.3 \times 10^{-10}$ and hypochlorous acid $HClO$, $K_a = 3.2 \times 10^{-8}$.

If $m = 1$ in the general formula, the acids are also weak, though considerably stronger than in the first case. The values of the dissociation constants range from 10^{-2} to 10^{-3}. This group of acids includes orthophosphoric acid $PO(OH)_3$, $K_{a1} = 6.3 \times 10^{-3}$, orthophosphorous acid $HPO(OH)_2$, $K_{a1} = 6.3 \times 10^{-3}$, sulphurous acid $SO(OH)_2$, $K_{a1} = 1.7 \times 10^{-2}$ and chlorous acid $ClO(OH)$ with $K_a = 10^{-2}$. An apparent exception is carbonic acid $CO(OH)_2$. The dissociation constant usually given in tables and corresponding to the pH values observed during titration is $K_{a1} = 4 \times 10^{-7}$, and thus is unexpectedly small. It turns out, however, that only 0.26% of the dissolved carbon dioxide in the solution is in the form of H_2CO_3. If we include this value in our calculations, then the true dissociation constant of H_2CO_3 will be $K_{a1} = 1.6 \times 10^{-4}$, i.e. it will be much closer to the expected value.

Acids with the general formula $XO_2(OH)_n$, such as sulphuric acid, nitric acid and chloric acid, are quite strong acids. The weakest in this group is iodic acid $IO_2(OH)$; its dissociation constant is 1.6×10^{-1}.

The last group, with the formula $XO_3(OH)_n$, consists of very strong acids, including the strongest known acid, perchloric acid $HClO_4$.

So far, we have considered only the first dissociation step of the oxyacid. Further dissociation steps of acids with one central atom are governed by another rule, which states that the successive dissociation constants of polyprotic oxyacids are in the ratio $1:10^{-5}:10^{-10}$. Examples obeying this simple rule can be found among all the groups of oxyacids previously discussed. Among them are germanic acid, orthophosphoric and orthophosphorous acids, arsenic and arsenious acids. Sulphuric acid and selenic acid ($m = 2$), which are strong in the first dissociation step, have second dissociation constants $K_{a2} = 1.3 \times 10^{-2}$ and 8×10^{-3}, respectively, which suggests that the rule is again satisfied.

The magnitude of the difference between the values of successive dissociation constants is affected by the interaction of OH groups connected to the central atom. The larger the distance of these groups from one another, the smaller the influence of that interaction, i.e. the smaller the ratio of the successive constants. An excellent illustration of this is pyrophosphoric acid with the structure $(OH)_2PO \cdot O \cdot PO(OH)_2$. The first dissociation constant K_{a1} ought to have a value intermediate between the predicted values for acids of type $XO(OH)_n$ and for acids of type $XO_2(OH)_n$ but actually it is 10^{-1}. Since the second proton dissociates from an oxygen atom attached to the second phosphorus atom, the electrostatic interaction is considerably smaller than, for example, in the case of orthophosphoric acid and the dissociation constant is $K_{a2} = 3.2 \times 10^{-3}$. The third proton dissociates from the OH group attached to a phosphorus atom where a dissociation has already taken place: hence $K_{a3} = 8 \times 10^{-7}$ and $K_{a1} : K_{a3}$

$= 1 : 8 \times 10^{-6}$, i.e. the value is close to the expected one for the ratio $K_{a1} : K_{a2}$ in the case of an oxyacid with one central atom. Similarly, $K_{a2} : K_{a4}$ $= 1 : 10^{-6}$, so again the ratio is relatively close to the expected one.

A good illustration of the weakening of acid group interaction is a homologous series of diprotic aliphatic acids. The ratio of the dissociation constants for the first two members of this series—oxalic acid, $COOHCOOH$, and malonic acid, $COOHCH_2COOH$—is about 1000. In succinic acid, in which the carboxylic groups are separated by two methylene groups, $COOH(CH_2)_2COOH$, the ratio is 20, and in sebacic acid, $COOH(CH_2)_8COOH$, it is 9. The theoretical limit to which this ratio should tend in the absence of electrostatic effects is 4. This limiting value results from the 50% smaller probability of proton loss from one than from two identical groups, and also a 50% smaller probability of proton attachment to one than to two basic groups (COO^-).

The rules given here are only approximate, since the values of the dissociation constants can be affected by many individual properties of the various acids. However, they can be useful when the appropriate data are lacking.

Such prediction based on evaluation of various effects are common for organic acid–base systems. An excellent and comprehensive treatment which allows for many theoretical and empirical relationships has been presented by Perrin *et al.* (see Bibliography).

CHAPTER 6

Other theories of acids and bases

6.1 SOLVENT THEORY OF ACIDS AND BASES

A fundamental assumption of Brønsted's theory is the autodissociation of water or other protic solvent (e.g. ammonia or acetic acid). In the auto-dissociation of a solvent, cations and anions characteristic of the solvent are formed. Cations arise from the attachment of protons to the mol-ecules of the solvent, and anions by the loss of a proton from the molecules of the solvent (Section 2.11). Acids are substances which, as a result of direct dissociation or a reaction with a molecule of the solvent, form in solution the cations characteristic of that solvent, (H_3O^+ in water, NH_4^+ in ammonia, $CH_3COOH_2^+$ in acetic acid). Bases, on the other hand, are substances which dissociate with formation of solvent anions or produce such anions in a reaction with the solvent. Thus for the examples given they are the anions OH^-, NH_2^- or CH_3COO^-.

It turns out, however, that such reactions can take place not only in protic solvents. Many solvents not containing hydrogen can undergo auto-dissociation to a small extent. The resulting cations and anions can also be regarded as acids and bases. Some such solvents are liquid sulphur dioxide SO_2, nitrogen dioxide NO_2, phosgene (carbon oxychloride) $COCl_2$, selenium oxychloride $SeOCl_2$, sulphuryl chloride SO_2Cl_2, bromine tri-fluoride BrF_3 and many others. Table 6.1 lists the reactions of autodis-sociation of some solvents, examples of substances which react as acids and as bases in a given solvent, and the reactions of neutralization. In many cases it has been possible to utilize such neutralizations in con-ductometric or potentiometric titrations. For example, in a BrF_3 solution it is possible to titrate the acid BrF_2SbF_6 with the base $AgBrF_4$. Another example of acid–base titration, this time in selenium oxychloride, is the titration of stannic chloride as an acid with a solution of chloride ions, and also the titration of pyridine as a base. When $SnCl_4$ is dissolved in

Table 6.1

Acid–base reactions in various solvents

Solvent	Cation of solvent	Anion of solvent	acid	Neutralization reactions		
				+ base	⇌ solvent	+ "salt"
H_2O	$H^+(H_3O^+)$	OH^-	HCl	$+ KOH$	$\rightleftharpoons H_2O$	$+ NaCl$
NH_3	$H^+(NH_4^+)$	NH_2^-	NH_4Cl	$+ NaNH_2$	$\rightleftharpoons NH_3$	$+ NaCl$
CH_3COOH	$H^+(CH_3COOH_2^+)$	CH_3COO^-	$HClO_4$	$+ CH_3COOK$	$\rightleftharpoons CH_3COOH$	$+ KClO_4$
SO_2	SO^{2+}	SO_3^{2-}	$SOCl_2$	$+ Cs_2SO_3$	$\rightleftharpoons 2SO_2$	$+ 2CsCl$
$SOCl_2$	$SOCl^+$	$Cl^-(SOCl_3^-)$	$SOClSbCl_6$	$+ HgCl_2$	$\rightleftharpoons SOCl_2$	$+ HgClSbCl_6$
$SeOCl_2$	$SeOCl^+$	$Cl^-(SeOCl_3^-)$	$SeOClAlCl_4$	$+ KCl$	$\rightleftharpoons SeOCl_2$	$+ KAlCl_4$
$COCl_2$	$COCl^+$	$Cl^-(COCl_3^-)$	$2COClAlCl_4$	$+ CaCl_2$	$\rightleftharpoons 2COCl_2$	$+ Ca(AlCl_4)_2$
N_2O_4	NO^+	NO_3^-	$NOCl$	$+ AgNO_3$	$\rightleftharpoons N_2O_4$	$+ AgCl$
BrF_3	BrF_2^+	BrF_4^-	BrF_2SbF_6	$+ AgBrF_4$	$\rightleftharpoons 2BrF_3$	$+ AgSbF_6$
$HgBr_2$	$HgBr^+$	$HgBr_3^-$	$(HgBr)_2SO_4$	$+ 2TlBr$	$\rightleftharpoons 2HgBr_2$	$+ Tl_2SO_4$

$SeOCl_2$ the solution obtained has acid properties (presence of the $SeOCl^+$ ion) according to the reaction

$$SnCl_4 + 2SeOCl_2 \rightleftharpoons SnCl_6^{2-} + 2SeOCl^+$$

Dissolution of pyridine results in a solution with basic properties

$$C_5H_5N + SeOCl_2 \rightleftharpoons C_5H_5NSeOCl^+ + Cl^-$$

This reaction can also be written so that, instead of the Cl^- ion, the product is solvated chloride ion expressed by the formula $SeOCl_3^-$, which in this solvent shows the properties of a base. As a consequence the acid solution obtained can be titrated with a solution of a base.

When we consider reactions in nitrogen dioxide (or rather in its dimer N_2O_4) we can perceive another similarity to reactions in aqueous solutions. In this case the acidic character is determined by the presence of the cation NO^+, which arises, for example, on dissolution of nitrosyl chloride $NOCl$. Such a solution dissolves metals with a reaction similar to the dissolution of metals in aqueous solutions of acids

$$Zn + 2H^+ \rightleftharpoons Zn^{2+} + H_2$$

$$Zn + 2NO^+ \rightleftharpoons Zn^{2+} + 2NO$$

$$Sn + 2H^+ + 4Cl^- \rightleftharpoons SnCl_4^{2-} + H_2$$

$$Sn + 4NO^+ + 6Cl^- \rightleftharpoons SnCl_6^{2-} + 4NO$$

In other solvents analogous reactions proceed in a similar way.

To illustrate the phenomenon of amphoterism, we consider the acid–base system in liquid sulphur dioxide, in which sulphite ions SO_3^{2-} are the bases and SO^{2+} ions the acids. If aluminium ions Al^{3+} are treated with sulphite solutions (e.g. tetramethylammonium sulphite $[(CH_3)_4N]_2SO_3$), aluminium sulphite is precipitated at first, but it then dissolves in excess of the reagent. The analogy to the amphoteric properties of aluminium hydroxide in water is shown by the following equations:

$$Al^{3+} + 3OH^- \rightleftharpoons Al(OH)_3$$

$$2Al^{3+} + 3SO_3^{2-} \rightleftharpoons Al_2(SO_3)_3$$

$$Al(OH)_3 + 3OH^- \rightleftharpoons [Al(OH)_6]^{3-}$$

$$Al_2(SO_3)_3 + 2SO_3^{2-} \rightleftharpoons 2[Al(SO_3)_3]^{3-}$$

The reaction can be reversed by addition of an acid, e.g. in the SO_2 solution by thionyl chloride $SOCl_2$.

Nowadays we know many reactions in different solvents which we can attempt to rationalize by theoretical considerations.

Different types of reactions have been described, i.e. acid–base reactions, redox reactions and complexation reactions. Some problems, however, are still unsolved. The question of ion solvation is one of these: there is a great deal of experimental evidence confirming that solvated ions exist; nevertheless, in many cases (e.g. in the case of water) the degree of solvation is not certain.

An unquestionable advantage of the solvent theory is the extension of Brønsted's theory to systems of non-protic solvents. However, it still concerns only solutions; and we know that reactions of a similar nature can take place not only in solutions and not only in ionic systems.

6.2 THE LEWIS ELECTRONIC THEORY

The Brønsted–Lowry theory and also the more general solvent theory are particularly concerned with the presence of the solvent—an important feature of both theories is the autodissociation of the molecules (of a protic solvent in the Brønsted–Lowry theory or of an aprotic solvent in solvent theory). The most general theory is the Lewis theory (1923), which is concerned not with investigating the behaviour of ions or molecules in solvents, but with studying their structure, which of course determines their properties. Thus, according to the Lewis theory a *Lewis acid* is a species with an incompletely filled outer electron shell. In the formation of a covalent coordinate bond, this acts as the acceptor of an electron pair. A *Lewis base* is a species with a free electron pair which it donates in formation of coordinate bond.

In a more modern formulation in terms of the theory of molecular orbitals, a Lewis base can be defined as a species which uses a doubly occupied orbital in a reaction. A Lewis acid uses an unoccupied orbital.

At first, the Lewis theory did not gain the wide and general appreciation which it was to have later. There were several reasons for this. First, the theory rejected the traditionally accepted connection of acid properties with labile hydrogen in a molecule. Moreover, it reduced the concepts of acid and base and their interactions to a common denominator, without differentiation between ionic and atomic bonds.

A significant advance in the propagation of the Lewis theory was the increased knowledge of chemical bond theory due to quantum mechanics, and also the formulation by Lewis (1938) of the following series of phenomenal criteria as a basis for his definition of acid–base systems.

1. Between an acid and a base, a neutralization reaction takes place, and this is a fast process. Since in this reaction, a compound with a co-valent coordinate bond is obtained, the reaction of neutralization in the Lewis theory is often called the *coordination reaction*, and the compound formed, the *coordination compound*. Examples of reactions which meet the requirements of the Lewis theory but are not reactions of neutralization in solvent theories are given below:

$$Al^{3+}(acid) + 6H_2O(base) \rightleftharpoons Al(H_2O)_6^{3+}$$

$$H^+(acid) + NH_3(base) \rightleftharpoons NH_4^+$$

$$BF_3(acid) + NH_3(base) \rightleftharpoons BF_3NH_3$$

$$FeCl_3(acid) + Cl^-(base) \rightleftharpoons FeCl_4^-$$

The products of these reactions are coordination compounds. They can be the final products of the reaction or they may undergo further re-arrangement or dissociation. Here are two examples of reactions of this kind:

$$SO^{2+}(acid) + SO_3^{2-}(base) \rightleftharpoons OS{:}SO_3 \rightarrow 2SO_2$$

$$SO_3(acid) + H_2O(base) \rightleftharpoons O_3S{:}OH_2 \rightarrow H_2SO_4$$

In the reactions we have quoted, the symbols Al^{3+} and H^+ cannot of course represent solvated ions, as they often do in the Brønsted–Lowry theory.

2. Reactions of displacement of a weak acid by a stronger one and a weaker base by a stronger one are also typical of Lewis acid–base reactions. There are many reactions in this group which in the Brønsted–Lowry theory are regarded as protolytic reactions because they consist of transfer of a proton as a particle, but the proton is also an acceptor of an electron pair, i.e. a Lewis acid

$$H_3O^+ + NH_3 \rightleftharpoons NH_4^+ + H_2O$$

$$HCN + OH^- \rightleftharpoons H_2O + CN^-$$

$$Al(H_2O)_6^{3+} + 6F^- \rightleftharpoons AlF_6^{3-} + 6H_2O$$

In those examples a stronger base (NH_3, OH^-, F^-) displaces a weaker one (H_2O, CN^-, H_2O), from the complex in which it was bound before the reaction. An example of a weak acid being displaced by a stronger one is the displacement of the silver ion (a weaker acid) by the hydrogen ion (a stronger acid) from the complex $Ag(NH_3)_2^+$

$$Ag(NH_3)_2^+ + 2H_3O^+ \rightleftharpoons Ag(H_2O)_2^+ + 2NH_4^+$$

3. Acids and bases can be used to titrate one another, with the use of chemical indicators. Such reactions are a particular case of the displacement reaction. Thus, for example, Methyl Violet may be used as an indicator both in aqueous solutions and in acid–base reactions in the sense of the Lewis theory, e.g. in the titration of BCl_3 (an acid) with pyridine (a base) in chlorobenzene solvent medium.

4. Catalytic properties are shown by acids (acid catalysis) or bases (base catalysis). These reactions are of great importance, particularly in organic chemistry, and their investigation in terms of the Lewis theory enables us to draw a number of conclusions of practical significance. Lewis acids—electron-pair acceptors—are thus *electrophilic reagents,* and Lewis bases—electron-pair donors—are *nucleophilic reagents.* For example in the Friedel–Crafts reaction we can use as catalysts $AlCl_3$, BF_3, H_2SO_4, P_2O_5, HF, which are Lewis acids; this makes it possible to explain the mechanism of their catalytic effect. Under the influence of $AlCl_3$ (a Lewis strong acid) an electrophilic acyl cation is formed from carboxylic acid chlorides; the cation reacts with the aromatic system (acting as a base) and finally forms a ketone

$$R-C\overset{O}{\underset{Cl}{\big\langle}} + AlCl_3 \rightleftharpoons R-C^+{=}O + AlCl_4^-$$

$$\text{C}_6\text{H}_6 + R-C^+{=}\dot{O} \longrightarrow \underset{\text{(benzene ring)}}{\overset{\overset{O}{\|}}{C-R}} + H^+$$

In many other organic reactions the formation of intermediate reactive products having the properties of Lewis acids can be a basis for explaining the processes taking place. Besides the acyl cation $R-C^+{=}O$ discussed above, we can mention carbonium ions $R-CH_2^+$, nitronium ions NO_2^+, or Br^+ cations. All of these are electrophiles, and play an important role in reactions of addition to double bonds or in nitration of aromatic compounds for example.

There are many types of particles which can be regarded as Lewis acids. The cations of metals are the first type. Some of them have comparatively weak acceptor properties, e.g. the cations of the alkali metals; other are considerably stronger acids, e.g. Al^{3+}, Fe^{3+}, Ag^+. In general, an increase in the positive charge on an ion and a reduction in its ionic radius lead to increased acid strength. Thus, the ions Co^{3+} and Fe^{3+} are stronger acids than the ions Co^{2+} and Fe^{2+}. In each group of the periodic system, acid strength increases as the atomic number decreases (K^+ <

$< \mathrm{Na^+} < \mathrm{Li^+}$) and among the ions of a given period, the acid strength considerably increases from left to right ($\mathrm{Na^+} < \mathrm{Mg^{2+}} < \mathrm{Al^{3+}}$). The ions of transition metals are strong acids. It is worth mentioning that acid strength also depends on the number of shielding electrons, which weaken the acid character. Below are some examples of reactions of this group of acids:

$$\mathrm{Cu^{2+}} + 4 \mathrm{:NH_3} \rightarrow \mathrm{Cu(:NH_3)_4^{2+}}$$
$$\mathrm{Al^{3+}} + 6 \mathrm{:OH_2} \rightarrow \mathrm{Al(:OH_2)_6^{3+}}$$
$$\mathrm{Fe^{3+}} + 4 \mathrm{:Cl^-} \rightarrow \mathrm{Fe(:Cl)_4^-}$$

Another group of Lewis acids comprises molecules in which one of the atoms has an incompletely filled valence-electron shell. They are mostly atoms which can form a stable eight-electron outer shell, e.g.

$$\mathrm{BF_3} + \mathrm{:NH_3} \rightarrow \mathrm{F_3B:NH_3}$$
$$\mathrm{SO_3} + \mathrm{:O^{2-}} \rightarrow \mathrm{O:SO_3^{2-}} \quad (\mathrm{SO_4^{2-}})$$
$$\mathrm{SO^{2+}} + \mathrm{:SO_3^{2-}} \rightarrow \mathrm{OS:SO_3} \rightarrow 2\,\mathrm{SO_2}$$

The coordination compound which is formed in the last reaction is unstable and decomposes into two molecules of $\mathrm{SO_2}$.

When one of the atoms in a molecule has an incompletely filled d-orbital, a stable system with more than an octet of electrons can be formed. We come across such reactions when the acids involved are, for example, $\mathrm{SnCl_4}$, $\mathrm{SiF_4}$, $\mathrm{TiCl_4}$, $\mathrm{SbF_5}$. In a reaction with halide ions as Lewis bases these form $\mathrm{SnCl_6^{2-}}$, $\mathrm{SiF_6^{-2}}$, $\mathrm{SbF_6^{2-}}$. Sometimes the reaction gives rise to an unstable product, which decomposes. For example, the reaction of $\mathrm{PCl_3}$ with water gives rise to $\mathrm{PCl_3 \cdot H_2O}$, which decomposes into HCl and $\mathrm{PCl_2OH}$. Further reaction of $\mathrm{PCl_2OH}$ with water leads to the formation of phosphorous acid $\mathrm{H_3PO_3}$, which is the final product.

A special type of reaction is one in which in the role of an acid is assumed by a molecule containing a proton. In this case a coordinate bond is formed with a strong base having a free electron pair. Thus the molecule of HCl reacts with water as a base, to form $\mathrm{H_3O^+}$ and $\mathrm{Cl^-}$ ions

$$\mathrm{ClH} + \mathrm{:OH_2} \rightarrow [\mathrm{ClH:OH_2}] \rightarrow \mathrm{Cl^-} + \mathrm{H_3O^+}$$

In a similar way sulphur trioxide $\mathrm{SO_3}$ as an acid reacts with water as a base to form $\mathrm{H_2SO_4}$

$$\mathrm{SO_3} + \mathrm{:OH_2} \rightarrow [\mathrm{O_3S:OH_2}] \quad (\mathrm{H_2SO_4})$$

We can also regard as Lewis acids the atoms of elements with six electrons in the outer shell. A good example is the sulphur atom, which reacts with the sulphite ion in a process which can be considered to be an acid–base reaction

$$S + :SO_3^{2-} \rightarrow [S:SO_3^{2-}] \quad (S_2O_3^{2-})$$

The oxygen atom can react similarly and is also regarded as an acid in such a reaction.

Finally we may regard as Lewis acids molecules with polarized bonds in which a more positive atom is the carrier of acid properties. Examples of such systems are:

It is worth mentioning that in such systems the more negative atom possesses basic properties. Finally the properties of a molecule will depend on the properties of the reaction partner and on the conditions in which the reaction takes place.

The Lewis theory has greatly extended the concept of acids and bases. A large number of the examples of Lewis type neutralizations are regarded in classical theories as complexation reactions. Metal ions are treated as acids, because they are acceptors, whereas ligands, as donors, are bases. The treatment is a very general approach to processes which we have customarily treated as reactions of different types. It is very valuable as a generalization of different phenomena, but certainly difficult to accept in terms of examples known from practice. This was probably the cause of the initial weak impact of the Lewis theory, in which the long-established concepts of acid and base were given a new sense. Perhaps, if the terms acceptor and donor or electrophilic and nucleophilic reagents had been used instead of 'acids' and 'bases', the Lewis theory would have gained general acceptance sooner. Also, its adversaries could no longer have argued that the terms acid and base, so widely exploited in the Lewis theory, lose their classifying power in chemistry.

Another difficulty is a certain duality in the treatment of acids—they can either combine directly with a base or just form a proton, which becomes attached to a free electron pair of the base. In many such cases the Lewis theory enables us to postulate a mechanism for the reaction, but the products of the neutralization reaction are often intermediate products only, and it is also necessary to consider the stability of the co-ordination compounds formed. However, if reactions in which an acid passes a proton to a base are treated, even only formally, as reactions of displacement of a weaker partner by a stronger one

$$H_3N: + H:Cl \rightleftharpoons H_3N:H^+ + Cl^-$$

it will be seen that they correspond to the Lewis theory without additional assumptions.

In some cases, the formation of a coordinate bond, which according to the Lewis theory results from neutralization, may be classically regarded as a redox reaction. An example is the reaction between sulphur and the sulphite ion. It is usually considered that the elemental sulphur atom is reduced and the sulphite-ion sulphur, conventionally regarded as quadrivalent, is oxidized to sexivalent sulphur in the thiosulphate ion, in a pair of two-electron half reactions. According to the Lewis theory, many similar reactions should also be regarded as acid–base reactions, and this is difficult to accept.

The Lewis theory is often condemned for its lack of a uniform quantitative scale of acid and base strengths. This results from the great diversity of the processes classified as acid–base reactions, and also from the conscious or unconscious disregard of situations in which, alongside the basic process of acid and base interaction, other reactions take place. These may arise, for example, from the interaction of one or both principal reagents with the other components of the systems, such as the solvent in which the reaction takes place. This fact shows again why it was not particularly difficult to introduce a uniform scale in the Brønsted–Lowry theory, because it considers systems for which the point of reference is the solvent and its protolytic properties.

Divergent evaluations of acid and base strength arise because of specific properties of the systems compared. If we compare, for example, two acid–base systems which have as common basic component the fluoride ion and as acids, beryllium (II) and copper (II) ions, the greater stability of the beryllium–fluoride complexes would suggest that the Be^{2+} ion is a stronger acid than the Cu^{2+} ion. On the other hand, if we compare the strengths of the same acids with an NH_3 molecule as the Lewis base, we arrive at just the opposite conclusion, because the copper–ammonia complexes are much stronger than those of beryllium.

Thus, in order to establish relative strengths of Lewis acids and bases, it is necessary to adopt appropriate reference systems, and the range of their application must be limited to a particular type of process.

An important group of typical donor–acceptor reactions is the interaction of substances with the solvent. Many solvents have donor properties, so that their molecules react with acceptor solutes. Examples of such interactions are the processes of solvation and ionic dissociation. These processes are of great importance, for example in dissolution processes and also in redox reactions. The more strongly the molecules of a solvent act upon the substances dissolved in it, the more limited is the possibility of occurrence of other reactions.

For estimation of the donor character (i.e. the basicity) of a solvent, Gutmann (1966) proposed a scale of *donor numbers* DN. Numerically, the value of the donor number of a given solvent X is determined by the negative value of the enthalpy (in kJ/mole) of formation of the adduct between the molecule X and antimony pentachloride, which is adopted as the standard acid

$$\text{X:} + \underset{\underset{Cl}{\diagup}\overset{\overset{Cl}{\diagdown}}{Sb}{\diagup}Cl \rightleftharpoons X - \underset{\underset{Cl}{\diagup}\overset{\overset{Cl}{\diagdown}}{Sb} - Cl$$

This reaction was chosen because it usually proceeds in 1:1 stoichiometric ratio; it is done in a dilute solution of 1,2-dichloroethane, which has negligible donor properties. Under such conditions it can be assumed that entropy changes can be disregarded and the interaction (bond formation) between $SbCl_5$ and X is predominantly determined by the value of the donor number DN.

Donor numbers have been determined for many solvents and some values are given in Table 6.2.

It has been found, moreover, that if we choose a different molecular or cationic acid as a standard, for example I_2 or phenol, then the sequence

Table 6.2

Empirical parameters of a solvent as an acceptor (E_T and AN) and as a donor (DN)

Solvent	E_T	AN	DN
Water, H_2O	63.1	54.8	18
Methanol, CH_3OH	55.5	41.3	19
Acetonitrile, CH_3CN	46.0	18.9	14.1
N,N-dimethylformamide, $HCON(CH_3)_2$	43.3	16.0	26.6
Acetone, $(CH_3)_2CO$	42.3	12.5	17.0
Hexamethylphosphoric amide, $C_6H_{18}N_3PO$	40.9	10.6	38.8
Pyridine, C_5H_5N	40.2	14.2	33.1
Chloroform, $CHCl_3$	39.1	23.1	—
Tetrahydrofuran, C_4H_8O	37.4	8.0	20.0
Benzene, C_6H_6	34.5	8.2	0.1

of the basicity of bases (represented by DN) is always the same. This permits, among other things, the prediction of the properties of many systems. It has also been discovered that some spectroscopic and electrochemical properties depend unequivocally on donor numbers; this enables us to use such measurements for empirical determination of donor numbers.

In addition to donor properties, many solvents show acceptor properties. To describe them, Meyer and Gutmann (1975) suggested an *acceptor number*, AN. Its determination is based on another standard reaction, the interaction of substance Y with triethylphosphine oxide as a standard base

$$\begin{array}{l} C_2H_5 \\ C_2H_5{-}P{=}O{:} \ + \ Y \ \rightleftharpoons \ C_2H_5{-}P{=}O{:}Y \\ C_2H_5 \end{array}$$

In this reaction, however, the thermodynamic quantity is not measured directly; instead the chemical shift of the ^{31}P nuclear magnetic resonance is determined. This is taken as 100 for the interaction of triethylphosphine oxide with antimony pentachloride as reference. Values of the acceptor number have been determined for many solvents (Table 6.2).

Several other methods of calculating the acceptor properties of solvents as Lewis acids have been suggested. These make use of the values of equilibrium constants, or the variability of the rate constants of reactions that take place in the solvents tested; they can also be connected with changes in the electronic absorption spectra of standard substances (these changes are defined as solvatochromic effects).

For the spectrum of a substance to be suitable for such an evaluation of acceptor properties there should (if possible) be no parallel interaction of the donor (basic) properties of the solvent with the substance. A suitable substance is *N*-phenolpyridine betaine (more precisely 2,4,6-triphenyl-*N*-(2,6-diphenyl-4-phenoxy)pyridine betaine), for which the excitation energy depends in a great measure on whether the phenol group with the free electron pair is solvated by a solvent with acceptor properties

The stronger the solvent acid (i.e. the stronger the solvation of the basic part of the betaine molecule) the greater is the energy ($h\nu$) needed for the charge transfer, i.e. for excitation. The numerical value of the Dimroth and Reichardt parameter E_T (1963) is calculated from the expression

$$E_T = h\nu cN = 2.86 \times 10^{-3}\nu$$

where h, c and N denote the Planck constant, the velocity of light and Avogadro's number respectively, and ν corresponds to the absorption maximum (in cm^{-1}). The structure of the betaine is such that its interaction with the solvent as a base can be disregarded. The data for several solvents are given in Table 6.2.

In reality, we hardly ever come across a solvent that reacts only as a donor or only as an acceptor. In other words, solvents are usually amphoteric in the sense of the Lewis theory. Thus in order to be able to describe both functions of a solvent and use them for the explanation of its behaviour in reactions we must make use of both the acceptor and the donor parameters. This was done by Krygowski and Fawcett (1975), who described the change of a quantity, Q, characteristic for a given process, in relation to the same quantity in a standard state, Q_0, by means of the expression

$$Q - Q_0 = \alpha E_T + \beta DN$$

where E_T denotes the Dimroth and Reichardt parameter, DN denotes the Gutmann donor number, and the coefficients α and β define the sensitivity of the physicochemical quantity Q to acidity and basicity changes of the solvent.

Perusal of the numerical data listed in Table 6.2, shows that the values which characterize the acidic properties of solvents (E_T and AN) determined on the basis of two different standard processes, have some discrepancies. These are, on the one hand, evidence for certain specific interactions not accounted for in the model process, and on the other hand a striking illustration of the difficulties involved in attempts to find a common state of reference for different systems, in order to achieve a uniform classification.

All these drawbacks of the Lewis theory, however, cannot obscure the fact that it is the most general theory and includes, as particular cases, many other theories of acids and bases (Fig. 6.1).

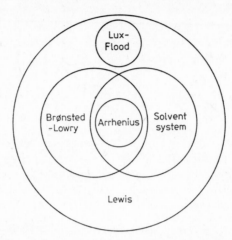

Fig. 6.1. Interdependence of various acid and base theories (from W. B. Jensen, *Chem. Rev.*, 1978, **78**, 1, by permission of the copyright holders).

6.3 PEARSON'S PRINCIPLE OF HARD AND SOFT ACIDS AND BASES

An attempt to introduce quantitative criteria into the Lewis theory and to connect the specific properties of acids and bases with their reactivity was made in 1963 by Pearson, who formulated the principle of hard and soft acids and bases (HSAB). The foundation of this concept is the classification of acids and bases as hard or soft, depending on observed preferences for bonding. Thus, the principle states that "hard acids prefer to combine with hard bases and soft acids prefer to combine with soft bases".

Like the Lewis theory, Pearson's principle concerns reactions which in more detailed classical formulations are defined as acid–base reactions and also as reactions of complexation of ions by different ligands, including the formation of solvates. However, Pearson's concept also concerns the behaviour of σ and π bonds in organic compounds, and situations in which a system of two atoms may conventionally be treated as a combination of an acid centre and a base centre. Thus, a molecule of an alcohol may be treated as a combination of an acid R^+ and a base OH^-, but also as a combination of a base RO^- and an acid H^+. Another characteristic case is the treatment of a C–H bond in hydrocarbons as a combination of an acid centre R^+ with a base centre H^-, or a base centre R^- with an acid centre H^+. An experimental proof of the validity of this approach is provided by the reactions of hydrocarbons (as bases) with metals and, on the other hand, their reactions (as acids) with halogens or nitric acid. Thus the HSAB concept allows us to explain the mechanisms of many important reactions in the chemistry of organic or organometallic compounds by assuming that the compounds considered are compounds of acids and bases.

The differentiation between hard and soft acids and bases depends on the polarizability of the species, i.e. their ability of their electron shells to be deformed under the influence of the electric field of the reaction partner. By a *hard acid* Pearson means a Lewis acid (acceptor) with valence electrons that are weakly polarizable and thus difficult to detach from small particles with a high positive charge. Examples of such acids are cations with a small ionic radius and a large charge, and molecules with substituents that induce on the central atom a high positive charge. *Soft acids*, on the other hand, have a small charge and a large radius, with filled outer orbitals or with an electron deficiency. Such acids are characterized by high polarizability.

The group of *hard bases* includes donors with high electronegativity, a large negative charge, and in which the donor atoms have a small radius. Thus they have strongly bonded electrons and their polarizability is small. *Soft bases* have high polarizability, and their donor atoms are characterized by lower electronegativity, for example sulphur or phosphorus. The division into hard and soft acids and bases is of course not sharp, and among both the acids and bases there exist intermediate groups, which can behave in various ways, depending on the medium and conditions (Table 6.3). Consequently, it is essential to compare the properties of the acids and bases under very definite conditions, for example in the gas phase, in which both the kinetic parameters (reaction rate constants) and the thermodynamic variables (equilibrium constants) provide data about the hardness or softness of the reagents. It is then possible to arrange the bases or acids in order of increasing hardness. The sequence may change if the reaction medium is changed and if the process takes place in a solvent in which solvation processes occur to a different degree.

The knowledge of the degree of hardness or softness of reagents is important from both the theoretical and practical points of view. Usually, the product of reaction between a hard base and a hard acid has ionic or polar bonding, whereas a soft acid and a soft base produce a covalently bound product. In addition to σ-bonding, π-bonding can also occur. Hard acids are usually π-bond acceptors, because they have empty outer orbitals which can accept π-electrons from a base. Only hard bases can donate electrons, via a π-bond, from filled outer orbitals, to a hard acid.

Table 6.3

Hard and soft acids and bases

	Acids	Bases
Hard	H^+, Na^+, K^+, Be^{2+}, Mg^{2+}, Ca^{2+}, Mn^{2+}, Al^{3+}, Cr^{3+}, Co^{3+}, Fe^{3+}, As^{3+}, Si^{4+}, Ti^{4+}, BF_3, $AlCl_3$, SO_3, $RC=O^{(+)}$, CO_2, I^{7+} I^{5+}, Cl^{7+}, Cr^{6+}	H_2O, OH^-, F^-, O^{2-}, CH_3COO^-, PO_4^{3-}, SO_4^{2-}, Cl^-, CO_3^{2-}, NO_3^-, ROH, RO^-, R_2O, NH_3, RNH_2
Intermediate	Fe^{2+}, Co^{2+}, Ni^{2+}, Cu^{2+}, Zn^{2+}, Pb^{2+}, Sn^{2+}, Sb^{3+}, $B(CH_3)_3$, SO_2, NO^+, R_3C^+	$C_6H_5NH_2$, C_5H_5N, N_3^- Br^-, SO_3^{2-}, N_2
Soft	Cu^+, Ag^+, Au^+, Mg^{2+}, Pd^{2+} Cd^{2+}, Pt^{2+}, BH_3, $GaCl_3$, I^+, Br^+, RO^+, I_2, Br_2, electron-deficient systems $[(CN)_2C=C(CN)_2$, polynitroaromatic compounds]	R_2S, RSH, RS^-, I^- SCN^-, $S_2O_3^{2-}$, R_3P, $(RO)_3P$, CN^-, CO, C_2H_4, C_6H_6, H^-, R^-, alkenes, carbanions

A standard reaction by means of which we can estimate the hardness of a base is a reaction in which a hard acid H^+ and a soft acid CH_3Hg^+ react with a base B

$$CH_3Hg^+ + BH^+ \rightleftharpoons CH_3HgB^+ + H^+$$

The harder the base B, the more easily it reacts with the hard acid H^+, and thus the greater is the shift of the reaction equilibrium to the left and the smaller is the value of the equilibrium constant

$$K = \frac{[CH_3HgB^+][H^+]}{[CH_3Hg^+][BH^+]}$$

If the base is soft, then it shows a more distinct tendency to react with the soft acid CH_3Hg^+ than with the hard acid H^+, so the equilibrium constant is greater than 1. Thus, from the values of the equilibrium constants we can place the bases in order of increasing hardness. An example of such a sequence of bases is

$$I^- < Br^- < Cl^- < S^{2-} < RS^- < CN^- < H_2O < NH_3 <$$
$$< F^- < OH^-$$

We can make a similar series for acids with respect to their relative reactivity with the very hard base OH^- and a soft base such as the chloride ion. A typical sequence in such a series is

$$Ag^+ < CH_3Hg^+ < Cd^{2+} < Zn^{2+} < Fe^{2+} < Cu^{2+} < In^{3+}$$
$$< Fe^{3+} < H^+$$

If these series are compared with the relative basicity measured by proton affinity, as is done in the proton theory of acids and bases, it will be found that good agreement is obtained with regard to hard acids and bases. It should be remembered, however, that this agreement is not always maintained in the study of properties of ions and molecules in solutions, because specific interactions with the solvent may alter the predicted sequence. Another factor that may complicate such a series of basicity or acidity is a reaction which proceeds in a different stoichiometric ratio. The acids H^+ and CH_3Hg^+ were chosen for evaluation of bases because they react with bases in 1:1 ratio.

As already mentioned, the HSAB theory allows prediction of the course of a reaction by consideration of the symmetry of the acid and base properties of the products and substrates of the reaction.

In the reactions of the thiocyanate ion, the potential centre of interaction with acids may be a sulphur atom (soft base) or a nitrogen atom (hard base). Thus if the reaction partner is a soft acid, for example the methyl carbonium ion CH_3^+, then the reaction will be

$$CH_3I + NCS^- \rightleftharpoons CH_3SCN + I^-$$

and the product of the reaction is a compound in which symmetry is maintained owing to the similarity in softness of both the acid and the base centres.

If, however, the reaction partner is an acyl cation, which is a hard acid, we shall have the reaction

$$CH_3COCl + NCS^- \rightleftharpoons CH_3CONCS + Cl^-$$

which will also result in a product with a higher acid–base symmetry than the substrate of the reaction.

Reactions with metal ions can be treated similarly; according to their hardness, they can form complexes in which the metal is bound through a nitrogen atom (e.g. Fe^{3+}–NCS^-) or through a sulphur atom (e.g. Hg^{2+}–SCN^-, Au^+–SCN^-). In the first case, that of hard reagents, the bonding has ionic character. In the second case, it is more covalent. Again it should be remembered that the absolute softness or hardness can refer only to reaction in the gas phase or possibly in non-solvating solvents, and moreover specific electronic or steric factors can cause some disturbances.

It should be mentioned that before Pearson developed this approach, there were numerous, often semi-empirical, rules concerning, for example, complexation reactions, which can now be successfully classified by the HSAB theory. For example, there was a well-known rule according to which metal cations capable of complexation can be divided into two groups. This division depends on which of the donor atoms of the ligands in the pairs N, P; O, S; F, Cl forms the more stable complexes with them. The first of each pair is the atom occurring in ligands which are hard bases. Thus, hard bases (ligands with donor atoms N, O, F) react with cations with typical properties of hard acids, and soft bases (ligands with donor atoms P, S, Cl) react with metals which form ions which are soft acids. Pearson's theory explains why the fluoride complexes of beryllium (hard base and acid) are more stable than the fluoride complexes of copper (hard base and soft acid), whereas the opposite order of stability is found in the case of the ammonia complexes (cf. Section 6.2).

6.4 USANOVICH'S THEORY

The idea of creating a theory capable of explaining many diverse processes led Usanovich (1939) to develop a new theory of acids and bases. The theory is based on the assumption that two substances with opposite properties can react with each other—neutralize each other—to produce a neutralization product called a *salt*. These substances belong to two opposite groups, one of which comprises acids and the other bases. There are differing criteria of assigning substances to a given group; they depend on the character of the reaction and the reactants, namely:

Acids	*Bases*
dissociate cations	associate with cations
associate with anions	dissociate anions
associate with electrons	dissociate electrons

To illustrate these properties, many chemical reactions can be quoted. Usanovich presents them according to a scheme similar to that of Brønsted and Lowry, in which from acid 1 and base 2, we obtain base 1 and acid 2. Thus, if the cation which dissociates of an acid is a proton, and that proton is attached to a base, this reaction, constituting the general case in the Brønsted–Lowry theory, becomes a particular case in Usanovich's theory. A number of different examples are given below:

$$Acid\ 1 + Base\ 2 \rightleftarrows Base\ 1 + Acid\ 2$$

proton transfer	$HCl + H_2O \rightarrow Cl^- + H_3O^+$
cation CH_3^+ transfer	$CH_3I + NH(CH_3)_2 \rightarrow I^- + NH(CH_3)_3^+$

anion O^{2-} transfer $SO_3 + Na_2O \rightleftharpoons SO_4^{2-} + 2Na^+$

anion CN^- transfer $Fe(CN)_3 + 3KCN \rightleftharpoons Fe(CN)_6^{3-} + 3K^+$

electron transfer $Cl_2 + 2Na \rightleftharpoons 2Cl^- + 2Na^+$

Thus Usanovich's theory includes not only acid–base reactions in the Brønsted sense but also many other types of reactions, including complexation and redox reactions.

A certain difficulty in Usanovich's theory seems to arise from the variety of criteria for evaluation of acid and base properties. However, all the criteria of acidity which have been mentioned give evidence of the presence of electropositive species, and the basicity criteria result from an excess of electronegative species. Hence all reactions in an acid–base system are accompanied by the transfer of the positive charge from acid to base or the transfer of the negative charge from base to acid, irrespective of the kind of species (electrons or ions) which are the carriers of the charges.

Usanovich's theory is much like the Lewis theory; it has virtually the same disadvantages and advantages, especially the versatility of the theoretical approach and the simultaneous limitation of practical applications, because of its too general character.

6.5 SOME REACTIONS OCCURRING IN MOLTEN SALTS

For aqueous solutions and other protic solvents (e.g. NH_3, CH_3COOH) the Brønsted–Lowry theory explains many ion-linking reactions which proceed with solvent participation. However, we can also have reactions in molten salts at an elevated temperature. It is impossible to apply the Brønsted–Lowry theory to such reactions, because of the crucial role attached in that theory to the transfer of the proton and its participation in the reactions. However, most reactions in molten salts are very well explained by the Lewis theory, and thus they are reactions which can be classified as acid–base reactions.

An essential difference between reactions occurring in water and in molten sodium chloride is that water is a weakly dissociable solvent (the ion product has a small value) but sodium chloride has an ionic structure, i.e. in a molten state it is fully dissociated and the cation concentration corresponds to the anion concentration and is constant

$$[Na^+] = [Cl^-] = \text{const}$$

Consequently, any reaction of dissolved substances which are donors or acceptors of one of the ions of the solvent is complete and the equilibrium state reached cannot be displaced by introducing into the molten salt other acceptors or donors.

However, in some molten-salt solvents one of the ions can dissociate with the formation of ions or molecules which do not appear directly when the relevant salt is dissolved in water. Thus in molten sodium carbonate, the anion dissociates according to the reaction

$$CO_3^{2-} \rightleftharpoons CO_2 + O^{2-}$$

for which the equilibrium is described by the "ion product"

$$K_i = [CO_2][O^{2-}]$$

Analogous reactions can occur in other molten salts, e.g.

in nitrates $\quad NO_3^- \rightleftharpoons NO_2^+ + O^{2-}, \qquad K_i = [NO_2^+][O^{2-}]$

in sulphates $\quad SO_4^{2-} \rightleftharpoons SO_3 + O^{2-}, \qquad K_i = [SO_3][O^{2-}]$

$\qquad\qquad\quad 2SO_4^{2-} \rightleftharpoons S_2O_7^{2-} + O^{2-}, \qquad K_i = [S_2O_7^{2-}][O^{2-}]$

in borates $\quad 2BO_2^- \rightleftharpoons B_2O_3 + O^{2-}, \qquad K_i = [B_2O_3][O^{2-}]$

in meta-
phosphates $\quad 2PO_3^- \rightleftharpoons P_2O_5 + O^{2-}, \qquad K_i = [P_2O_5][O^{2-}]$

Such reactions are not limited to molten oxygen-containing salts, because, for example, in molten cryolite, Na_3AlF_6, a similar type of reaction takes place

$$AlF_6^{3-} \rightleftharpoons AlF_4^- + 2F^-, \quad K_i = [AlF_4^-][F^-]^2$$

Under the given conditions the ions O^{2-} or F^- act as an indicator of the basicity of the system. Attachment of the O^{2-} ion is characteristic for acid entities and its detachment for basic entities. Thus in molten carbonates, CO_3^{2-} can be treated as a base and CO_2 as an acid. In nitrates NO_3^- is the base and NO_2^+ (or N_2O_5) the acid and in sulphates SO_4^{2-} is the base and SO_3 the acid. To make a clear distinction between protic acids and the substances mentioned above, the latter, as suggested by Bjerrum, are sometimes called *antibases*.

Thus, if we introduce into a molten-salt system a compound with donor or acceptor properties, the reactions which can take place will be of the types: acid–base or antibase–base.

The description of such reactions as acid–base systems was introduced by Lux (1939) and developed by Flood and his collaborators (1947) and thus it is sometimes called the *Lux–Flood theory*. The structure of this theory is similar to the Brønsted–Lowry theory; it should be carefully noted, however, that while in the Brønsted–Lowry theory the acid was the donor of a particle which was the carrier of the principal properties of a system, i.e. the proton, in the Lux–Flood oxotropic theory the donor of the particle characteristic for the system—the O^{2-} ion—is a base. This is an obvious consequence of the fact that in the first case the proton and in the second case the oxide ion with a free electron pair are the operative species in the reactions. In this respect both prototropy and oxotropy are particular cases of the Lewis theory.

For the oxotropic reaction it is possible to form a scale of acidity (or basicity) similar to the pH scale in protic solvents. It is based on the value of $-\log(O^{2-}) = pO^{2-}$ (or of pF^- in cryolite). Thus, if we introduce into a molten alkali metal carbonate an acceptor stronger than CO_2, e.g. silicon dioxide SiO_2 or aluminium oxide Al_2O_3, it will behave as a strong acid, i.e. it will react quantitatively according to the equations

$$CO_3^{2-} + SiO_2 \rightarrow CO_2 + SiO_3^{2-}$$
$$CO_3^{2-} + Al_2O_3 \rightarrow CO_2 + 2AlO_2^-$$

If the acceptor properties of the substance introduced are weaker than those of the CO_3^{2-} ion, the reaction will take place only to a small extent, analogously to aqueous solutions containing a weak acid. Examples of such weaker acceptors are the ions BO_2^-, SiO_3^{2-}—hence the reactions

$$CO_3^{2-} + BO_2^- \rightleftharpoons CO_2 + BO_3^{3-}$$
$$CO_3^{2-} + SiO_3^{2-} \rightleftharpoons CO_2 + SiO_4^{4-}$$

proceed reversibly and the equilibrium state can be shifted according to the pO^{2-} value of the molten-salt system. Particular attention should be paid to silicon dioxide SiO_2 because of the two-stage reaction; it behaves similarly to, say, sulphuric acid which, as a strong acid in aqueous solution reacts completely with the solvent, producing HSO_4^-, which in turn undergoes only a partial reaction resulting in the formation of the SO_4^{2-} ion.

In molten oxyanion salts, many metal oxides behave as O^{2-} donors, and therefore are readily soluble in such media. Such reactions have been utilized in industry and analytical practice for a very long time, though their explanation, based on the Lewis theory, is comparatively recent.

Thus by use of molten oxides, hydroxides and alkali metal carbonate, it is easy to dissolve aluminium, boron and silicon oxides, which under these conditions have an acid character and produce soluble aluminates, borates and silicates. These processes, apart from their application to the dissolution of substances that are only sparingly soluble in aqueous solutions, are used, for example, in the formation of glass and cement and also in metallurgy during the production of slag. Molten sulphates or pyro-sulphates are used for dissolving ores containing sparingly soluble metal oxides, such as Fe_2O_3, which are stronger donors of oxygen than the $S_2O_7^{2-}$ ion.

In analytical chemistry we are familiar with the identification of sparingly soluble oxides by means of the so-called borax bead or microcosmic salt bead tests. These reactions are also acid–base reactions proceeding in molten borax or sodium diammonium phosphate. When borax ($Na_2B_4O_7 \cdot$ $\cdot 10H_2O$) is heated, it is dehydrated and then melts. In molten borax the following reaction takes place:

$$B_4O_7^{2-} \rightleftharpoons 2BO_2^- + B_2O_3$$

Molten metaborates and boric oxide are acceptors of the O^{2-} ion

$$B_2O_3 + O^{2-} \rightleftharpoons 2BO_2^-$$
$$BO_2^- + O^{2-} \rightleftharpoons BO_3^{3-}$$

which causes metal oxides to appear as bases in them (donors of O^{2-}) and hence readily dissolve, often with production of characteristic colours. The BO_2^- ion may show, besides acceptor (acid) properties, donor (basic) properties with respect to stronger acceptors. These can be observed, for example, in the presence of chromium trioxide

$$CrO_3 + 2BO_2^- \rightleftharpoons CrO_4^{2-} + B_2O_3$$

Similar reactions to those occurring in the borax bead take place in the microcosmic salt bead. When heated, sodium diammonium phosphate decomposes

$$Na(NH_4)_2PO_4 \rightarrow NaPO_3 + 2NH_3 \uparrow + H_2O \uparrow$$

Molten metaphosphate has strong acceptor properties

$$PO_3^- + O^{2-} \rightleftharpoons PO_4^{3-}$$

and very weak donor properties

$$2PO_3^- \rightleftharpoons P_2O_5 + O^{2-}$$

Hence the "ion product"

$$K_i = [P_2O_5][PO_4^{3-}]$$

characterizes the behaviour of metaphosphate as a rather acidic solvent, in which oxides that which are donors will dissolve.

Numerous oxides, such as CrO_3, V_2O_5, Fe_2O_3, may undergo acid–base reactions in this solvent, and the colour developed permits their use as indicators in exchange reactions of the O^{2-} ion.

The reactions which we have mentioned as examples are of great practical importance; it is possible to explain them on the basis of the theory of acid–base reactions and hence also to predict their course. However, the reaction schemes we have given are greatly simplified, and further investigations are needed to explain them in a more comprehensive manner.

Finally it is worth mentioning that many exchange reactions of the O^{2-} ion can be done in molten-salt solvents which do not contain oxygen ions. Such reactions, for example in molten alkali metal halides (KCl, NaCl), occur between the dissolved substances, without participation of the solvent, similarly to the proton exchange reactions which can take place in aprotic solvents. Thus, in molten metal halides the reaction of binding of the O^{2-} is possible.

For example, the cations of alkali metals react to a small extent with the O^{2-} ion, whereas the aluminium cation (in molten $AlCl_3$), showing strong acceptor tendencies, causes the decomposition of many compounds with donor properties

$$CO_3^{2-} + Al^{3+} \rightleftharpoons CO_2 + AlO^+$$
$$ZnO + Al^{3+} \rightleftharpoons Zn^{2+} + AlO^+$$
$$UO_3 + Al^{3+} \rightleftharpoons UO_2^{2+} + AlO^+$$

Appendix

Acid dissociation constants of inorganic compounds in aqueous solution at 25°C

Acid	Formula	pK_a $I = 0$	$I = 0.1$
1	2	3	4
Ammonium ion	NH_4^+	9.24	9.37
Antimonic	$HSb(OH)_6$	4.4	4.3
Arsenic	H_3AsO_4	2.2	2.1
	$H_2AsO_4^-$	7.0	6.7
	$HAsO_4^{2-}$	11.4	11.2
Arsenious	H_3AsO_3	9.2	9.1
	$H_2AsO_3^-$		12.1
	$HAsO_3^{2-}$		13.4
Boric	H_3BO_3	9.2	9.1
Carbonic*	$CO_2 + H_2CO_3$	6.4	6.3
	H_2CO_3	3.8	
	HCO_3^-	10.3	10.1
Chromic	H_2CrO_4	0.8	0.7
	$HCrO_4^-$	6.5	6.2
	$2\,HCrO_4^- \rightleftharpoons Cr_2O_7^{2-}$		
	$\quad + H_2O; \quad \log K = 1.6$		1.5
Cyanic	$HOCN$	3.7	3.6
Germanic	H_4GeO_4	9.4	9.1
	$H_3GeO_4^-$		12.7
Hydrogen hexacyanoferrate (II)	$H_4Fe(CN)_6$	< 1	
	$H_3Fe(CN)_6^-$	< 1	
	$H_2Fe(CN)_6^{2-}$	2.2	
	$HFe(CN)_6^{3-}$	4.2	
Hydrogen hexacyanoferrate (III)	$H_3Fe(CN)_6$	< 1	
Hydrazinium ion	$N_2H_6^{2+}$	−0.9	−0.6
	$N_2H_5^+$	8.0	8.1
Hydrazoic	HN_3	4.7	4.6
Hydrocyanic	HCN	9.36	9.2

continued

1	2	3	4
Hydrofluoric	HF	3.2	3.1
	$HF + F^- \rightleftharpoons HF_2^-$; $\log K = 0.6$		0.6
Hydrogen peroxide	H_2O_2	11.8	11.6
Hydrogen selenide	H_2Se	3.9	3.8
	HSe^-	11.0	10.7
Hydrogen sulphide	H_2S	7.2	7.0
	HS^-	13.2	12.9
Hydrogen telluride	H_2Te	2.6	2.5
	HTe^-	11.0	10.7
Hydroxylammonium ion	NH_3OH^+	6.1	6.2
Hypochlorous	HClO	7.5	7.4
Hypoiodous	I^+	4.5	
	HIO	10.4	
Iodic	HIO_3	0.8	0.7
Molybdic	H_2MoO_4		1.8
	$HMoO_4^-$		4.1
Nitramide	H_2NNO_2	6.6	6.5
Nitrous	HNO_2	3.3	3.2
Orthophosphoric	H_3PO_4	2.12	2.0
	$H_2PO_4^-$	7.20	6.9
	HPO_4^{2-}	11.90	11.7
Orthophosphorous	H_3PO_3	2.2	2.0
	$H_2PO_3^-$	6.7	6.4
Periodic	H_5IO_6	1.6	1.5
	$H_4IO_6^-$	8.4	8.2
	$H_3IO_6^{2-}$	15.0	14.5
Pyrophosphoric	$H_4P_2O_7$		1.0
	$H_3P_2O_7^-$		2.5
	$H_2P_2O_7^{2-}$		6.1
	$HP_2O_7^{3-}$		8.5
Selenic	H_2SeO_4	—	—
	$HSeO_4^-$	2.1	2.0
Selenious	H_2SeO_3	2.6	2.5
	$HSeO_3^-$	8.3	8.1
Silicic	H_4SiO_4		9.6
	$H_3SiO_4^-$		12.7
Sulphuric	H_2SO_4	—	—
	HSO_4^-	1.90	1.80
Sulphurous	H_2SO_3	1.76	1.7
	HSO_3^-	7.20	6.9
Telluric	H_6TeO_6	7.6	7.5
	$H_5TeO_6^-$	10.4	10.2
	$H_4TeO_6^{2-}$	15.0	14.5

continued

1	2	3	4
Thiocyanic	HSCN	0.9	0.8
Thiosulphuric	$H_2S_2O_3$	0.6	0.5
	$HS_2O_3^-$	1.72	1.4

* See Section 5.2.

Acid dissociation constants of organic compounds in aqueous solutions at 25°C

Compound	Symbol for the compound	Symbol for the acid form	pK_a	
			$I = 0$	$I = 0.1$
1	2	3	4	5
Acetic acid CH_3COOH	HA	HA	4.88	4.76
Acetylacetone $CH_3COCH_2COCH_3$	HA	HA	9.0	8.9
Aminoacetic acid CH_2NH_2COOH	HA	H_2A^+	2.35	2.5
		HA	9.77	9.7
Aniline $C_6H_5NH_2$	A	HA^+	4.6	4.7
Ascorbic acid $C_6H_8O_6$	H_2A	H_2A	4.17	4.1
		HA^-	11.56	11.3
Benzoic acid C_6H_5COOH	HA	HA	4.20	4.12
2,2′-Bipyridyl $C_{10}H_8N_2$	A	HA^+		4.4
Chloroacetic acid $CH_2ClCOOH$	HA	HA	2.9	2.7
Citric acid $C_3H_4OH(COOH)_3$	H_4A	H_4A	3.1	3.0
		H_3A^-	4.8	4.4
		H_2A^{2-}	6.4	6.1
		HA^{3-}		(16.0)
1,2-Diaminopropane $NH_2CH_2CHNH_2CH_3$	A	H_2A^{2+}		6.9
		HA^+		10.0
1,3-Diaminopropane $NH_2(CH_2)_3NH_2$	A	H_2A^{2+}		9.0
		HA^+		10.7
Dichloroacetic acid $CHCl_2COOH$	HA	HA	1.3	1.1
Dimethylamine $(CH_3)_2NH$	A	HA^+	11.0	11.1
Ethanolamine $HOCH_2CH_2NH_2$	A	HA^+	9.4	9.6
Ethylamine $C_2N_5NH_2$	A	HA^+	10.7	10.8
Ethylenediamine $NH_2CH_2CH_2NH_2$	A	H_2A^{2+}	7.5	7.3
		HA^+	10.6	10.1
Ethylenediaminetetra-acetic acid (EDTA)	H_4A	H_4A		2.1
$(HOOCCH_2)_2N(CH_2)_2N(CH_2COOH)_2$		H_3A^-		2.8
		H_2A^{2-}		6.2
		HA^{3-}		10.3
Formic acid $HCOOH$	HA	HA	3.75	3.65
Fumaric acid (trans) $C_2H_2(COOH)_2$	H_2A	H_2A	3.0	2.9
		HA^-	4.4	4.1
Hexamethylenetetramine $C_6H_{12}N_4$	A	HA^+	5.1	5.3
8-Hydroxyquinoline C_9H_6NOH	HA	H_2A^+	5.0	5.1
		HA	9.9	9.7
8-Hydroxyquinoline-5-sulphonic acid				
$C_9H_5NOHSO_3H$	H_2A	H_3A^+		1.3
		H_2A		3.8
		HA^-		8.4
Iminodiacetic acid $NH(CH_2COOH)_2$	H_2A	H_2A		2.7
		HA^-		9.5

continued

1	2	3	4	5
Lactic acid $CH_3CHOHCOOH$	HA	HA	3.9	3.8
Maleic acid (*cis*) $C_2H_2(COOH)_2$	H_2A	H_2A	1.9	1.8
		HA^-	6.2	5.9
Malonic acid $CH_2(COOH)_2$	H_2A	H_2A	2.9	2.7
		HA^-	5.7	7.4
Methylamine CH_3NH_2	A	HA^+	10.6	10.7
Nitrilotriacetic acid $N(CH_2COOH)_3$	H_3A	H_3A		2.0
		H_2A		2.6
		HA^{2-}		9.8
Oxalic acid $(COOH)_2$	H_2A	H_2A	1.3	1.1
		HA^-	4.3	4.0
1,10-Phenanthroline $C_{12}H_8N_2$	A	HA^+		5.0
Phenol C_6H_5OH	HA	HA	10.0	9.8
o-Phthalic acid $o\text{-}C_6H_4(COOH)_2$	H_2A	H_2A	2.9	2.8
		HA^-	5.4	5.1
Picolinic acid $C_6H_4N(COOH)$	HA	H_2A^+	1.4	1.5
		HA	5.5	5.4
Pyridine C_5H_5N	A	HA^+	5.3	5.3
Salicylic acid $C_6H_4OHCOOH$	H_2A	H_2A	3.0	2.9
		HA^-	12.8	13.1
Sulphosalicylic acid $C_6H_3OHSO_3HCOOH$	H_3A	H_2A^-		2.6
		HA^{2-}		11.6
Tartaric acid $COOH(CHOH)_2COOH$	H_2A	H_2A	3.0	2.9
		HA^-	4.3	4.1
Trichloroacetic acid CCl_3COOH	HA	HA	0.7	0.5
Trimethylamine $(CH_3)_3N$	A	HA^+	9.8	9.9

Logarithms of the equilibrium constants of protolytic reactions of certain metal ions in aqueous solutions at 25°C at $I = 0.1$. Column $*\beta_{qp}$ gives the data for polynuclear complexes with the formula $M_p(OH)_q$, the corresponding stoichiometric coefficients are given in brackets (q, p) preceding the value of $\log *\beta$

Metal ion	$*\beta_1$	$*\beta_2$	$*\beta_3$	$*\beta_4$	$*\beta_5$	$*K_{s0}$	$*\beta_{qp}$
Ag^+	−11.9	−23.8				6.3	
Al^{3+}	−5.4	−10.0	−15.7	−23.6		9.2	(2, 2) −7.7, (4, 3) −13.7, (32, 13) −102.8
Be^{2+}	−5.7	−13.9	−23.8	−37.8		6.9	(1, 2) −3.2, (3, 3) −8.9, (8, 6) −27.5
Bi^{3+}	−1.4	−4.6	−9.4	−22.2		4.0	(12, 6) 0.3
Ca^{2+}	−12.6					23.0	
Cd^{2+}	−10.3	−20.6	−33.8	−46.9		13.7	(1, 2) −9.2, (4, 4) −32.4
Co^{2+}	−9.9	−18.8	−31.5			12.8	(1, 2) −11.0, (4, 4) −30.1
Cu^{2+}	−8.2	−17.5	−27.8	−39.1		8.9	(2, 2) −10.6
Fe^{2+}	−9.7	−20.8	−31	−46		13.1	
Fe^{3+}	−2.6	−6.2	−10	−21.9		2.5	(2, 2) −2.9, (4, 3) −6.1
Hg^{2+}	−3.8	−6.2	−21.1			2.5	(1, 2) −2.7, (3, 4) −6.4
La^{3+}	−8.5	−17.2	−25.9	−36.9		19.7	
Mg^{2+}	−12	−27				18	(4, 4) −39.8
Mn^{2+}	−10.8	−22.4	−34.8	−47.9		15.4	(3, 3) −24.4
Ni^{2+}	−10.2	−19.2	−30	−44		13.3	(1, 2) −10.7, (4, 4) −27.3
Pb^{2+}	−7.9	−17.3	−28.0			13.0	(1, 2) −6.2, (4, 3) −24.0, (4, 4) −20,3 (8, 6) −43,3
Sc^{3+}	−4.9	−10.7	−17.3	−26.6		10.5	(2, 2) −5.8, (5, 3) −17.2
Sn^{2+}	−3.6	−7.3	−16.6			2.0	(2, 2) −5.0, (4, 3) −7.3
Th^{4+}	−4.2	−7.7	−12.4	−17.7		11.4	(2, 2) −4.7, (15, 6) −40.0
Zn^{2+}	−9.2	−17.1	−28.4	−40.7		12.1	(1, 2) −8.9, (6, 2) −57.5
Zr^{4+}	−0.6	−2.1	−6.9	−11.2	−17	−0.4	(4, 3) 5.1, (5, 3) 5.5, (8, 4) 8.0

Solutions of problems

CHAPTER 1

1. *a.* 0.15; *b.* 0.10; *c.* 0.80; *d.* 1.40; *e.* 1.00; *f.* 1.20; *g.* 0.10; *h.* 0.10; *i.* 0.10; *j.* 0.10.

2. *a.* 0.30M; *b.* 0.10M; *c.* 0.075M; *d.* 0.02M; *e.* 0.10M; *f.* because the acid is only slightly dissociated it is not possible to prepare a solution with an ionic strength of 0.30 without addition of strong electrolytes.

3. *a.* 1; *b.* 4; *c.* 9; *d.* 3; *e.* 15.

4. *a.* 0.90; *b.* 0.86; *c.* 0.82; *d.* 0.75; *e.* the value 0.56 obtained from Eq. (1.15) does not correspond to the true activity coefficient because the ionic strength is outside the range of applicability of this equation.

5. *a.* 0.82; *b.* 0.89; *c.* 0.91.

6. *a.* 3.02×10^{-5}; *b.* 5.75×10^{-10}; *c.* 9.33×10^{-2}; *d.* 1.62×10^{-4}; *e.* 1.74×10^{-14}.

7. *a.* $pK = p_aK$; *b.* $pK = p_aK + \dfrac{\sqrt{I}}{1 + \sqrt{I}}$; *c.* $pK = p_aK + \dfrac{2\sqrt{I}}{1 + \sqrt{I}}$; *d.* $pK = p_aK$.

8. *a.* 18.09 kJ/mole; *b.* -53.62 kJ/mole; *c.* 79.88 kJ/mole; *d.* 12.67 kJ/mole; *e.* -12.67 kJ/mole.

9. *a.* 7.8×10^1; *b.* 1.8×10^{-4}; *c.* 2.0×10^4; *d.* 2.0×10^1; *e.* 4.0×10^{-10}; *f.* 1.08×10^2; *g.* 1.6×10^2; *h.* 1.59×10^{-9}.

CHAPTER 2

1. *a.* 1.26×10^{-8}; *b.* 2.5×10^{-3}; *c.* 4.0×10^{-12}; *d.* 2.0×10^{-6}; *e.* 5.0×10^{-10}; *f.* 2.3×10^{-5}; *g.* 1.6×10^{-7}; *h.* 1.6×10^{-10}.

2. *a.* 5.0×10^{-6}; *b.* 2.5×10^{-5}; *c.* 1.6×10^{-10}; *d.* 3.1×10^{-8}; *e.* 2.5×10^{-11}; *f.* 4.0×10^{-5}.

3. *a.* 1.68; *b.* 5.70; *c.* 6.16; *d.* 9.00.

4. *a.* 0.010M; *b.* $7.08 \times 10^{-6}M$; *c.* $1.0 \times 10^{-7}M$; *d.* $1.0 \times 10^{-14}M$.

5. *a.* $2.0 \times 10^{-14}M$; *b.* $1.0 \times 10^{-7}M$; *c.* $2.0 \times 10^{-2}M$; *d.* $5.0 \times 10^{-3}M$; *e.* 3.2M; *f.* 1.26M.

6. *a.* 7.0; *b.* 9.45; *c.* 8.35; *d.* 1.45.

CHAPTER 3

1. *a.* 1.37; *b.* 3.94; *c.* 5.02; *d.* 0.70.

2. *a.* $2.05 \times 10^{-3} M$, pH 2.69; *b.* $1.28 \times 10^{-7} M$, pH 6.89; *c.* $0.185 M$, pH 0.73; *d.* $1.01 \times \times 10^{-7} M$, pH 7.00; *e.* $1.82 \times 10^{-11} M$, pH 10.74; *f.* $1.00 \times 10^{-4} M$, pH 4.00; *g.* $2.00 \times \times 10^{-10} M$, pH 9.70; *h.* $1.00 \times 10^{-11} M$, pH 11.00.

3. *a.* $2.22 \times 10^{-2} M$, pH 12.35; *b.* $1.72 \times 10^{-2} M$, pH 10.76; *c.* $3.98 \times 10^{-9} M$, pH 5.60; *d.* $6.30 \times 10^{-4} M$, pH 10.80; *e.* $9.01 \times 10^{-8} M$, pH 6.95; *f.* $1.50 \times 10^{-3} M$, pH 11.18.

4. *a.* $4.60 \times 10^{-4} M$, pH 3.37; *b.* $2.93 \times 10^{-3} M$, pH 2.53; *c.* $1.19 \times 10^{-3} M$, pH 2.92; *d.* $6.93 \times 10^{-7} M$, pH 6.16; *e.* $1.00 \times 10^{-6} M$, pH 6.00; *f.* $3.08 \times 10^{-6} M$, pH 5.51; *g.* $1.58 \times \times 10^{-4} M$, pH 3.80; *h.* $1.28 \times 10^{-6} M$, pH 5.89; *i.* $5.71 \times 10^{-6} M$, pH 5.24; *j.* $1.80 \times 10^{-6} M$, pH 5.74.

5. *a.* 1.46×10^{-4}; *b.* 1.11×10^{-5}; *c.* 1.03×10^{-6}; *d.* 1.00×10^{-7}.

6. *a.* $3.5 \times 10^{-2} M$, pH 1.46; *b.* $4.3 \times 10^{-2} M$, pH 1.37; *c.* $1.6 \times 10^{-2} M$, pH 1.78; *d.* $9.4 \times 10^{-3} M$, pH 2.02; *e.* $3.3 \times 10^{-2} M$, pH 1.48; *f.* $3.03 \times 10^{-13} M$, pH 12.52; *g.* $2.23 \times 10^{-12} M$, pH 11.65.

7. *a.* $1.12 \times 10^{-3} M$, pH 2.95; *b.* $1.12 \times 10^{-5} M$, pH 4.95; *c.* $1.41 \times 10^{-4} M$, pH 3.85; *d.* $3.16 \times 10^{-6} M$, pH 5.50.

8. *a.* 3.39; *b.* 11.0; *c.* 7.90; *d.* 6.15; *e.* 8.28; *f.* 8.85; *g.* 8.51; *h.* 8.06.

9. *a.* 3.6%; *b.* 5.0%; *c.* 10.8%; *d.* 30.3%; *e.* 0.02%; *f.* 0.04%; *g.* 1.58%; *h.* 4.9%; *i.* 27.0%.

10. *a.* $1.9 \times 10^{-7} M$; *b.* $1.2 \times 10^{-5} M$; *c.* $4.78 \times 10^{-3} M$; *d.* $4.78 \times 10^{-2} M$; *e.* $2.40 M$; *f.* $19 M$.

11. *a.* 7.48; *b.* 7.66; *c.* 6.48; *d.* 6.85; *e.* 6.80; *f.* 6.48; *g.* 4.71; *h.* 9.44; *i.* 10.12.

12. *a.* 3.1; *b.* 2.8; *c.* 3.4; *d.* 9.2; *e.* 9.2; *f.* 3.1; *g.* 5.0; *h.* 1.18; *i.* 1.43; *j.* 2.72; *k.* 11.1; *l.* 11.75; *m.* 11.47; *n.* 12.07.

13. *a.* 63; *b.* 6.3; *c.* 0.63.

14. *a.* 69.1 ml; *b.* 14.8 ml; *c.* 43.1 ml.

15. *a.* 98.9 ml; *b.* 71.5 ml; *c.* 50.0 ml; *d.* 20.1 ml.

16. *a.* 0.090; *b.* 0.067; *c.* 0.048; *d.* 0.020.

17. *a.* 3.61; *b.* 3.22; *c.* 3.79; *d.* 4.18.

18. *a.* $2.51 M$; *b.* $0.0032 M$; *c.* $0.032 M$; *d.* $0.063 M$.

19. *a.* 0.12; *b.* 0.058; *c.* 0.058; *d.* 0.102; *e.* 0.185.

20. *a.* 0.4; *b.* 0.9; *c.* 0.9; *d.* 0.5; *e.* 0.3.

22. *a.* $0.174 M$ of the acidic form and $0.174 M$ of the basic form.

23. *a.* $0.23 M$ NH_4Cl + $0.09 M$ NH_3; *b.* $0.10 M$ HCOOH + $0.20 M$ HCOONa; *c.* $0.15 M$ C_6H_5COOH + $0.12 M$ C_6H_5COONa; *d.* $0.08 M$ $(CH_3)_2NH$ + $0.33 M$ $(CH_3)_2NH \cdot HCl$.

24. *a.* 2.70; *b.* 2.26; *c.* 4.80; *d.* 2.45; *e.* 1.91; *f.* 2.15; *g.* 1.70; *h.* 1.51; *i.* 2.00; *j.* 2.50; *k.* 2.85; *l.* 2.73; *m.* 4.15; *n.* 1.64; *o.* 1.57.

25. *a.* 9.86; *b.* 11.3; *c.* 7.9; *d.* 11.2; *e.* 10.8; *f.* 10.3.

26. *a.* $[H_3O^+] = 6.97 \times 10^{-3}M$, $[HSO_3^-] = 6.97 \times 10^{-3}M$, $[H_2SO_3] = 3.03 \times 10^{-3}M$, $[SO_3^{2-}] = 6.23 \times 10^{-8}M$;

b. $[H_3O^+] = 6.3 \times 10^{-6}M$, $[H_2CO_3] = 9.4 \times 10^{-5}M$, $[HCO_3^-] = 6.3 \times 10^{-6}M$, $[CO_3^{2-}] = 4.7 \times 10^{-11}M$;

c. $[H_3O^+] \ 1.7 \times 10^{-3}M$, $[H_3PO_4] = 2.9 \times 10^{-4}M$, $[H_2PO_4^-] = 1.7 \times 10^{-3}M$, $[HPO_4^{2-}] = 9.9 \times 10^{-8}M$, $[PO_4^{3-}] = 9.2 \times 10^{-17}M$;

d. $[H_3O^+] = 1.83 \times 10^{-3}M$, $[H_2A] = 3.2 \times 10^{-3}M$, $[HA^-] = 1.76 \times 10^{-3}M$, $[A^{2-}] = 3.8 \times 10^{-5}M$;

e. $[OH^-] = 7.73 \times 10^{-4}M$, $[H_3O^+] = 1.29 \times 10^{-11}M$, $[A] = 7.27 \times 10^{-4}M$, $[HA^+] = 7.73 \times 10^{-4}M$, $[H_2A^{2+}] = 1.53 \times 10^{-7}M$;

f. $[OH^-] = 1.78 \times 10^{-4}M$, $[H_3O^+] = 5.62 \times 10^{-11}M$, $[A^{2-}] = 8.22 \times 10^{-4}M$, $[HA^-] = 1.78 \times 10^{-4}M$, $[H_2A] = 4.11 \times 10^{-12}M$.

27. *a.* $[H_3O^+] = 0.10M$, $[HSO_3^-] = 1.36 \times 10^{-3}M$, $[H_2SO_3] = 8.64 \times 10^{-3}M$, $[SO_3^{2-}] = 8.64 \times 10^{-10}M$;

b. $[H_3O^+] = 0.10M$, $[HCO_3^-] = 4.0 \times 10^{-10}M$, $[H_2CO_3] = 1 \times 10^{-4}M$, $[CO_3^{2-}] = 2 \times 10^{-19}M$;

c. $[H_3O^+] = 0.1M$, $[H_3PO_4] = 1.82 \times 10^{-3}M$, $[H_2PO_4^-] = 1.82 \times 10^{-4}M$, $[HPO_4^{2-}] = 1.82 \times 10^{-9}M$, $[PO_4^{3-}] = 2.9 \times 10^{-21}M$;

d. $[H_3O^+] = 0.1M$, $[H_2A] = 4.95 \times 10^{-3}M$, $[HA^-] = 4.95 \times 10^{-5}M$, $[A^{2-}] = 1.97 \times 10^{-8}M$;

e. $[H_3O^+] = 0.1M$, $[H_2A^{2+}] = 5 \times 10^{-3}M$, $[HA^+] = 1.6 \times 10^{-9}M$, $[A] = 4.0 \times 10^{-19}M$;

f. $[H_3O^+] = 0.1M$, $[H_2A] = 1.0 \times 10^{-3}M$, $[HA^-] = 2.0 \times 10^{-5}M$, $[A^{2-}] = 6.3 \times 10^{-14}M$.

For more accurate results the change of ionic strength should be taken into account.

28. *a.* Starting at pH 0–12.2; 10.2; 8.3; 6.7; 5.5; 4.5; 3.5; 2.5; 1.5; 0.6; 0.1; at pH $\geqslant$ 11–0.

b. Starting at pH 0–14.4; 11.4; 8.7; 7.0; 5.8; 4.8; 3.8; 2.8; 1.8; 0.9; 0.2; at pH $\geqslant$ 11–0.

c. Starting at pH 0–16.0; 14.0; 12.1; 9.1; 8.1; 7.1; 6.1; 5.1; 4.1; 3.1; 2.1; 1.1; 0.3; 0.1.

d. Starting at pH 0–6.9; 5.0; 3.5; 2.4; 1.4; 0.5; 0.1; at pH $\geqslant$ 7–0.

e. Starting at pH 0–23.7; 20.7; 17.7; 14.7; 11.8; 9.3; 7.2; 5.2; 3.3; 1.5; 0.4; at pH $\geqslant$ 11–0.

f. Starting at pH 0–29.4; 25.4; 21.5; 17.8; 14.1; 11.0; 8.1; 5.5; 3.3; 1.5; 0.3; at pH $\geqslant$ 11–0.

g. Starting at pH 0–35.4; 29.4; 23.6; 18.4; 14.2; 10.9; 8.0; 5.7; 3.2; 1.8; 0.3; 0.1; at pH $\geqslant$ 12–0.

29. *a.* $1.26 \times 10^{-5}M$, $1.6 \times 10^{-6}M$, $2.5 \times 10^{-7}M$, $5.0 \times 10^{-10}M$;

b. $1.0 \times 10^{-12}M$, $1.0 \times 10^{-13}M$;

c. $6.3 \times 10^{-9}M$, $5.0 \times 10^{-10}M$, $5.0 \times 10^{-11}M$, $5.0 \times 10^{-12}M$;

d. $2.0 \times 10^{-6}M$, $6.3 \times 10^{-9}M$, $1.26 \times 10^{-10}M$, $1.0 \times 10^{-14}M$.

30. *a.* From pH $\leqslant$ 9–0, 0.2, 1.5, 3.4, 5.7, 9.2.

b. From pH $\leqslant$ 7–0, 0.2, 1.0, 1.8, 5.4, 9.0, 12.9, 16.9.

c. From pH $\leqslant$ 2–0, 0.3, 1.8, 3.8, 5.8, 7.8, 9.8, 11.8, 13.8, 15.8, 17.8, 19.8, 21.9.

d. From pH $\leqslant$ 6–0, 0.1, 0.4, 1.3, 2.9, 5.2, 8.0, 11.0, 14.0.

31. *a.* 10^{-25}; *b.* 10^{-19}; *c.* 10^{-13}; *d.* 10^{-7}.

32. *a.* $10M$; *b.* $4 \times 10^{-13}M$; *c.* $3.2 \times 10^{-7}M$.

33. *a.* pH = 1.99; *b.* pH = 1.35; *c.* 2.19.

34. 4.17.

35. 2.82.

36. From 3.95 to 3.86.

37. 2.29.

38. 11.8.

39. *a.* 7.2; *b.* 6.55; *c.* 6.05; *d.* 5.0; *e.* 4.3; *f.* 7.55.

40. *a.* $2.5 \times 10^{-5}M$; *b.* $1.26 \times 10^{-3}M$; *c.* $6.3 \times 10^{-11}M$; *d.* $2.5 \times 10^{-9}M$.

41. *a.* 9.7; *b.* 4.2; *c.* 3.7; *d.* 5.0.

42. *a.* 4.8; *b.* 6.1; *c.* 7.3; *d.* 8.7; *e.* 10.1.

43. *a.* 6.20; *b.* 3.55; *c.* 3.75; *d.* 4.4; *e.* 9.3; *f.* 3.0; *g.* 3.5; *h.* 4.1; *i.* 4.5.

44. *a.* 2.3 ml; *b.* 4.3 ml; *c.* 8.2 ml.

CHAPTER 4

1. *a.* 11.5, 11.2, 10.7; *b.* 11.7, 11.4, 11.0; *c.* 10.7, 10.5, 10.0; *d.* 4.3, 4.5, 5.0.

2. *a.* 4.3, 4.8, 5.1, 5.8, 11.3; *b.* 4.3, 4.8, 5.1, 5.8, 9.7; *c.* 2.8, 3.0, 3.3, 3.7, 9.7; *d.* 4.8, 5.2, 5.8, 6.2, 11.9.

3. *a.* 5.2, 4.8, 4.3, 2.1; *b.* 9.8, 9.4, 8.8, 2.1, *c.* 9.2, 8.8, 8.2, 2.1; *d.* 6.1, 5.7, 5.1, 2.8.

4. *a.* 2.1, 2.4, 2.7, 2.8, 4.3, 4.8, 8.1, 11.6;   *b.* 11.2, 10.6, 10.1, 9.1, 7.5, 7.0, 4.9, 2.7; *c.* 10.5, 10.1, 9.8, 9.1, 7.5, 7.0, 5.4, 3.6; *d.* 11.3, 11.3, 10.3, 9.3, 6.8, 6.3, 5.0, 3.1; *e.* 2.5, 3.1, 3.5, 3.6, 4.3, 4.8, 5.9, 11.6; *f.* 3.1, 3.5, 3.8, 6.1, 9.5, 10.0, 10.2, 11.0; *g.* 3.3, 4.2, 4.7, 7.9, 11.1, 11.2, 11.4, 12.1.

5. *a.* pH 5.30–8.70; *b.* pH 6.30–7.70; *c.* pH 6.00–8.00.

6. *a.* -0.02%; *b.* -0.2%; *c.* -2%; *d.* -1%.

7. *a.* -0.6%; *b.* -0.6%; *c.* $-2 \times 10^{-4}\%$; *d.* -100%; *e.* -0.6%; *f.* -0.1%.

8. *a.* $+0.014\%$; *b.* $+0.010\%$; *c.* $+0.11\%$; *d.* -91%; *e.* $+0.18\%$; *f.* $+0.55\%$.

9. *a.* pH 7.5–9.0; *b.* pH 4.7–6.7; *c.* pH 8.3–9.8; *d.* pH 4.3–3.7.

10. *a.* 6.0; *b.* 4.0; *c.* 2.2; *d.* 2.2.

11. *a.* 1.1×10^4; *b.* 2.0×10^4; *c.* 2.0×10^3; *d.* 2.15×10^3.

12. *a.* 5.4×10^3; *b.* 1.9×10^2; *c.* 9.9×10^2; *d.* 3.5×10^1.

13. *a.* $2.5 \times 10^{-3}\%$; *b.* $2.3 \times 10^{-2}\%$; *c.* $2.5 \times 10^{-2}\%$; *d.* $7.6 \times 10^{-2}\%$; *e.* 0.11%; *f.* 5.0%.

14. *a,* I—3.3%; II—$2.6 \times 10^{-3}\%$; *b.* I—$3.7 \times 10^{-2}\%$, II—1.5%; *c.* I—18%, II—13%, III—$2.7 \times 10^{-2}\%$.

Bibliography

The textbooks and monographs mentioned below deal at least partially with topics that are similar or related to the subject of this book. They are recommended fo a broader and deeper understanding of the problemr discussed here.

A. Albert and E. P. Serjeant, *The Determination of Ionization Constants*, 3rd Ed., Chapman and Hall, London, 1984.

C. F. Baes Jr. and R. E. Mesmer, *The Hydrolysis of Cations*, Wiley, New York, 1976.

R. G. Bates, *Determination of* pH, *Theory and Practice*, Wiley, New York, 1973.

R. P. Bell, *Acids and Bases*, Methuen, London, 1952.

R. P. Bell, *The Proton in Chemistry*, Chapman and Hall, London, 1973.

C. Bliefert, *pH Wert Berechnungen*, Verlag Chemie, Weinheim, 1978.

J. N. Butler, *Ionic Equilibrium, A Mathematical Approach*, Addison Wesley, Reading, Mass., 1964.

J. N. Butler, *Solubility and* pH *Calculations*, Addison Wesley, Reading, Mass., 1964.

G. Charlot and B. Trémillon, *Les Réactions Chimiques dans les Solvants et les Sels Fondus*, Gauthier-Villars Editeur, Paris, 1963.

F. Dietze, E. Hoyer, F. Lorenz, W. Seifert and D. Wagler, *Säuren und Basen*, Akademische Verlagsgesellschaft Geest & Portig, Leipzig, 1971.

H. Freiser and Q. Fernando, *Ionic Equilibria in Analytical Chemistry*, Wiley, New York, 1963.

A. M. Huntz, *Equilibres Acido–Basiques, pH*, Masson, Paris, 1975.

J. Inczédy, *Analytical Applications of Complex Equilibria*, Horwood, Chichester, 1973.

E. J. King, *Acid–Base Equilibria*, Pergamon Press, Oxford, 1965.

J. Kragten, *Atlas of Metal–Ligand Equilibria in Aqueous Solution*, Horwood, Chichester, 1978.

A. P. Kreshkov, *Analiticheskaya Khimia Nevodnykh Rastvorov*, Khimia, Moscow, 1982.

D. D. Perrin and B. Dempsey, *Buffers for pH and Metal-Ion Control*, Chapman and Hall, London, 1974.

D. D. Perrin, B. Dempsey and E. P. Serjeant, pK_a *Prediction for Organic Acids and Bases*, Chapman and Hall, London, 1981.

J. E. Prue, *Ionic Equilibria*, Pergamon Press, Oxford, 1966.

A. Ringbom, *Complexation in Analytical Chemistry*, Wiley, New York, 1963.

H. Rossotti, *The Study of Ionic Equilibria*, Longmans, London, 1978.

L. Rougeot, *Acides et Bases*, Presses Universitaires de France, Paris, 1970.

K. Schwabe, *pH-Messung*, Akademie Verlag, Berlin, 1980.

F. Seel, *Grundlagen der analytischen Chemie*, 6th Ed., Verlag Chemie, Weinheim, 1976.

H. H. Sisler, *Chemistry in Non-Aqueous Solvents*, Reinhold, New York, 1961.

L. Šucha and S. Kotrly, *Solution Equilibria in Analytical Chemistry*, Van Nostrand-Reinhold, London, 1972.

B. Trémillon, *Chemistry in Non-Aqueous Solvents*, D. Reidel Publ. Co., Dordrecht, 1974.

Subject Index